Mathematics Study Resources

Volume 13

Series Editors

Kolja Knauer, Departament de Matemàtiques Informàtic, Universitat de Barcelona, Barcelona, Barcelona, Spain

Elijah Liflyand, Department of Mathematics, Bar-Ilan University, Ramat-Gan, Israel

This series comprises direct translations of successful foreign language titles, especially from the German language.

Powered by advances in automated translation, these books draw on global teaching excellence to provide students and lecturers with diverse materials for teaching and study.

Joachim Hilgert

Mathematical Structures

From Linear Algebra over Rings
to Geometry with Sheaves

Joachim Hilgert
Institut für Mathematik
Universität Paderborn
Paderborn, Germany

ISSN 2731-3824 ISSN 2731-3832 (electronic)
Mathematics Study Resources
ISBN 978-3-662-69411-4 ISBN 978-3-662-69412-1 (eBook)
https://doi.org/10.1007/978-3-662-69412-1

English translation of the 2nd original German edition published by Springer-Verlag Heidelberg, 2024
The original submitted manuscript has been translated into English. The translation was done using artificial intelligence. A subsequent revision was performed by the author(s) to further refine the work and to ensure that the translation is appropriate concerning content and scientific correctness. It may, however, read stylistically different from a conventional translation.
Translation from the German language edition: "Mathematische Strukturen" by Joachim Hilgert, © Springer-Verlag Berlin Heidelberg 2024. Published by Springer Berlin Heidelberg. All Rights Reserved.

© The Editor(s) (if applicable) and The Author(s), under exclusive license to Springer-Verlag GmbH, DE, part of Springer Nature 2024

This work is subject to copyright. All rights are solely and exclusively licensed by the Publisher, whether the whole or part of the material is concerned, specifically the rights of translation, reprinting, reuse of illustrations, recitation, broadcasting, reproduction on microfilms or in any other physical way, and transmission or information storage and retrieval, electronic adaptation, computer software, or by similar or dissimilar methodology now known or hereafter developed.
The use of general descriptive names, registered names, trademarks, service marks, etc. in this publication does not imply, even in the absence of a specific statement, that such names are exempt from the relevant protective laws and regulations and therefore free for general use.
The publisher, the authors and the editors are safe to assume that the advice and information in this book are believed to be true and accurate at the date of publication. Neither the publisher nor the authors or the editors give a warranty, expressed or implied, with respect to the material contained herein or for any errors or omissions that may have been made. The publisher remains neutral with regard to jurisdictional claims in published maps and institutional affiliations.

This Springer imprint is published by the registered company Springer-Verlag GmbH, DE, part of Springer Nature.
The registered company address is: Heidelberger Platz 3, 14197 Berlin, Germany

If disposing of this product, please recycle the paper.

Preface

Mathematics as a science fulfills certain criteria for both natural science and humanities science. A significant reason for this dual position is that it cannot be determined whether the subject of mathematical research is naturally given or man-made. A commonly taken conceptual way out is to refer to mathematics as a *structural science*. However, in mathematics, there is no formal definition of what a structure is. If we use the working definition "A structure is a collection of objects between which relationships exist that are subject to certain rules", the term structural science is very aptly chosen for mathematics. Starting with simple number systems, modern mathematics works with an abundant variety of such structures. Even for very simple questions, such as those about the solvability of certain equations, these structures provide the framework within which the questions can be answered.

Not all mathematical structures are equal, there is a certain hierarchy in the significance of such structures. Some can rightly be called fundamental, while others must be considered very specific. Still others lead, at least currently, a niche existence.

The focus of this book is on those mathematical structures that I consider fundamental and that I believe every professional mathematician should know. The structures are presented in the context of concrete and significant mathematical results, so that readers can more easily form their own opinion on the relevance of these structures. The selection of both the structures and the results is of course influenced by my own mathematical background. However, I am convinced that it provides a good platform from which to delve deeper into any area of mathematics currently covered in research and teaching.

I am addressing readers who are familiar with the standard content of the first two years of a solid mathematics undergraduate program. Specifically, I assume elementary knowledge of the following concepts:

(1) Prime factorization of natural numbers
(2) Real numbers
(3) Vector spaces and linear mappings
(4) Groups and their homomorphisms
(5) Topological spaces and continuous mappings
(6) Differential and integral calculus in one and several real variables

(7) σ-algebras and elementary measure or probability theory
(8) Zorn's lemma and the axiom of choice

This material can be found in many standard textbooks. See for instance [La67] for (1), (4) and (8), [Ap75] for (2), (3) and (6), [Mu18] for (5), and [CK05] for (7).

Particular attention is paid to the motivation of the structures and the explanation of the connections between different structures. A separate treatment of the fields of algebra, geometry, and analysis is deliberately avoided, although the curricula of most courses in the context of undergraduate education suggest such a separation. The organization of the material is not based on mathematical disciplines but on the nature of the described structures. This allows the connections and parallels between the mentioned fields to be made clear. The same purpose is served by the explicit discussion of examples that cover several fields and the outlooks on later applications.

The aim of this book is to give the reader an understanding of the architecture of a typical undergraduate mathematics curriculum. The detailed discussion of the concepts and the various examples are intended to encourage active engagement with the structural definitions and arguments that often pose difficulties for students. In order to better highlight the cross-connections between different contents of typical undergraduate mathematics programs, we also introduce some concepts that are often only presented in detail in graduate programs.

From the elementary courses, most mathematics majors are familiar with three types of structures. The first type is the *algebraic structure* such as fields, groups, and vector spaces, in which a set is provided with one or more operations that must obey certain rules. The second type is the *comparative structure* such as order or equivalence relations on a given set, with the help of which one can compare elements of this set in terms of relevant properties. The third type is the *subset structure*, such as topologies or σ-algebras, in which a family of subsets of a given set is distinguished, which must also obey certain rules. Most of the basic mathematical structures are obtained by combining such types of structures.

In this book, we deal with exemplary algebraic structures in Part I. The starting point of the presentation is basic concepts and constructions from (multi)linear algebra over rings. In the process, various analogies between definitions and constructions become apparent, which I take as an opportunity to introduce concepts from category theory for the systematic comparison of structures. In Part II, the insights from Part I are combined with subset structures, especially topologies. The focus is on local structures, i.e., those that are completely determined by their properties in (small) neighborhoods of points. The mathematical concept that clarifies this vague description is that of a sheaf. Sheaves are discussed at the beginning of Part II and illustrated with numerous examples. Exemplary for local structures, introductions to differentiable manifolds and algebraic varieties follow.

Sheaves not only provide a practical conceptual framework for describing local mathematical structures. They are also a powerful tool for systematically investigating global properties of local and mixed structures. However, their use requires additional technical tools from algebra and topology, the provision of which would have exceeded the scope of this book. Instead, the short Part III contains an outlook on further structures that result from combination, modification, or enrichment of already described structures, with particular emphasis on the motivation for the introduction and investigation of the respective structures. The aim is to convey a sense that the structural considerations of the first two parts allow us to appreciate a multitude of mathematical concepts without much extra effort. As an additional test, I recommend reading [Go11], for example the contribution IV.8 on moduli spaces.

Paderborn, Germany Joachim Hilgert
April 2024

Contents

Part I Algebraic Structures

1 Rings 3
   1.1 Elementary Definitions and Examples 3
   1.2 Some Structure Theory for Rings 9
   1.3 Special Classes of Rings 13

2 Modules 25
   2.1 Structure Theory of Modules 25
   2.2 Applications to Linear Mappings 43

3 Multilinear Algebra 55
   3.1 Tensor Products 55
   3.2 Tensor Algebras 71
   3.3 Symmetric and Exterior Algebras 75

4 Pattern Recognition 91
   4.1 Universal Algebra 91
   4.2 Naive Category Theory 97
   4.3 Categorical Constructions: Limits 108
   4.4 Adjoint Functors 121

Part II Local Structures

5 Sheaves 139
   5.1 Presheaves and Sheaves 139
   5.2 Étalé Spaces 153
   5.3 Ringed Spaces 160
   5.4 Module Sheaves 166

6 Manifolds 171
   6.1 Charts and Parametrizations 172
   6.2 Tangent Spaces and Derivatives 182
   6.3 Tangent and Tensor Bundles 193
   6.4 Differential Forms 209
   6.5 Integration on Real Manifolds 216
   6.6 Applications to Complex Differentiability 229

7 Algebraic Varieties ... 239
- 7.1 Algebraic Sets ... 240
- 7.2 Algebraic Varieties ... 254
- 7.3 Schemes ... 266

Part III Outlook

8 Transfer of Arguments and Structures ... 279
- 8.1 Technologies of Structural Comparison ... 279
- 8.2 Group Objects and Group Actions ... 283

9 Specialization, Generalization and Unification of Structures ... 289
- 9.1 Special Tensors ... 289
- 9.2 Connections and Fiber Bundles ... 302
- 9.3 Structured Analysis? ... 313

References ... 319

Index ... 323

Part I
Algebraic Structures

We begin the part on algebraic structures with an introductory chapter on rings. In Chap. 2 we then present a rich algebraic structure, the *module* over a *ring*. We describe several examples, prove a number of basic facts about this structure and also a prototypical theorem about a special case, the finitely generated modules over Euclidean rings. We also show some non-trivial applications, especially to the linear mappings known from elementary linear algebra. Later we will also see applications to the integration of functions in several variables.

In Chap. 3 we explain several constructions in the context of modules. Also in this chapter, the selection of material follows the two basic criteria: relevance in mathematical practice and the model character for general mathematical procedures. My choice fell on multilinear algebra, which is treated rather stepmotherly in the common curricula despite its omnipresence in all parts of mathematics.

To make the model character of the presented structures and constructions clear to the readers, we introduce basic concepts of universal algebra and category theory in Chap. 4. Using many examples, we sketch how the use of the categorical language in particular unifies algebraic structure considerations. In the last two sections, we then show that the constructions presented in Chaps. 2 and 3 are natural in a precise sense, and even inevitable in a certain way.

Rings

We begin with a brief introduction to general ring theory, in which we mainly discuss some basic definitions and many examples. We then delve a little deeper into ring theory and discuss some interesting and fundamental structural properties of rings that we will need in later applications. At the same time, these considerations illustrate how to deal with algebraic structures.

1.1 Elementary Definitions and Examples

The following definition of a ring should be considered as a weakening of the definition of a field, which captures the following three significant examples: the integers, the square matrices of a (arbitrarily) given size, and the mappings of a set into a field with the pointwise defined additions and multiplications.

Definition 1.1 (Rings) Let $R \neq \emptyset$ be a set and $+ : R \times R \to R$ (*addition*) and $\cdot : R \times R \to R$ (*multiplication*) two mappings. $(R, +, \cdot)$ is called a *ring*, if the following properties hold:

(i) $(R, +)$ is an abelian group.
(ii) The multiplication on R is associative.
(iii) *Right-distributivity*:

$$\forall x, y, z \in R : \quad (x + y) \cdot z = x \cdot z + y \cdot z.$$

(iii′) *Left-distributivity*:

$$\forall x, y, z \in R : \quad z \cdot (x + y) = z \cdot x + z \cdot y.$$

In the distributive laws, the multiplications are done before the additions. A ring $(R, +, \cdot)$ is called *commutative*, if the multiplication is commutative, i.e.,

$$\forall x, y \in R: \quad x \cdot y = y \cdot x.$$

If there is an element $e \in R$ in a ring $(R, +, \cdot)$ with

$$\forall x \in R: \quad e \cdot x = x \cdot e = x \tag{1.1}$$

then $(R, +, \cdot, e)$ is called a *ring with identity*. In this case, an $x \in R$ is called a *unit*, if there is an element $y \in R$ with $x \cdot y = y \cdot x = e$. We denote the set of units of R with $\mathrm{Unit}(R)$.

As with fields, the multiplication symbol is often omitted in rings, i.e., one writes xy instead of $x \cdot y$. For the following examples, the ring axioms are easy to verify directly from the definitions. Here, as in all further examples, the verification of the not explicitly proven assertions is to be understood as an exercise.

Example 1.2 (Rings)

(i) Every field is a commutative ring with identity.
(ii) $(\mathbb{Z}, +, \cdot, 1)$ is a commutative ring with identity.
(iii) The $n \times n$ matrices $\mathrm{Mat}(n \times n, \mathbb{K})$ over a field $\mathbb{K}$ form a ring with respect to the usual matrix addition and multiplication with the identity matrix as the identity element. This ring is not commutative for $n > 1$.
(iii') The subset $T \subseteq \mathrm{Mat}(n \times n, \mathbb{K})$ of strictly upper triangular matrices (all elements on or below the diagonal are 0) form a ring with respect to matrix addition and multiplication. This ring does not contain an identity, which can be seen, e.g., from the fact that the n-th power $A^n = A \cdot \ldots \cdot A$ for each $A \in T$ is equal to 0.
(iv) Let $\mathbb{K}$ be a field and V a $\mathbb{K}$-vector space. Then the set $\mathrm{End}_\mathbb{K}(V) = \mathrm{Hom}_\mathbb{K}(V, V)$ of linear self-mappings of V is a ring with respect to pointwise addition and composition as multiplication. We describe these two operations in more detail: For $\varphi, \psi \in \mathrm{End}_\mathbb{K}(V)$ and $v \in V$ we have:
 (Add) $(\varphi + \psi)(v) = \varphi(v) + \psi(v)$, where the addition on the right side is that of V.
 (Mult) $(\varphi \circ \psi)(v) = \varphi(\psi(v))$.
 The ring $\mathrm{End}_\mathbb{K}(V)$ has an identity, namely the identity id_V.
(v) Let M be a set and R a ring with an identity. Then the set $F(M, R) := \{f : M \to R\}$ of R-valued mappings with pointwise addition and pointwise multiplication is a ring with the constant function 1 as the identity element. Again, we describe these two operations in more detail: For $f, g \in F(M, R)$ and $x \in M$ we have:
 (Add) $(f + g)(x) = f(x) + g(x)$, where the addition on the right side is that of R.

1.1 Elementary Definitions and Examples

(Mult) $(f \cdot g)(x) = f(x) \cdot g(x)$, where the multiplication on the right side is that of R.

(v') Various subsets of function spaces of the form $F(M, R)$ are commutative rings with respect to pointwise operations. This is the case, e.g., for $C^k(U, \mathbb{R})$, the space of k-times continuously differentiable real-valued functions on an open subset U of $\mathbb{R}^n$.

(vii) $\mathbb{Z}[i] := \mathbb{Z} + i\mathbb{Z} := \{c \in \mathbb{C} \mid c = a + ib; \ a, b \in \mathbb{Z}\}$ is a ring with respect to complex addition and complex multiplication. It is called the ring of *Gaussian integers*.

□

With Examples 1.2(ii)–(iv), all the rings mentioned before the definition are covered. However, the list of rings that appear even in elementary courses is far from exhausted. The following class of examples is encountered, e.g., in elementary number theory.

Example 1.3 (Residue Class Rings) Let $n \in \mathbb{Z}$ and $n\mathbb{Z} := \{nz \in \mathbb{Z} \mid z \in \mathbb{Z}\}$. Then $n\mathbb{Z}$ is a subgroup of the additive abelian group $(\mathbb{Z}, +)$. The subsets of $\mathbb{Z}$ of the form $k + n\mathbb{Z} := \{k + nz \in \mathbb{Z} \mid z \in \mathbb{Z}\}$ for $k \in \mathbb{Z}$ are called the *cosets* of $n\mathbb{Z}$. Because for $n \in \mathbb{N}$ and $k = 0, \ldots, n-1$ the set $k + n\mathbb{Z}$ can also be considered as the set of all integers that give the remainder k when divided by n, we also speak of the *residue classes* modulo n. The set $\mathbb{Z}/n\mathbb{Z} := \{k + n\mathbb{Z} \mid k \in \mathbb{Z}\}$ of all *cosets* of $n\mathbb{Z}$ forms an abelian group with respect to the addition

(Add) $(k + n\mathbb{Z}) + (k' + n\mathbb{Z}) = (k + k') + n\mathbb{Z}$,

which is shown to be well-defined by a short calculation. If we additionally equip $\mathbb{Z}/n\mathbb{Z}$ with the multiplication

(Mult) $(k + n\mathbb{Z}) \cdot (k' + n\mathbb{Z}) = (k \cdot k') + n\mathbb{Z}$,

which is also easily recognizable as well-defined, one obtains a commutative ring with identity, which is called the *residue class ring* modulo n. The identity element is given by $1 + n\mathbb{Z}$, the zero by $0 + n\mathbb{Z} = n\mathbb{Z}$.

□

From additively written abelian groups, we know that they have exactly one neutral element, i.e., exactly one element that can be added to any other element without changing this element. This element is then called the *zero* of the group. For rings, there is always exactly one zero, which is usually denoted by 0. For multiplicatively written abelian groups, the neutral element fulfills exactly the condition from (1.1), and it is called the *identity* of the group. As can be seen from Example 1.2(iii'), a ring does not have to contain an identity. But there can also not be more than one identity, as the following proposition shows.

Proposition 1.4 (Zero and Identity in Rings) *Let $(R, +, \cdot)$ be a ring. Then the following holds:*

(i) *There is at most one identity element.*
(ii) *If there is an identity element e, then for each $x \in R$ there is at most one multiplicative inverse, i.e., there is at most one element $y \in R$ with $xy = e = yx$.*
(iii) *$\forall x \in R : \quad 0 \cdot x = x \cdot 0 = 0$.*
(iv) *If the additive inverse of $x \in R$ is denoted by $-x$, then*

$$\forall x, y \in R : \quad x(-y) = -xy = (-x)y.$$

Proof

(i) Let e and e' be identity elements. Then $e = e \cdot e' = e'$.
(ii) Let y and y' be multiplicative inverses of $x \in R$. Then $y = ey = (y'x)y = y'(xy) = y'e = y'$.
(iii) $0 \cdot x + 0 \cdot x = (0+0) \cdot x = 0 \cdot x$. But if $y + y = y$, then follows $y = 0 + y = (-y + y) + y = -y + (y + y) = -y + y = 0$. Analogously, one obtains $x \cdot 0 = 0$, because $x \cdot 0 + x \cdot 0 = x \cdot (0+0) = x \cdot 0$ holds.
(iv) This follows from $x(-y) + xy = x(-y + y) = x \cdot 0 = 0$ and $(-x)y + xy = (-x + x)y = 0 \cdot y = 0$.

$\square$

There are also rings with identity, for which every non-zero element has a multiplicative inverse as in a field, but which are not commutative. Such rings are called *division rings* or *skew-fields*.

Example 1.5 (Quaternions) Let $\mathbb{H} = \mathbb{R}^4$ with a given basis, which is denoted by $\{1, i, j, k\}$. If one extends the multiplication table

$\cdot$	1	i	j	k
1	1	i	j	k
i	i	-1	k	$-j$
j	j	$-k$	-1	i
k	k	j	$-i$	-1

1.1 Elementary Definitions and Examples

real bilinearly, one obtains a multiplication, with respect to which $(\mathbb{H}, +, \cdot)$ becomes a division ring, the *quaternions*. The multiplicative inverse of an element $z = r + is + ju + kv \neq 0$ is given by

$$z^{-1} = \frac{r - is - ju - kv}{r^2 + s^2 + u^2 + v^2}.$$

The similarity with the construction of complex numbers as $\mathbb{R}^2$ with a new multiplication is evident. Indeed, the quaternions were found in 1843 by William Rowan Hamilton after he had spent a long time unsuccessfully trying to find a multiplication as for $\mathbb{R}^1 = \mathbb{R}$ and $\mathbb{R}^2 = \mathbb{C}$ also on $\mathbb{R}^3$ that makes $(\mathbb{R}^3, +)$ a field. □

The next family of examples for rings is of fundamental importance in algebra, geometry, and analysis. It includes as special cases the polynomial functions on $\mathbb{R}$, but also the Taylor series of infinitely differentiable functions in one or more variables.

Example 1.6 (Formal Power Series and Polynomials) Let R be a ring with identity and $X_1, \ldots, X_k$ symbols. Elements of $\mathbb{N}_0^k$ we call *multi-indices*. They are denoted by $\alpha = (\alpha_1, \ldots, \alpha_k)$. The sum of two multi-indices is given component-wise.

(i) A *formal power series* in $X_1, \ldots, X_k$ with coefficients in R is a formal sum

$$\sum_{\alpha \in \mathbb{N}_0^k} a_\alpha X^\alpha := \sum_{\alpha \in \mathbb{N}_0^k} a_\alpha X_1^{\alpha_1} \cdots X_k^{\alpha_k},$$

where the $a_\alpha \in R$. This is initially nothing more than a suggestive notation for a family $(a_\alpha)_{\alpha \in \mathbb{N}_0^k}$ of elements in R. The point of this notation is that it makes the following multiplication on the set $R[[X_1, \ldots, X_k]]$ of all formal power series in $X_1, \ldots, X_k$ appear "natural" because it mimics the "expansion" of products of finite sums (this is called the *Cauchy product*):

(Mult) $\left(\sum_{\alpha \in \mathbb{N}_0^k} a_\alpha X^\alpha \right) \left(\sum_{\alpha \in \mathbb{N}_0^k} b_\alpha X^\alpha \right) := \sum_{\alpha \in \mathbb{N}_0^k} \left(\sum_{\beta + \gamma = \alpha} a_\beta b_\gamma \right) X^\alpha.$

In addition to this product, there is the usual coefficient-wise addition, which corresponds to point-wise addition when the elements of $R[[X_1, \ldots, X_k]]$ are considered as functions $\mathbb{N}_0^k \to R$:

(Add) $\left(\sum_{\alpha \in \mathbb{N}_0^k} a_\alpha X^\alpha \right) + \left(\sum_{\alpha \in \mathbb{N}_0^k} b_\alpha X^\alpha \right) := \sum_{\alpha \in \mathbb{N}_0^k} (a_\alpha + b_\alpha) X^\alpha.$

Together with this addition and this multiplication, the set $R[[X_1, \ldots, X_k]]$ is a ring with identity, which is commutative exactly when the ring R is commutative. The identity is the element X^0, which has $a_0 = 1$ at the position $0 = (0,\ldots,0)$ as the only non-zero coefficient. The multiplication is constructed so that in a term of the form $X^\alpha = X_1^{\alpha_1} \cdots X_k^{\alpha_k}$ all factors can simply be omitted for which $\alpha_j = 0$. In the extreme case, instead of $a_0 X^0$ we simply write a_0 and, for $a_0 = 1$, instead of X^0 only 1.

(ii) A formal power series $\sum_{\alpha \in \mathbb{N}_0^k} a_\alpha X^\alpha$ is called a *polynomial*, if only finitely many a_α are different from zero. The *degree* of a polynomial $0 \neq f = \sum_{\alpha \in \mathbb{N}_0^k} a_\alpha X^\alpha$ is given by

$$\deg(f) := \max\{|\alpha| \mid a_\alpha \neq 0\},$$

where $|\alpha| := \alpha_1 + \ldots + \alpha_k$. Also, we set $\deg(0) := -\infty$. If $a_\alpha \neq 0$ only for $|\alpha| = d$, then f is called *homogeneous* of degree d. The set of polynomials in $X_1, \ldots, X_k$ over R is denoted by $R[X_1, \ldots, X_k]$. The subset of homogeneous polynomials of degree d is denoted by $R[X_1, \ldots, X_k]_d$. If $k = 1$, then $a_{\deg f}$ is called the *leading coefficient* of f. If the leading coefficient is equal to 1, then f is called *monic* or *normalized*.

The subset $R[X_1, \ldots, X_k]$ of $R[[X_1, \ldots, X_k]]$ is closed with respect to the addition and multiplication from (i) and forms a ring with identity with these operations. □

In the description of the addition of formal power series in Example 1.6(i), there is a phrase that often signals that something interesting or surprising is happening: "When one [...] considers as [...]." It is used when one looks at a known object from a new or at least unexpected perspective for the given context.

While one usually encounters the objects from Example 1.6 in the study of algebra or function theory, the following example is essential for the theory of linear partial differential equations and is at least implicitly introduced in pertinent courses.

Example 1.7 (Differential Operators) Let $U \subseteq \mathbb{R}^k$ be an open subset and $C^\infty(U)$ the real vector space of infinitely differentiable functions $f: U \to R$. We write ∂_j for the partial derivatives $\frac{\partial}{\partial x_j}$, i.e., $\partial_j f$ instead of $\frac{\partial f}{\partial x_j}$. For a multi-index $\alpha = (\alpha_1, \ldots, \alpha_k) \in \mathbb{N}_0^k$ and $x = (x_1, \ldots, x_k) \in \mathbb{R}^k$ we set

$$\partial^\alpha := \left(\frac{\partial}{\partial x_1}\right)^{\alpha_1} \cdots \left(\frac{\partial}{\partial x_n}\right)^{\alpha_n} = \partial_1^{\alpha_1} \cdots \partial_k^{\alpha_k}.$$

Similar to Example 1.6, we set ∂_j^0 equal to 1. We consider self-mappings D of $C^\infty(U)$ of the following type: For each $\alpha \in \mathbb{N}_0^k$, let $c_\alpha \in C^\infty(U)$, where only finitely many of these c_α are different from zero. Then we set for $f \in C^\infty(U)$

$$D(f) := \sum_{\alpha \in \mathbb{N}_0^k} c_\alpha \partial^\alpha f. \tag{1.2}$$

1.2 Some Structure Theory for Rings

These self-mappings are called *differential operators* with smooth coefficients on U. Using the known rules of differentiation, it is easy to calculate that every such differential operator is an $\mathbb{R}$-linear self-mapping of $C^\infty(U)$. Another such calculation shows that the set $\mathcal{D}(U) \subseteq \text{End}_\mathbb{R}\left(C^\infty(U)\right)$ of all differential operators with smooth coefficients on U is closed under the addition and multiplication from Example 1.2(iv) for $V = C^\infty(U)$. Moreover, $\mathcal{D}(U)$ is a ring with respect to the restricted operations. To verify this, one only needs to check that if D is in $\mathcal{D}(U)$, then so is $-D$. The associative and distributive laws are automatically satisfied because they hold for all elements of $\text{End}_\mathbb{R}\left(C^\infty(U)\right)$. The ring $\mathcal{D}(U)$ has an identity of the form (1.2): one chooses c_0 to be constantly 1 and all other c_α to be constantly 0.

For every differential operator $D \in \mathcal{D}(U)$ and every function $h \in C^\infty(U)$, there is a differential equation, namely $D(f) = h$, where one seeks an $f \in C^\infty(U)$ that satisfies this equation. If one could invert the mapping D, then with $f = D^{-1}(h)$ one would have a solution to the differential equation. The definition of ring multiplication shows that the invertibility of D as a mapping is equivalent to the (multiplicative) invertibility of D as a ring element (see Proposition 1.4). □

Exercise 1.1 (Weyl-Algebra) Example 1.7 can be varied in different ways. Show:

(i) The differential operators of the form (1.2) with constant functions c_α form, with the same operations, also a non-commutative ring with identity. This is referred to as the algebra of differential operators with *constant coefficients*.
(ii) The differential operators of the form (1.2) with polynomial functions c_α form, with the same operations, also a non-commutative ring with identity. If $U = \mathbb{R}^k$, this ring is called the *Weyl-algebra*.

1.2 Some Structure Theory for Rings

Mappings between sets often play the role of a "size comparison" in mathematics. Does one set fit into another? Can one set be covered by another? If so, how often? If the sets carry additional structures, the same questions can be asked, but specified in such a way that the structures should be preserved in the embeddings or coverings. What exactly is meant by "preserved" depends on the type of structure. In the case of algebraic structures, this can be formulated relatively easily.

In the case of rings, we concretize the idea of structure-preserving mapping in the following definition, which is completely analogous to the definition of a group homomorphism as a structure-preserving mapping of a group and the linear mapping as a structure-preserving mapping of a vector space. Later, this idea will be incorporated into the concept of a *morphism* of a category.

Definition 1.8 (Ring Homomorphism) Let R and S be rings. A mapping $\varphi \colon R \to S$ is called a *ring homomorphism*, if the following holds:

(a) $\forall x, y \in R: \quad \varphi(x+y) = \varphi(x) + \varphi(y)$
(b) $\forall x, y \in R: \quad \varphi(xy) = \varphi(x)\varphi(y)$

We denote the image of φ with im (φ) and the *kernel* $\varphi^{-1}(0_S)$ of φ with ker (φ). If φ is bijective, φ is called a *isomorphism*. In this case, it is easy to verify that the inverse mapping φ^{-1} is also a ring homomorphism (see also Definition 2.6).

If R and S are rings with identity elements 1_R and 1_S respectively, a ring homomorphism $\varphi: R \to S$ is called a homomorphism of rings with identity, if additionally

(c) $\varphi(1_R) = 1_S$

holds. The definitions of kernel, image, and isomorphism are also transferred verbatim for rings with identity.

Rings of very different formation can be isomorphic, i.e., isomorphisms are not necessarily just renamings of elements. Nevertheless, one usually does not distinguish between isomorphic rings as long as one is only interested in properties of the ring that can be expressed by addition and multiplication, because such properties can be transferred from one ring to an isomorphic ring. A corresponding remark also applies to all other algebraic structures in which one speaks of isomorphism.

Example 1.9 (Ring homomorphism) Let $R = \mathbb{R}$ and $R' = \mathrm{Mat}(2 \times 2, \mathbb{R})$. Then the mapping defined by $\varphi(r) = \begin{pmatrix} r & 0 \\ 0 & r \end{pmatrix}$ is $\varphi: R \to R'$ a homomorphism of rings with identity. On the other hand, the mapping defined by $\psi(r) = \begin{pmatrix} r & 0 \\ 0 & 0 \end{pmatrix}$ is indeed a ring homomorphism, but not a homomorphism of rings with identity. □

The following definition of an *ideal* in a ring has its historical origin in the effort to transfer the concept of a prime decomposition of integers to more general rings in which there is no analogue of the fundamental theorem of arithmetic, and thus to remedy a weak point of a failed attempt to prove Fermat's conjecture (see [Ko97, § 1.3]). For the structural theory of rings, the concept is central because it enables the description of quotient rings in analogy to quotient vector spaces.

Definition 1.10 (Ideal) Let R be a ring. A subset $I \subseteq R$ is called an *ideal* of R, if

$$I - I \subseteq I, \quad RI \subseteq I, \quad IR \subseteq I.$$

We write $I \trianglelefteq R$, if I is an ideal in R.

The condition $I - I \subseteq I$ simply says that I is a subgroup of R with respect to addition.

1.2 Some Structure Theory for Rings

Example 1.11 (Ideals)

(i) Let $\varphi : R \to S$ be a ring homomorphism. Then $\ker(\varphi)$ is an ideal in R, because

$$\varphi(xr) = \varphi(x)\varphi(r) = 0 \cdot \varphi(r) = 0 = \varphi(r) \cdot 0 = \varphi(r)\varphi(x) = \varphi(rx)$$

for $r \in R$ and $x \in \ker(\varphi)$. That $\ker(\varphi)$ is an additive subgroup of R is clear anyway, because φ is in particular a group homomorphism with respect to addition.

(ii) The evaluation of a function at a point x provides a ring homomorphism $\mathrm{ev}_x : C^k(]a, b[) \to \mathbb{R}, f \mapsto f(x)$. More generally, one can also restrict functions to subsets, e.g.,

$$\mathrm{rest}_{]c,d[} : C^k(]a, b[) \to C^k(]c, d[), \quad f \mapsto f|_{]c,d[}$$

for a subinterval $]c, d[\subseteq]a, b[$. The associated kernels consist of the functions that vanish on $\{x\}$ respectively $]c, d[$.

(iii) The ring $R = \mathrm{Mat}(n \times n, \mathbb{K})$ contains no ideals other than $\{0\}$ and R. Indeed, by suitable multiplication from the left and right with matrices that have only one entry different from 0, one sees that every ideal different from $\{0\}$ contains all such matrices and then the whole R.

(iv) Let R be a commutative ring with identity. Then $aR := \{ar \mid r \in R\}$ for every $a \in R$ is an ideal, which is called the *principal ideal* generated by a.

(v) Let R be a commutative ring with identity and $A \subseteq R$. Then $\{\sum_{j=1}^n a_j r_j \mid r_j \in R, a_j \in A, n \in \mathbb{N}\}$ is an ideal, which is called the *ideal generated* by A.

(vi) Let R be a field. Then $\{0\}$ and R are the only ideals in R. If $0 \neq r$ is contained in an ideal $I \triangleleft R$, then for every $s \in R$, $s = 1 \cdot s = r(r^{-1}s) \in I$, so $I = R$. □

In the case of the ring of integers, ideals and subgroups with respect to addition are one and the same objects.

Example 1.12 (Subgroups and Ideals in $\mathbb{Z}$) Every subgroup I of $(\mathbb{Z}, +)$ is of the form $d\mathbb{Z}$ with $d \in \mathbb{N}_0$ and thus an ideal in $\mathbb{Z}$. In particular, every ideal in $\mathbb{Z}$ is of this form. To see this, we first consider the case that $I = \{0\}$. In this case, we choose $d = 0$. Otherwise, we find a non-zero element $n \in I$. Because $I - I \subseteq I$, also $-n \in I$, and we can assume that $n > 0$. Let d be the *smallest* positive number in I and $k \in I$. Dividing with remainder yields $k = md + r$ for a $m \in \mathbb{Z}$ and $0 \leq r < d$. But since $\mathbb{N}I \subseteq I$ and $I = -I$ hold, we have $\mathbb{Z}I \subseteq I$. Thus, $r = k - md \in I$, so due to the minimality of d, the equality $r = 0$ holds, and thus $k \in d\mathbb{Z}$ follows. Since conversely with d also $d\mathbb{Z}$ is in I, $I = d\mathbb{Z}$ holds. □

We now show how to form the quotient ring to an ideal. If one only considers the addition, the construction is a special case of forming the quotient group of an abelian group with respect to a subgroup.

Proposition 1.13 (Quotient Rings) *Let R be a ring and $I \subseteq R$ an ideal.*

(i) *Through $x \sim y$ for $x - y \in I$ an equivalence relation on R is defined, whose equivalence classes are given by*
$$[x] := x + I := \{x + i \mid i \in I\}$$

(ii) *The set R/I of equivalence classes under $\sim$ is a ring with respect to the addition and multiplication on R/I defined for $x, y \in R$ by*
(Add) $[x] + [y] := [x + y]$,
(Mult) $[x] \cdot [y] := [x \cdot y]$
This ring is called the quotient ring *of R modulo I.*

(iii) *The mapping $\varphi \colon R \to R/I, r \mapsto r + I$ is a surjective ring homomorphism with kernel I.*

Proof The assertions about addition are known from the case of quotient groups of abelian groups. This can also be compared with the construction of quotient vector spaces. The conditions $RI \subseteq I$ and $IR \subseteq I$ are used to prove the well-definedness of the multiplication on R/I. The details of the routine verifications that arise are left as an exercise for the reader. □

The following lemma about quotient rings originates from a method known in China as early as the third century for finding a natural number with given remainders when divided by several coprime numbers. We discuss it here because we will need it later for an application to normal forms of linear mappings.

Lemma 1.14 (Chinese Remainder Theorem) *Let R be a ring with identity and I and J ideals in R with $I + J = R$. Then the following holds:*

(i) *The set $R/I \times R/J := \{(r + I, s + J) \mid r, s \in R\}$ is a ring with respect to component-wise addition and multiplication, and the mapping*
$$\bar{\varphi} \colon R/(I \cap J) \to R/I \times R/J, \quad r + I \cap J \mapsto (r + I, r + J)$$

is a ring isomorphism.

(ii) *If R is commutative, then*
$$I \cap J = IJ := \left\{ \sum_{\text{finite}} i_k j_k \,\middle|\, i_k \in I, j_k \in J \right\}.$$

Proof The calculation rules for rings on $R/I \times R/J$ are easily verified by applying the corresponding rules for R/I or R/J in the components (exercise!). It is also easy to verify that
$$\varphi \colon R \to R/I \times R/J, \quad r \mapsto (r + I, r + J)$$

is a ring homomorphism with kernel $I \cap J$. This shows that $\bar{\varphi}$ is a well-defined injective ring homomorphism (exercise! Hint: In construction 2.9(iii) a suitable argument is presented in detail). For (i), it remains only to show that φ is surjective. To this end, we consider $(x_1 + I, x_2 + J) \in R/I \times R/J$. Since by assumption $1 = y_1 + y_2$ for suitable $y_1 \in I$ and $y_2 \in J$, we can write with $z = x_1 y_2 + x_2 y_1$

$$z + I = x_1 y_2 + I = x_1(y_2 + y_1) + I = x_1 + I,$$
$$z + J = x_2 y_1 + J = x_2(y_1 + y_2) + J = x_2 + J,$$

so $\varphi(z) = (x_1 + I, x_2 + J)$. This shows the surjectivity of φ.

Now let R be commutative and $z \in I \cap J$. We again write $1 = y_1 + y_2$ for suitable $y_1 \in I$ and $y_2 \in J$. Then we have

$$z = zy_1 + zy_2 = y_1 z + zy_2 \in IJ.$$

Conversely, $IJ \subseteq I \cap J$ follows immediately from the definition. □

Exercise 1.2 (Chinese Remainder Theorem) Let m and n be coprime natural numbers and $p \in \{0, 1, \ldots, m - 1\}$ and $q \in \{0, 1, \ldots, n - 1\}$. Show using Lemma 1.14 that there is a natural number k that gives the remainder p when divided by m and the remainder q when divided by n.

It is not necessary to limit oneself to two ideals in the Chinese remainder theorem, as the following exercise shows. We use the standard notation $A \cong B$ for the fact that A and B are isomorphic (here as rings).

Exercise 1.3 (Chinese Remainder Theorem) Let R be a commutative ring with identity and $I_1, \ldots, I_n$ ideals in R with $I_i + I_j = R$ for $i \neq j$. Let

$$I_1 \cdots I_n := \left\{ \sum_{\text{fin.}} r_1 \ldots r_n \,\bigg|\, r_i \in I_i \right\}.$$

Then the following holds:

(i) $(I_1 \cdots I_{n-1}) + I_n = R$.
(ii) $R / \left(\bigcap_{i=1}^{n} I_i \right) \cong \left(R / \left(\bigcap_{i=1}^{n-1} I_i \right) \right) \times R/I_n$.
(iii) $R / \left(\bigcap_{i=1}^{n} I_i \right) \cong R/I_1 \times \ldots \times R/I_n$.
(iv) $I_1 \cdots I_n = \bigcap_{i=1}^{n} I_i$.

1.3 Special Classes of Rings

In this section, we consider special classes of rings. Our selection follows two basic criteria: Each of the discussed classes is of fundamental importance, and each discussed result will be used at least once in a crucial way later on.

The question of the solvability of linear differential equations of the form $Df = h$ (see Example 1.7), as well as the question of the solvability of equations of the form $ax = b$ for integer coefficients a and b, motivates a general question: Is there a multiplicative inverse for a given ring element? If so, how can it be found? If one has the multiplicative inverses D^{-1} or a^{-1}, one can solve the equations $f = D^{-1}h$ and $x = a^{-1}b$. In general, however, such multiplicative inverses do not exist, and even if they do, they are difficult to find. The question can be modified: Can one embed a ring into a (skew) field in such a way that the addition and multiplication on the ring are the restrictions of the respective operation on the field? In this case, one would know for each non-zero element of the ring (this property is easy to test) that there is an inverse at least in the (skew) field, and one could find *weak solutions* $f = D^{-1}h$ and $x = a^{-1}b$ for the equations in question. "Weak" would these solutions be in the following sense: one can no longer immediately guarantee that they lie in the desired range. For example, in the equation $2x = b$, it depends on b whether the solution $x = \frac{b}{2}$ is an integer or not. This can be tested, e.g., by the last digit of b in decimal notation.

The approach just described is typical for structural thinking in mathematics: One creates a structural framework in which one can solve a problem, possibly the solution is not automatically a solution to the original problem. Then one develops further tools with which one tests whether the "weak" solution actually also provides a solution to the original problem.

For the integers $\mathbb{Z}$, one indeed finds an embedding of the desired kind, because $\mathbb{Z}$ lies in the field of rational numbers $\mathbb{Q}$. In general, this will not be possible. For non-commutative rings, the answer to the question of embedding is somewhat complicated, so we restrict ourselves here to the commutative case, i.e., to the case of embedding into a field. Then the ring must not have two non-zero elements x, y whose product is zero. Otherwise, according to Proposition 1.4(iii) for the 1 of the field, one would have

$$y = 1 \cdot y = (x^{-1}x) \cdot y = x^{-1} \cdot 0 = 0.$$

Assuming that R is a commutative ring with identity, there are no further restrictions. This leads to the following definition.

Definition 1.15 (Integral Domains) Let R be a commutative ring. An element $r \in R \setminus \{0\}$ is called a *zero divisor* if $0 \in r(R \setminus \{0\}) \cup (R \setminus \{0\})r$. An *integral domain* is a commutative ring R with identity element 1, in which $1 \neq 0$ and there are no zero divisors.

The following example shows in particular that $\mathbb{Z} \subseteq \mathbb{Q}$ is an integral domain.

Example 1.16 (Integral Domains)

(i) Every additive subgroup R of a field $\mathbb{K}$, which is closed under multiplication and contains identity, is an integral domain: If $R \ni r \neq 0$ and $sr = 0$ for $s \in R$, then it follows $0 = 0 \cdot r^{-1} = srr^{-1} = s$. For $rs = 0$ one argues analogously.

1.3 Special Classes of Rings

(ii) The residue class ring $\mathbb{Z}/4\mathbb{Z}$ (see Example 1.3) is not an integral domain, because $[2] \cdot [2] = [0]$ holds. On the other hand, $\mathbb{Z}/2\mathbb{Z}$ is even a field. It can be shown that $\mathbb{Z}/n\mathbb{Z}$ is an integral domain exactly when it is a field, and this is exactly the case when n (or $-n$) is a prime number.

□

In preparation for proving that every integral domain can be embedded in a field in the desired way, we provide a characterization of non-divisibility by a cancellation rule.

Proposition 1.17 (Characterization of Integral Domains) *For a commutative ring $R \neq \{0\}$ with identity, the following statements are equivalent:*

(1) *R is an integral domain.*
(2) *From $ra = rb$ with $r \neq 0$ it follows $a = b$.*

Proof For the implication (1) $\Rightarrow$ (2) one concludes

$$r \neq 0, \; ra = rb \implies r(a-b) = 0 \stackrel{(1)}{\implies} a - b = 0 \implies a = b,$$

and the reversal is seen with

$$r \neq 0, \; ra = 0 \implies ra = r \cdot 0 \stackrel{(2)}{\implies} a = 0.$$

□

The embedding of an integral domain into a field works completely analogous to the embedding of the integers into the rational numbers. One can regard the construction as an abstract variant of fraction calculation. In particular, exactly the same calculation rules apply.

Theorem 1.18 (Field of Fractions) *Let R be an integral domain and $S = R \times (R \setminus \{0\})$.*

(i) *$(a, b) \sim (c, d) :\Leftrightarrow ad = bc$ defines an equivalence relation on S.*
(ii) *Denote the equivalence class of (a, b) with $\frac{a}{b}$ and the set of equivalence classes with $Q(R)$. Then define*

$$\text{(Add)} \quad \frac{a}{b} + \frac{c}{d} := \frac{ad + bc}{bd},$$
$$\text{(Mult)} \quad \frac{a}{b} \cdot \frac{c}{d} := \frac{ac}{bd}$$

an addition and a multiplication on $Q(R)$, which make $(Q(R), +, \cdot)$ a field with zero element $\frac{0}{1}$ and one element $\frac{1}{1}$.

(iii) *The additive inverse of $\frac{a}{b}$ is $\frac{-a}{b}$, and the multiplicative inverse of $\frac{a}{b}$ with $a \neq 0$ is $\frac{b}{a}$.*

Proof Due to the commutativity of R, the relation $\sim$ is symmetric. The reflexivity is obvious. Now let $(a, b) \sim (c, d)$ and $(c, d) \sim (e, f)$. Then

$$adf = bcf = bde$$

and therefore $d(af - be) = 0$. Because $d \neq 0$, the non-existence of zero divisors now shows that $af = be$, i.e., $(a, b) \sim (e, f)$. This shows the transitivity of $\sim$. The argument also shows that

$$\frac{ad}{bd} = \frac{a}{b}$$

for $d \neq 0$. So one can cancel non-zero elements in fractions. The rest of the proof is routine, starting with the proof of the well-definedness of the operations which must be ensured (exercise). □

Note that the mapping $\varphi : R \to Q(R)$, $r \mapsto \frac{r}{1}$ is injective and *structure preserving* in the following sense:

$$\forall x, y \in R: \quad \varphi(x+y) = \frac{x+y}{1} = \frac{x}{1} + \frac{y}{1} = \varphi(x) + \varphi(y),$$

$$\forall x, y \in R: \quad \varphi(xy) = \frac{xy}{1} = \frac{x}{1} \cdot \frac{y}{1} = \varphi(x)\varphi(y),$$

$$\varphi(1) = \frac{1}{1} = 1_{Q(R)},$$

i.e., φ is a homomorphism of rings with identity. This allows us to view R as a subset of $Q(R)$ and the operations on R as restrictions of the respective operations on $Q(R)$.

The field constructed in Theorem 1.18 is called the *field of fractions* of the integral domain R.

Even if a commutative ring with identity is not an integral domain, one can find weakened variants of the construction of a field of fractions. Exercise 1.4 deals with the question: Given a commutative ring R with identity and a suitable subset $S \subseteq R$. How do you find the smallest possible ring R' that contains R and in which the elements from S are units? The name "localization" for the resulting construction will only become clear later, but it actually has to do with the fact that one wants to consider objects only in a certain vicinity of a given point (see e.g., Construction 7.25).

1.3 Special Classes of Rings

Exercise 1.4 (Localization) Let R be a commutative ring with identity and S a subset of R that satisfies:

(a) $1 \in S$.
(b) $a, b \in S$ implies $ab \in S$.

On $R \times S$ we define the relation $\sim$ by

$$(r_1, s_1) \sim (r_2, s_2) \quad :\Leftrightarrow \quad \exists s \in S \text{ with } (r_1 s_2 - r_2 s_1)s = 0.$$

(i) Show that $\sim$ is an equivalence relation that has only one equivalence class if $0 \in S$.
(ii) For $0 \notin S$, let $S^{-1}R$ be the set of equivalence classes of the relation $\sim$ in $R \times S$. The equivalence class of $(r, s) \in R \times S$ is denoted by $\frac{r}{s}$. Show: By

(Add) $\dfrac{r_1}{s_1} + \dfrac{r_2}{s_2} := \dfrac{r_1 s_2 + r_2 s_1}{s_1 s_2},$

(Mult) $\dfrac{r_1}{s_1} \cdot \dfrac{r_2}{s_2} := \dfrac{r_1 r_2}{s_1 s_2}$

for $r_1, r_2 \in R$ and $s_1, s_2 \in S$, operations are defined on $S^{-1}R$ for which $(S^{-1}R, +, \cdot)$ is a commutative ring, which is called the *localization* of R in S.
(iii) Consider for the case $0 \notin S$ the mapping $\varphi : R \to S^{-1}R$, $r \mapsto \frac{r}{1}$ and show: φ is a ring homomorphism, and φ is injective if and only if S contains no zero divisors. The elements $\frac{s_1}{s_2}$ in $S^{-1}R$ with $s_1, s_2 \in S$ are units.

The following definitions of special types of ideals are derived from the example of integers. In particular, an ideal $d\mathbb{Z} \trianglelefteq \mathbb{Z}$ (see example 1.12) will be a prime ideal if and only if d is zero or a prime number. $d\mathbb{Z} \trianglelefteq \mathbb{Z}$ is maximal if and only if d is prime (see Remark 1.32).

Definition 1.19 (Maximal and Prime Ideals) An ideal I in a commutative ring R with identity is called *prime*, if $1 \notin I$ and from $xy \in I$ with $x, y \in R$ it follows: $x \in I$ or $y \in I$. The ideal I is called *maximal*, if $1 \notin I$ and for every $r \in R \setminus I$ there exists an $s \in R$ with $1 \in sr + I$.

One can immediately see from the definition that the maximal ideals are precisely those ideals I that are not contained in any ideal different from R and I (an ideal I is equal to R exactly when $1 \in I$).

The following proposition shows that the property of an ideal to be maximal or prime translates into already known properties of the associated quotient ring. This additional perspective on the properties opens up new possibilities for investigation.

Proposition 1.20 (Characterization of Maximal and Prime Ideals) *Let R be a commutative ring with one and $I \subseteq R$ an ideal.*

(i) *R/I is an integral domain if and only if I is prime.*
(ii) *R/I is a field if and only if I is maximal.*
(iii) *If I is maximal, then I is prime.*

Proof

(i) Let R/I be an integral domain. If $xy \in I$ for $x, y \in R$, then
$$(x + I)(y + I) = xy + I = 0 + I,$$
so $x + I = 0 + I$ or $y + I = 0 + I$, i.e., $x \in I$ or $y \in I$. Thus, I is prime.

Conversely, assume that I is prime. If now $(x+I)(y+I) = 0+I$, then this means $xy \in I$, so $x \in I$ or $y \in I$. This implies $x + I = 0+I$ or $y+I = 0+I$, and R/I is free of zero divisors.

(ii) Let R/I be a field. If $r \in R \setminus I$, then $r + I \in (R/I) \setminus \{0 + I\}$, so there exists an $s + I \in R/I$ with
$$(s + I)(r + I) = 1 + I.$$
But this immediately implies $1 \in sr + I$, i.e., I is maximal.

Conversely, assume that I is maximal and $r + I \in (R/I) \setminus \{0 + I\}$. Then $r \in R \setminus I$, and there is an $s \in R$ with $sr \in 1+I$. But this means $(s+I)(r+I) = 1 + I$, so $r + I$ has a multiplicative inverse in R/I. Thus, R/I is a field.

(iii) This follows immediately from (i) and (ii), because every field is an integral domain.

□

Next, we describe an analogue to the Euclidean algorithm for polynomial rings. The existence of such an analogue then motivates the definition of a Euclidean ring. This approach is typical in mathematics. When there are multiple types of examples that all share a certain property E, one abstracts from the examples and considers all objects with property E. In a next step, one then explores which further properties of the examples can be derived from property E and therefore apply to all objects with property E.

Proposition 1.21 (Polynomial Division) *Let R be a commutative ring with identity and $f, g \in R[X] \setminus \{0\}$ (see Example 1.6).*

(i) *If the highest (non-zero) coefficients of f and g do not multiply to zero, then*
$$\deg(fg) = \deg(f) + \deg(g).$$

(ii) *If the highest coefficient of g is a unit, then there exist uniquely determined polynomials $q, r \in R[X]$ with $f = qg + r$, where either $r = 0$ or $\deg(r) < \deg(g)$.*

Proof Let $f = \sum_{i=0}^{m} a_i X^i$ and $g = \sum_{i=0}^{n} b_i X^i$ with $a_m \neq 0 \neq b_n$. Then $\deg(f) = m$ and $\deg(g) = n$.

1.3 Special Classes of Rings

(i) $fg = a_m b_n X^{m+n} + \sum_{i=0}^{n+m-1} \left(\sum_{l+m=i} a_l b_m \right) X^i$. But since by assumption $a_m b_n \neq 0$, it follows that $\deg(fg) = m+n$.

(ii) Existence of q and r: If $m < n$, then choose $q = 0$ and $r = f$. We can therefore assume $n \leq m$, so that $f = (a_m b_n^{-1} X^{m-n}) g + \tilde{f}$, where either $\tilde{f} = 0$ or $\deg(\tilde{f}) < m$. If $\tilde{f} = 0$, then we choose $r = 0$ and $q = a_m b_n^{-1} X^{m-n}$. Otherwise, by induction over the degree, we find elements $\tilde{q}, \tilde{r} \in R[X]$ as stated in the theorem, in particular with $\tilde{f} = \tilde{q}g + \tilde{r}$. It then holds

$$f = (a_m b_n^{-1} X^{m-n} + \tilde{q}) g + \tilde{r},$$

which proves the existence of q and r.

For uniqueness, we assume two decompositions $f = qg + r = \tilde{q}g + \tilde{r}$ as stated. Then $(\tilde{q} - q)g = r - \tilde{r}$ and with (i)

$$\deg\bigl((\tilde{q}-q)g\bigr) = \deg(q - \tilde{q}) + \deg(g) > \deg(r - \tilde{r}),$$

if $\tilde{q} \neq q$. Thus, $q = \tilde{q}$ and then also $r = \tilde{r}$.

□

The second part of the following definition is the announced abstraction of the Euclidean algorithm. The first part is also an abstraction of a property of $\mathbb{Z}$, namely the description of the ideals in Example 1.12.

Definition 1.22 (Principal Ideal and Euclidean Rings) Let R be an integral domain.

(i) R is called a *principal ideal domain*, if every ideal I in R is of the form $I = xR$ with $x \in R$.
(ii) R is called a *Euclidean ring*, if there is a function $d \colon R \setminus \{0\} \to \mathbb{N}_0$ that has the following properties:
   (a) $d(ab) \geq d(a)$ for all $a, b \in R \setminus \{0\}$.
   (b) If $a \in R \setminus \{0\}$ and $b \in R$, then there are elements $q, r \in R$ with

$$b = qa + r, \quad \text{where } r = 0 \text{ or } d(r) < d(a) \quad \text{(division with remainder)}.$$

The function d with properties (a) and (b) is called a *degree function* for R.

Using the Euclidean algorithm and Proposition 1.21, we see that $\mathbb{Z}$ and polynomial rings over fields are indeed Euclidean rings.

Example 1.23 (Euclidean Rings)

(i) $\mathbb{Z}$ is a Euclidean ring with $d(n) = |n|$.

(ii) Let $\mathbb{K}$ be a field. Then $\mathbb{K}[X]$ is a Euclidean ring with degree function deg. This follows immediately from Proposition 1.21, because in a field every non-zero element is a unit. □

We illustrate the above-described procedure of deriving a property from an abstracted property E: One of the standard applications of the Euclidean algorithm is the calculation of the greatest common divisor (gcd) of two integers. If we define the concept of divisor appropriately for commutative rings, we can also algorithmically determine the gcd in Euclidean rings. However, it must be noted that the number of examples of Euclidean rings beyond those just described is rather small. This explains why the concept does not play an important role in modern algebra.

Definition 1.24 (Divisors and Prime Elements) Let R be a commutative ring and $a, b \in R$. We say *a divides b* and write $a \mid b$, if there is an $r \in R$ with $ra = b$. In this case, a is also called a *divisor* of b in R. A *greatest common divisor* (gcd) of $a_1, \ldots, a_k \in R$ is then a common divisor of the a_j, which is divided by every other common divisor. If R has an identity, two elements a and b in R are called *coprime*, if 1 is a gcd of a and b.

In generalization of the concept of a prime number, an element $d \neq 0$ in $R \setminus \text{Unit}(R)$ is called *prime*, if from $d \mid ab$ for $a, b \in R$ it follows: $d \mid a$ or $d \mid b$. Two elements $p, q \in R$ are called *associated*, if there is a unit $u \in \text{Unit}(R)$ with $p = uq$.

Prime elements in commutative rings do not play nearly the role that prime numbers play for number theory. Far more important are the prime ideals. Since prime principal ideals in integral domains are generated by prime elements (see Proposition 1.29), there is a connection.

Note that in this definition every unit is a divisor of any ring element. In particular, in $\mathbb{Z}$ the gcd is not unique, but only determined up to the sign. More generally, in an integral domain the gcd is only determined up to units (exercise; see Proposition 1.17).

Exercise 1.5 (gcd in $\mathbb{Z}$) Considering the definition of a gcd for $R = \mathbb{Z}$ given in Definition 1.24, it is noticeable that in middle school for two natural numbers the *greatest* common divisor is usually really determined with respect to the order relation. Show that this leads to the same gcd as with Definition 1.24.

Proposition 1.25 (gcd in Principal Ideal Domains)

(i) *Every Euclidean ring is a principal ideal domain.*
(ii) *Let R be a principal ideal domain. Then two elements $a, b \in R$ have a greatest common divisor (gcd) i.e., unique up to multiplication by a unit, and is contained in the set $\{ma + nb \mid n, m \in R\}$.*

1.3 Special Classes of Rings

Proof

(i) Let I be an ideal in R. If $I = \{0\}$, then $I = 0 \cdot R$. Otherwise, we choose an element $d \in I$ with minimal $d(d)$. For each $i \in I$, we then find $q, r \in R$ with $i = qd + r$ and $r = 0$ or $d(r) < d(d)$. But since $r = i - qd \in I$, the second case cannot occur, so $i = dq \in dR$. Conversely, with d, dR is also in I, so $I = dR$.

(ii) Set
$$\langle a, b \rangle := \{na + mb \mid n, m \in R\}$$

for $a, b \in R$. Then one immediately checks that $\langle a, b \rangle$ is an ideal in R, so by assumption it is of the form dR. Then d is a common divisor of a and b. But since $d \in \langle a, b \rangle$, there are $n, m \in R$ with $d = na + mb$. Therefore, every common divisor of a and b is also a divisor of d. So d is a gcd of a and b.

To show the uniqueness statement, we assume that d and d' are each a gcd of a and b. Then there are $r, r' \in R$ with $dr = d'$ and $d = d'r'$, so $d = drr'$. Due to Proposition 1.17, this gives $1 = rr'$ (R is in particular an integral domain), i.e., r and r' are units in R. □

Note that from the existence of a gcd of two elements, one can immediately conclude the existence of a gcd of finitely many elements.

Exercise 1.6 (Euclidean Algorithm) Formulate a Euclidean algorithm for Euclidean rings and show that the result of this algorithm is the gcd of two given ring elements.

Using the Euclidean algorithm, one can derive the fundamental theorem of arithmetic, which states that every natural number can be written in a unique way as a product of prime numbers, up to the order. This insight can also be transferred to any Euclidean ring, if one has a suitable definition of prime factor decomposition.

Definition 1.26 (Factorial Ring) Let R be an integral domain. Then R is called a *factorial ring*, if every non-unit $0 \neq r \in R \setminus \text{Unit}(R)$ can be written as a product of prime elements.

We will use the following lemma to prove that Euclidean rings are factorial.

Lemma 1.27 *Let R be a Euclidean ring with degree function d. Let $a, b \in R \setminus \{0\}$. If $b \mid a$, but not $a \mid b$, then $d(b) < d(a)$.*

Proof The assumptions show: On the one hand, $a = qb + r$ with $r \neq 0$ and $d(r) < d(a)$, on the other hand, we have $a = cb$. Therefore, we calculate

$$r = a - qb = (q - c)b$$

and find $d(a) > d(r) \geq d(b)$. □

Theorem 1.28 (Euclidean Implies Factorial) *Let R be a Euclidean ring. Then R is factorial.*

Proof Let $0 \neq r \in R$. We want to show that r is either a unit or can be written as a product $r = p_1 \cdots p_k$ of prime elements $p_1, \ldots, p_k \in R$. For this, we perform an induction over the degree $d(r)$ of r, i.e., we assume that every element of smaller degree is either a unit or can be written as a product of prime elements.

If r is a unit or a prime element, there is nothing more to show. Therefore, we can assume that $0 \neq r \in R \setminus \text{Unit}(R)$ is not a prime element. Then there are $a, b \in R$ with $r|ab$, but not $r|a$ or $r|b$.

One can choose a and b such that $d(a), d(b) < d(r)$. To see this, we divide a and b by r with remainder and find

$$a = cr + a', \quad b = dr + b'$$

with $a' \neq 0 \neq b'$ and $d(a'), d(b') < d(r)$. Also, $r|a'b'$, but not $r|a'$ or $r|b'$. In other words, a' and b' have the desired properties.

We now choose a and b with the mentioned properties such that $d(a)$ is minimal. Because a and b are not units (otherwise $r|b$ or $r|a$ would hold), it follows from $d(a), d(b) < d(r)$ that a and b can be written as products of prime elements:

$$a = p_1 \cdots p_k, \quad b = q_1 \cdots q_l.$$

Now write $ab = rs$ with $s \in R$. Because p_1 is prime, it follows $p_1|r$ or $p_1|s$. We show that the case $p_1|s$ cannot occur: For this, write $a = p_1 a'$ and $s = p_1 s'$. Then $rp_1 s' = rs = ab = p_1 a' b$, so $rs' = a'b$. Note that a cannot be a divisor of a', because $ar' = a'$ would result in the equation $p_1 r' a' = a'$, thus $p_1 r' = 1$. Then p_1 would be a unit, contradicting the assumption that p_1 is prime. So we have $a'|a$ and $a \nmid a'$, so Lemma 1.27 gives the inequality $d(a') < d(a)$. But since $r|a'b$ and $r \nmid b$ and $r \nmid a'$ (otherwise $r|a$ would hold), this is a contradiction to the minimality of $d(a)$.

So we have now shown that $p_1|r$, i.e., there is an $r_1 \in R$ with $r = p_1 r_1$. Then $r|r_1$, and as before, one sees that $r_1 \nmid r$, because p_1 is not a unit. So Lemma 1.27 this time gives that $d(r) > d(r_1)$. Therefore r_1 is either a unit or the product of prime elements. But then $r = p_1 r_1$ is in any case a product of prime elements. □

The concepts of divisors and prime elements introduced in the discussion of Euclidean rings allow us to show the announced connection between prime numbers and prime ideals or maximal ideals in greater generality than just for $\mathbb{Z}$.

Proposition 1.29 (Prime Elements and Prime Ideals) *Let R be an integral domain.*

(i) *If a prime element $p \in R$ is of the form $p = ab$ with $a, b \in R$ and $p \mid a$, then b is a unit.*

1.3 Special Classes of Rings

(ii) *If $p, q \in R$ are prime with $p \mid q$, then q is of the form up with $u \in \text{Unit}(R)$.*
(iii) *$0 \neq d \in R$ is prime if and only if dR is a prime ideal.*

Proof

(i) From $p = ab$ and $a = pc$ it follows $p = pcb$, i.e., $1 = cb$.
(ii) From $pr = q$ it follows $q \mid p$, because $q \mid r$ would lead to p being a unit according to (i). But (i) also shows that because $q \mid p$ the element r is a unit.
(iii) If p is prime and $ab \in pR$ holds, it follows that $p \mid ab$ and therefore $p \mid a$ or $p \mid b$. However, this means that $a \in pR$ or $b \in pR$, so pR is prime. Conversely, if pR is prime and $p \mid ab$ holds, then it follows that $ab \in pR$, so $a \in pR$ or $b \in pR$, which means $p \mid a$ or $p \mid b$. Therefore, p is prime. □

Remark 1.30 (Prime Elements and Prime Ideals) Note that in the proof of Proposition 1.29(iii) the absence of zero divisors was not used. This means that for commutative rings with identity, a non-zero element is prime if and only if the corresponding principal ideal is a prime ideal.

Proposition 1.31 (Maximality of Prime Ideals) *Let R be a principal ideal domain. Then every prime ideal $I \neq \{0\}$ is maximal.*

Proof Let $I \neq \{0\}$ be prime and $J \trianglelefteq R$ an ideal that contains I. Since R is a principal ideal domain, there are $a, b \in R$ with $I = (a)$ and $J = (b)$. According to Proposition 1.29(iii), $a \in R$ is prime. Because $I \subset J$, $a = br$ holds with $r \in R$. Now Proposition 1.29(i) shows that r or b is a unit. In the first case, $I = J$ holds and in the second $J = R$. This proves the maximality of I. □

Remark 1.32 (Prime Numbers and Prime Ideals) If one combines Propositions 1.29 and 1.31, it follows for $R = \mathbb{Z}$, that for every prime number p the ideal $p\mathbb{Z}$ is prime and thus maximal. Conversely, if $\{0\} \neq d\mathbb{Z}$ is maximal, then $d\mathbb{Z}$ is prime according to Proposition 1.20(iii), so d is a prime element according to Proposition 1.29.

Exercise 1.7 (Prime Ideals) Let $\varphi R \to R'$ be a homomorphism between two commutative rings with identity that preserves the identity, and $I' \trianglelefteq R'$ a prime ideal. Show that $\varphi^{-1}(I')$ is a prime ideal in R.

Exercise 1.8 (Irreducible Elements of an Integral Domain) In an integral domain R, an element $r \in R \setminus \text{Unit}(R)$ is called *irreducible*, if r cannot be decomposed into the product of two non-units, i.e., $r = ab$ with $a, b \in R$ implies $a \in \text{Unit}(R)$ or $b \in \text{Unit}(R)$. Show:

(i) If $p \in R$ is prime, then p is irreducible.
(ii) If R is factorial, then the converse of (i) also holds.

Exercise 1.9 (Irreducible Elements in a Principal Ideal Domain) Let R be a principal ideal domain and $0 \neq r \in R$. Show that the following statements are equivalent:

(1) r is irreducible.
(2) (r) is a maximal ideal.
(3) The congruence $ax \equiv b \mod r$, i.e., $ax - b \in rR$, has a solution for all b in R and for all a in R that are not multiples of r.
(4) r is prime.

Exercise 1.10 (Principal Ideal Domains are Factorial) Let R be a principal ideal domain. Show:

(i) For any two elements $a, b \in R$ there exists a gcd $d \in R$ (see Definition 1.24).
(ii) Every non-empty set of principal ideals in R has a maximal element.
(iii) R is factorial.

Literature: The material presented in this chapter is covered in all books on algebra, whether they are designed as elementary introductions like [Ar23, Kn06], as concise presentations of central results like [Ke95], or as reference works like [La93].

Modules 2

The concept of a module is simultaneously a generalization of the concepts of "abelian group", "ring", and "vector space". The module structure is so broad that it can form the framework for the relevant functions of almost all central areas of mathematics. On the other hand, the module structure is also so rich that it allows a significant theory that has extremely interesting applications. A key goal of this chapter is to provide convincing evidence for these two claims.

We assume that the reader is familiar with the concept of an abelian group. Modules, like vector spaces, are abelian groups with an additional operation. In the case of vector spaces, this operation consists of being able to multiply its elements with the elements of a field. This scalar multiplication then fulfills various rules such as two distributive laws. For modules, more general scalars are allowed. These no longer need to be elements of a field, but only elements of a ring. Thus, in the case of modules, several characteristics of a field are dispensed with for the scalars, such as the invertibility of the non-zero elements, the existence of an identity element, and also the commutativity of multiplication. Non-commutative rings are rarely dealt with in first year courses, but they, like their modules, play a very significant role, e.g., in the study of differential equations and in functional analysis.

2.1 Structure Theory of Modules

In this section, we present a prototypical algebraic structure theory. We start with the definition of a module, give a series of examples, and introduce homomorphisms, i.e., structure-preserving mappings, between modules. With this, we can then compare modules over a fixed ring and encounter submodules and quotient modules. After that, we study the submodules generated by subsets—a concept that is found in all algebraic structures with appropriate variations. Somewhat more specific is the concept of a free module, but it also has a correspondence in a number of algebraic structures. Finally, we study the possibility of composing modules from smaller modules—an objective that is considered for all algebraic structures.

Elementary Definitions and Examples

The following definition of a module generalizes in an obvious way both the definition of a ring and that of a vector space. Less obvious is that every abelian group is also a module.

Definition 2.1 (Modules) Let R be a ring. A *left-R-module* (or simply R-module) M is an abelian group with a mapping

$$R \times M \to M, \ (r, m) \mapsto rm,$$

that satisfies the following conditions:

(i) $\forall r_1, r_2 \in R, \ \forall m \in M: \quad (r_1 r_2)m = r_1(r_2 m)$.
(ii) $\forall r_1, r_2 \in R, \ \forall m \in M: \quad (r_1 + r_2)m = r_1 m + r_2 m$.
(iii) $\forall r \in R, \ \forall m_1, m_2 \in M: \quad r(m_1 + m_2) = rm_1 + rm_2$.
(iv) If R has an identity element $1 \in R$, then $1m = m$ for all $m \in M$.

Our first set of examples is of a rather abstract nature. We start with rings or vector spaces and naturally find associated modules.

Example 2.2 (Modules)

(i) Let R be a ring, then the ring multiplication makes R an R-module.
(i') Let R be a ring and $I \subseteq R$ an additive subgroup. If the ring multiplication can be restricted to a mapping $R \times I \to I$, i.e., if

$$\forall r \in R, \ \forall x \in I: \quad rx \in I,$$

then this mapping makes $(I, +)$ into an R-module. I is then called a *left ideal* in R.
(i'') Let R be a ring and I a left ideal in R. Then the set $R/I = \{r + I \mid r \in R\}$ of cosets of I with respect to addition
(Add) $(r + I) + (r' + I) = (r + r') + I$
is an abelian group, the *quotient group*. With respect to multiplication
(Mult) $r(s + I) := rs + I$
$(R/I, +)$ then becomes an R-module. The R-modules of this form are called *cyclic*.
(ii) Let $\mathbb{K}$ be a field and V a $\mathbb{K}$-vector space. Then scalar multiplication $\mathbb{K} \times V \to V$ makes the abelian group $(V, +)$ into a $\mathbb{K}$-module.

2.1 Structure Theory of Modules

(iii) Let $(M, +)$ be an abelian group. Then the mapping given by

$$na := \begin{cases} \underbrace{a + \ldots + a}_{n-\text{times}} & n \in \mathbb{N} \\ 0 & n = 0 \\ -\underbrace{(a + \ldots + a)}_{(-n)-\text{times}} & -n \in \mathbb{N} \end{cases}$$

makes the group $(M, +)$ into a $\mathbb{Z}$-module.

(iv) Let V be a $\mathbb{K}$-vector space. Then V is an $\mathrm{End}(V)$-module with respect to $\varphi v := \varphi(v)$.

$\square$

The next example is of a much more specific nature, but already suggests that module structures can also be of interest for the treatment of differential equations.

Example 2.3 (Vector Fields) Let $U \subseteq \mathbb{R}^k$ be an open subset and $C^\infty(U, \mathbb{R}^m)$ the $\mathbb{R}$-vector space of all smooth mappings from U to $\mathbb{R}^m$. Then $C^\infty(U, \mathbb{R})$ is a ring with respect to pointwise addition and multiplication and $C^\infty(U, \mathbb{R}^m)$ is a $C^\infty(U, \mathbb{R})$-module with respect to pointwise addition and scalar multiplication (see Example 1.2). Again, we describe these two operations in more detail: For $f, g \in C^\infty(U, \mathbb{R}^m), s \in C^\infty(U, \mathbb{R})$ and $x \in U$ we have:

(Add) $(f + g)(x) = f(x) + g(x)$, where the addition on the right side is that of $\mathbb{R}^m$.
(Mult) $(sf)(x) = s(x) \cdot f(x)$, where the multiplication on the right side is the scalar multiplication $\mathbb{R} \times \mathbb{R}^m \to \mathbb{R}^m$.

In the case that $k = m$, the elements of $C^\infty(U, \mathbb{R}^m)$ can be interpreted as *vector fields*: The function value $f(x) \in \mathbb{R}^k$ at the point $x \in \mathbb{R}^k$ is a vector, which is thought of as being "attached" at the point x. Such vector fields define ordinary differential equations: One seeks differentiable *solution curves* $\gamma : I \to \mathbb{R}^k$, where $I \subseteq \mathbb{R}$ is as large an interval as possible. γ is supposed to solve the differential equation

$$\gamma'(t) = f(\gamma(t)).$$

Note that (Mult) already defines a second module structure on $C^\infty(U, \mathbb{R}^m)$. After all, $C^\infty(U, \mathbb{R}^m)$ is also a real vector space, i.e., an $\mathbb{R}$-module. If one interprets a real number as a constant function on U, one finds that the scalar multiplication $\mathbb{R} \times C^\infty(U, \mathbb{R}^m) \to C^\infty(U, \mathbb{R}^m)$ is a restriction of the scalar multiplication $C^\infty(U, \mathbb{R}) \times C^\infty(U, \mathbb{R}^m) \to C^\infty(U, \mathbb{R}^m)$. Thus, the $C^\infty(U, \mathbb{R})$-module structure is a refinement of the vector space structure.

$\square$

The rings of differential operators from Example 1.7 and Exercise 1.1 suggest the following example. It is a simple prototype for a so-called *D*-module (see [Co95]).

Example 2.4 (*D*-Modules) The abelian group $C^\infty(U, \mathbb{R})$ is a $\mathcal{D}(U)$-module with respect to the mapping defined by

$$\forall D \in \mathcal{D}(U), f \in C^\infty(U, \mathbb{R}): \quad Df := D(f)$$

$\mathcal{D}(U) \times C^\infty(U, \mathbb{R}) \to C^\infty(U, \mathbb{R})$, $(D, f) \mapsto Df$. The same equation also defines a Weyl algebra module structure on the space $\mathbb{R}[x_1, \ldots, x_k]$ of polynomial functions on $\mathbb{R}^n$. This algebraic variant of the *D*-module can be formed for any field of characteristic zero, if one considers the formal derivatives of the polynomials given by the same formulas instead of the derivative of polynomial functions. □

The following example is the key to an extremely elegant treatment of various normal form problems from linear algebra (see Section 2.2).

Example 2.5 (Vector Space Endomorphisms) Let V be a $\mathbb{K}$-vector space and $\varphi \in \text{End}_\mathbb{K}(V) = \text{Hom}_\mathbb{K}(V, V)$. Then V is a $\mathbb{K}[X]$-module (see Example 1.6) via

$$\left(\sum a_j X^j\right) v := \sum a_j \varphi^j(v).$$

□

Analogous to Definition 2.1, one defines *right-R-modules* via a scalar multiplication $M \times R \to M$, $(m, r) \mapsto mr$. The associative law then takes the form $m(r_1 r_2) = (m r_1) r_2$, and the distributive laws are written as $m(r_1 + r_2) = m r_1 + m r_2$ and $(m_1 + m_2) r = m_1 r + m_2 r$. One sees that for commutative rings, every left module becomes a right module if one simply moves the ring element to the other side. For commutative rings, the abbreviation from left modules to modules is therefore harmless. If the ring is non-commutative, one should explicitly note when considering right modules.

Exercise 2.1 (Right Modules) Find examples of right modules for non-commutative rings.

Exercise 2.2 (Adjoining a Ring Identity)

(i) Show that every ring R can be embedded ring-homomorphically into a ring with an identity. *Hint:* Define component-wise addition on $\widetilde{R} := R \times \mathbb{Z}$ and define a multiplication

$$(r, n)(r', n') := (rr' + r \cdot n' + n \cdot r', nn'),$$

where $n \cdot r := r \cdot n := \text{sign}(n)(r + \ldots + r)$ is the $|n|$-fold sum of $\text{sign}(n)r$. Then one can easily verify that $\widetilde{R}$ with these operations is a ring, in which $(0, 1) \in \widetilde{R}$ is the unit. Furthermore, one sees that the embedding $R \to \widetilde{R}, r \mapsto (r, 0)$ preserves addition and multiplication.

(ii) Show: If M is an R-module, then one can define a $\widetilde{R}$-module structure on M by $(r, n) \cdot x := r \cdot x + n \cdot x$ for $(r, n) \in \widetilde{R}$ and $x \in M$. Here, $n \cdot x$ is again the $|n|$-fold sum of $\text{sign}(n)x$.

2.1 Structure Theory of Modules

Homomorphisms, Submodules, and Quotient Modules

We begin our systematic structure theory of modules with the definition of the appropriate structure-preserving mappings.

Definition 2.6 (Module Homomorphisms) Let R be a ring and M, N left-R-modules. A mapping $\varphi\colon M \to N$ is called an *R-module homomorphism*, if for all $r_1, r_2 \in R$ and $m_1, m_2 \in M$

$$\varphi(r_1 m_1 + r_2 m_2) = r_1 \varphi(m_1) + r_2 \varphi(m_2).$$

If φ is bijective, then φ is called an *R-module isomorphism* or simply *isomorphism*. The set of R-module homomorphisms $M \to N$ is denoted by $\operatorname{Hom}_R(M, N)$.

It is easy to verify (exercise!) that the inverse of a bijective module homomorphism is itself a module homomorphism. This fact justifies the name *isomorphism*. If the inverse mapping were not automatically structure-preserving, one would only call a bijective structure-preserving mapping an isomorphism if its inverse is also structure-preserving.

The size comparison between two modules through a homomorphism is particularly simple when one module is a subset of the other module. In this case, one would like to have the inclusion mapping $\iota\colon M \to N$ as a homomorphism. This only works if

$$\forall m_1, m_2 \in M, r_1, r_2 \in R: \quad r_1 m_1 + r_2 m_2 \in M$$

holds. In this case, M is called a *submodule* of N.

Exercise 2.3 (Submodules) Let R be a ring, N a left-R-module and $M \subseteq N$ a subset. Show that M is a submodule of N if and only if $M - M \subseteq M$ and $RM \subseteq M$ hold.

Analogous to the concept of the submodule just introduced, one could also have formed the concept of the *subring* in Ch. 1. According to Definition 1.8, a subring S of a ring R is a subset for which

$$\forall x, y \in S: \quad x + y \in S \text{ and } xy \in S$$

hold, as well as $1 \in S$, if R is a ring with identity. The fact that we did not do so has two reasons. First, we did not even attempt to develop a systematic structure theory of rings. Second, for historical reasons, the phraseology for the fact that S is a subring of R is rather "R is a *ring extension* of S". The same language convention exists for fields, where one normally does not speak of $\mathbb{R}$ as a subfield of $\mathbb{C}$, but of $\mathbb{C}$ as a field extension of $\mathbb{R}$.

From the definitions, it is immediately apparent that the terms *submodule* and *module homomorphism* reduce to *subspace* and *linear mapping* when R is a field.

This immediately provides a large class of examples for submodules. One could even say that module theory is linear algebra over rings.

Another class of examples is obtained from Example 2.2(iii): Every subgroup M of an abelian group N is, considered as a $\mathbb{Z}$-module, a submodule of the $\mathbb{Z}$-module N.

Example 2.7 (Module Homomorphisms and Submodules)

(i) $C^k(\mathbb{R})$ is a $C^\infty(\mathbb{R})$-submodule of $C(\mathbb{R})$.
(ii) Let R be a ring and $R^n := \{(r_1, \ldots, r_n) \mid r_j \in R\}$. Then R^n is an R-module with respect to

$$r(r_1, \ldots, r_n) := (rr_1, \ldots, rr_n)$$

and $\{(r_1, \ldots, r_k, 0, \ldots, 0) \mid r_j \in R\}$ is a submodule of R^n.
(iii) If we consider R as a left R-module as in Example 2.2(i), then $\rho_r : R \to R$, $s \mapsto s \cdot r$ is a module homomorphism for any $r \in R$.
(iv) The derivative

$$D : C^\infty(\mathbb{R}) \to C^\infty(\mathbb{R}), \quad f \mapsto \frac{df}{dt}$$

is an $\mathbb{R}$-module homomorphism, but not a $C^\infty(\mathbb{R})$-module homomorphism.
(v) $I := \left\{ \begin{pmatrix} a & 0 \\ c & 0 \end{pmatrix} \,\middle|\, a, c \in R \right\}$ is a left ideal in $\mathrm{Mat}(2 \times 2, R)$ in the sense of Example 2.2(i'), but not a *right ideal*, i.e., it does not hold that

$$\forall r \in \mathrm{Mat}(2 \times 2, R), \forall x \in I : \quad xr \in I.$$

(vi) We generalize the example from (v): Let R be a ring. If we consider R as a left or right R-module and $I \subseteq R$ is a submodule, then I is called a *left* or *right ideal*. I is exactly an ideal of R if it is both a left and a right ideal. □

The submodules that we have found as images of inclusion mappings are special cases of a very general phenomenon: images of module homomorphisms are always submodules of the range. Conversely, pre-images of submodules under module homomorphisms are always submodules.

Example 2.8 (Module Homomorphisms and Submodules) Let $\varphi : M \to N$ be a module homomorphism and $M' \subseteq M$ and $N' \subseteq N$ submodules. Then $\varphi^{-1}(N')$ is a submodule of M and $\varphi(M')$ is a submodule of N. In particular, the *kernel* $\ker(\varphi) := \varphi^{-1}(0) = \varphi^{-1}(\{0\})$ is a submodule of M and the *image* $\mathrm{im}(\varphi) := \varphi(M)$ is a submodule of N. □

2.1 Structure Theory of Modules

The composition of module homomorphisms is a module homomorphism. Since the composition of mappings is associative, in particular the *module endomorphisms* $\varphi \colon M \to M$ of a fixed module M form a semigroup. The set $\operatorname{Hom}_R(M, M)$ of such endomorphisms is also denoted by $\operatorname{End}_R(M)$.

Exercise 2.4 (Automorphism Group) Let M be an R-module. Show:

$$\operatorname{Aut}_R(M) := \{\varphi \in \operatorname{Hom}_R(M, M) \mid \varphi \text{ is bijective}\}$$

is a group with the composition of mappings as multiplication.

Exercise 2.5 (Module Structures on $\operatorname{Hom}_R(M, N)$) Let R be a commutative ring with identity and M, N two R-modules. Show that by

$$\forall r \in R, \varphi \in \operatorname{Hom}_R(M, N), m \in M: \quad (r \cdot \varphi)(m) := r\bigl(\varphi(m)\bigr) = \varphi(r \cdot m)$$

a R-module structure is defined on $\operatorname{Hom}_R(M, N)$.

Just as with the construction of quotient vector spaces and quotient rings (see Proposition 1.13), quotient modules can also be found.

Construction 2.9 (Quotient Modules) Let M be an R-module and N a submodule of M. Furthermore, let

$$M/N := \{m + N \mid m \in M\}$$

be the set of all additive N-cosets.

(i) M/N is an R-module via

$$r(m + N) := rm + N, \quad (m_1 + N) + (m_2 + N) := (m_1 + m_2) + N$$

for $r \in R$ and $m, m_1, m_2 \in M$. The well-definedness of the two operations follows from the submodule properties $RN \subseteq N$ and $N + N \subseteq N$, the calculation rules are then immediate consequences of the corresponding calculation rules for M. The R-module M/N is called the *quotient module* or *factor module* of M over N.

(ii) The mapping $\pi \colon M \to M/N$, $m \mapsto m + N$ is a surjective R-module homomorphism with kernel N. The surjectivity and homomorphism property follow immediately from the definition. The fact that the kernel is equal to N is due to the fact that the coset $N = 0 + N$ is the zero of M/N and $m + N = N$ holds exactly when m is in $N - N = N$.

(iii) Let $\varphi \colon M \to L$ be an R-module homomorphism with kernel N. Then $\operatorname{im}(\varphi) = \varphi(M)$ is isomorphic to M/N. For this, consider the mapping $\bar{\varphi} \colon M/N \to \operatorname{im}(\varphi)$, which is defined by $\bar{\varphi}(m + N) := \varphi(m)$. It is well-defined because with $m + N = m' + N$ also $m - m' \in N$ holds, so $\varphi(m) - \varphi(m') = \varphi(m - m') = 0$. It is then easy to verify that $\bar{\varphi}$ is a module homomorphism. The surjectivity of $\bar{\varphi}$ is an immediate consequence of the definition. As in (ii), one sees that $\bar{\varphi}(m + N) = \varphi(m) = 0$ holds exactly when $m \in N$, i.e., when $m + N$ is the zero in M/N. Thus, $\ker(\bar{\varphi}) = \{0\}$, and as in the case of linear mappings, this implies the injectivity of $\bar{\varphi}$:

$$\bar{\varphi}(m + N) = \bar{\varphi}(m' + N) \Rightarrow \bar{\varphi}(m - m' + N) = 0 \Rightarrow m - m' \in N$$
$$\Rightarrow m + N = m' + N.$$

Together, this shows that $\bar{\varphi}$ is a module isomorphism. This statement is also called the *first isomorphism theorem for modules*.

□

The parallelism of the definitions of subring and quotient ring as well as of submodule and quotient module is as striking as the similarities to the concepts of subspace and quotient space known from linear algebra. These similarities can be made precise and thus the constructions can also be transferred to other structures. For this, a framework must be created within which one can talk about different algebraic structures at the same time. Such a framework is provided by the field of *universal algebra*. We will return to this in Ch. 4.

Generated and Free Modules

A particularly important tool for studying vector spaces in linear algebra is the concept of a basis, which can be used to characterize abstract objects by tuples of numbers. We will examine in the following to what extent these ideas from linear algebra can also be implemented for modules.

We start with a generalization of the *linear hull* of a subset E of a vector space V. The linear hull can be described in different ways. Particularly easy to generalize is the view that the linear hull of E in V is the smallest subspace of V that contains E.

Definition 2.10 (Generated Submodule) Let M be a left R-module and $E \subseteq M$ a subset. Then

$$\langle E \rangle := \bigcap \{N \subseteq M \mid E \subseteq N, \ N \text{ submodule}\}$$

is called the left R-module *generated* by E (see Example 1.11).

In linear algebra, it is shown that every element of the linear hull of E can be written as a linear combination of elements from E. This perspective also gives rise to the alternative name *linear span* for the linear hull. This perspective can also be transferred to modules: For this, we refer to finite sums of the form $\sum r_j m_j$ with $r_j \in R$ and $m_j \in M$ as *R-linear combinations*.

2.1 Structure Theory of Modules

Proposition 2.11 (Characterization of the Generated Submodule) *Let R be a ring with identity and M a left R-module. For $E \subseteq M$ it holds*

$$\langle E \rangle = \Big\{ \sum_{\text{finite}} r_j e_j \;\Big|\; r_j \in R, \; e_j \in E \Big\}.$$

Proof The right side is obviously a submodule that contains E (because $1 \in R$). But then the definition implies that $\langle E \rangle$ is contained in the right side. Conversely, every submodule containing E also contains all R-linear combinations of E. □

When R is a field, Definition 2.10 coincides with the definition of the linear hull and Proposition 2.11 with its characterization as a set of linear combinations. We have thus found a generalization of the linear hull for *all* modules.

The second defining element of a basis, the *linear independence*, can also be formulated in a very general way. However, it will turn out that a module does not generally have a linearly independent generating system.

Definition 2.12 (Independence, Bases and Free Modules) Let R be a ring with identity, M a left R-module and $E \subseteq M$. We say E generates M or *spans* M, when $\langle E \rangle = M$. If M is spanned by a finite subset, then M is *finitely generated*. The set E is called R-*independent*, if for all $n \in \mathbb{N}, r_1, \ldots, r_n \in R$ and pairwise different $m_1, \ldots, m_n \in E$ it holds:

$$r_1 m_1 + \ldots + r_n m_n = 0 \quad \Rightarrow \quad r_1 = \ldots = r_n = 0.$$

If E generates the module M and is R-independent, then E is called an R-*basis* for M. This is equivalent to the fact that every element of M can be written in exactly one way (up to the order) as an R-linear combination of the basis elements. Finally, M is called a *free* (left-)R-module, if there is an R-basis for M.

When $R = \mathbb{K}$ is a field, then the concept of R-*independence* reduces to the concept of *linear independence* from the theory of vector spaces. From the above discussion of the linear hull, it follows that an R-basis in this case is the same as a vector space basis. Since by Zorn's lemma every $\mathbb{K}$-vector space has a basis, every $\mathbb{K}$-module is therefore free.

Example 2.13 (Free Modules)

(i) Let R be a ring with identity. Then R^n with the R-module structure from Example 2.7(ii) is free with basis

$$\{(1, 0, \ldots, 0), \; (0, 1, 0, \ldots, 0), \ldots, (0, \ldots, 0, 1)\}.$$

(ii) $R = \text{Mat}(2 \times 2, \mathbb{R})$, $M = \left\{ \begin{pmatrix} a & 0 \\ b & 0 \end{pmatrix} \mid a, b \in \mathbb{R} \right\}$. If there were an R-basis E for M, then $2 = \dim_{\mathbb{R}} M \geq |E| \cdot \dim_{\mathbb{R}} R = 4|E|$. Therefore, M is not free.

(iii) A non-zero free R-module has at least as many elements as R. Therefore, e.g., the $\mathbb{Z}$-module $\mathbb{Z}/n\mathbb{Z}$ is not free, meaning it has no basis.

□

Torsion phenomena like the one described in Example 2.13(iii), i.e., the existence of $m \in M$ and $r \in R$ with $rm = 0$, lead to the fact that modules are usually not free. Nevertheless, free modules are very important for structure theory. This is because all modules can be written as quotient modules of free modules. The proof of this fact will occupy us for a while, but it is extremely instructive, as all individual steps are fundamental constructions and thought figures of algebraic structure theories. We start with a characterization of the freedom of a module by a so-called *universal property*.

Theorem 2.14 (Universal Property of Free Modules) *Let R be a ring with identity, M a left-R-module and $E \subset M$. Then the following statements are equivalent:*

(1) *M is free with basis E.*
(2) *For every left-R-module V and for every mapping $\varphi : E \to V$ there exists exactly one R-module homomorphism $\overline{\varphi} : M \to V$ with $\overline{\varphi}|_E = \varphi$.*

Proof

(1) $\Rightarrow$ (2) : $\overline{\varphi}(\Sigma r_j m_j) := \Sigma r_j \varphi(m_j)$ for $m_j \in E$.

(2) $\Rightarrow$ (1) : We first show the R-independence of E: For $m_1, \ldots, m_n \in E$, $r_1, \ldots, r_n \in R$ with $\Sigma r_j m_j = 0$ choose $V = R$ and $\varphi_j : E \to V$ with

$$\varphi_j(m) = \begin{cases} 1 & m = m_j \\ 0 & \text{otherwise.} \end{cases}$$

Then one calculates $\overline{\varphi}_i(\Sigma r_j m_j) = \Sigma r_j \varphi_i(m_j) = r_i$, and this provides with $\overline{\varphi}_i(0) = 0$ the R-independence of E. To show $\langle E \rangle = M$, set $N := \langle E \rangle$ and consider the factor module $M/N = \{m + N \mid m \in M\}$. The R-module structure of M/N is given by $r \cdot (m + N) = (r \cdot m) + N$. Apply (2) to

$$\varphi : E \to M/N, \ m \mapsto [0] = 0 + N$$

The mappings

$$\left. \begin{array}{l} \overline{\varphi}_1 : M \to M/N \\ m \mapsto [0] \end{array} \right\} \quad \text{and} \quad \left. \begin{array}{l} \overline{\varphi}_2 : M \to M/N \\ m \mapsto m + N \end{array} \right\}$$

2.1 Structure Theory of Modules

are both R-module homomorphisms, which extend φ (because $E \subseteq N$). Therefore, (2) provides that $\overline{\varphi}_1 = \overline{\varphi}_2$, and this shows $M = N$. □

It is condition (2) from Theorem 2.14 i.e., called the universal property of free modules. This is because a statement is made for *arbitrary* mappings $\varphi \colon E \to V$. Universal properties play an important role in mathematics, and we will see a lot more of such properties in the course of this text. The advantage of universal properties is usually that one can work easily with objects whose construction was complicated, because only a few properties, including the characterizing universal property, are used. This is in some way analogous to the situation one creates when characterizing the real numbers as (the only) complete ordered field and then working with these properties without having to remember the constructions via Cauchy sequences or Dedekind cuts.

Universal properties can often be represented in a clear way by *commutative diagrams*. Here, "commutative" means that the mappings that can be composed from the diagram by composition of arrows coincide as soon as the starting and ending points coincide. The universal property of free modules is represented by the following commutative diagram:

$$\begin{array}{ccc} E & \xrightarrow{\forall \varphi} & V \\ \downarrow & \nearrow_{\exists ! \overline{\varphi}} & \\ M & & \end{array}$$

For vector spaces, there is the principle of "invariance of basis length", which states that every basis of a finite-dimensional vector space has the same number of elements. For commutative rings with identity, this principle can also be transferred to free modules.

Proposition 2.15 (Invariance of Basis Length) *Let R be a commutative ring with identity. Any two finite bases of a free R-module M have the same number of elements. In particular, n is determined by $M = R^n$.*

Proof If $\{m_1, \ldots, m_k\}$ is a basis for M, then

$$\eta \colon R^k \to M, \quad (r_1, \ldots, r_k) \mapsto \sum_{j=1}^{k} r_j m_j$$

is a module isomorphism. According to Zorn's lemma, there is a maximal ideal $I \trianglelefteq R$ (the union of a chain of proper ideals is a proper ideal). The set $IM := \{\sum_{\text{finite}} i_\alpha m_\alpha \mid i_\alpha \in I, m_\alpha \in M\}$ is an R-submodule of M, and the induced mapping $\varphi \colon M/IM \to R^k/IR^k$, $m + IM \mapsto \eta^{-1}(m) + IR^k$ is a bijective R-module homomorphism. Since multiplication with elements from I always yields zero on both sides, φ can also be interpreted as an R/I-module homomorphism. The maximality of I shows after Proposition 1.20 that R/I is a field, i.e., φ is a vector space isomorphism. But also $\psi_k : R^k/IR^k \to (R/I)^k$, $(r_1,\ldots,r_k) + IR^k \mapsto (r_1+I,\ldots,r_k+I)$ is an R/I-module isomorphism, i.e., a vector space isomorphism. This gives $\dim_{R/I}(M/IM) = k$, and k is determined by M and I. Therefore, the lengths of two finite bases coincide. $\square$

For a commutative ring with identity, the number n of elements of a finite basis of a free R-module is called the *rank* of the module. In the case of vector spaces, i.e., when R is a field, the rank is nothing other than the dimension of the vector space.

Exercise 2.6 (A Ring R that is Isomorphic to R^2 as a Module) Let $\mathbb{K}$ be a field, V an infinitely-dimensional $\mathbb{K}$-vector space. Show:

(i) V is isomorphic to V^2 as a $\mathbb{K}$-vector space.
(ii) The ring $R := \mathrm{Hom}_{\mathbb{K}}(V, V)$ is isomorphic to $\mathrm{Hom}_{\mathbb{K}}(V, V^2) \cong \mathrm{Hom}_{\mathbb{K}}(V, V)^2 = R^2$ as an $\mathbb{R}$-module.

Excursion: Direct Sums and Products

In the next chapter, we will get to know various ways of constructing new modules from given ones. However, we already here consider one such construction, namely the direct sum of modules, because we need it for the proof of the existence of free modules with a given basis.

Construction 2.16 (Direct Sums and Products of Modules) Let R be a ring with unity and M_λ, with $\lambda \in \Lambda$ a family of left R-modules. Then it is easy to verify that

$$\prod_{\lambda \in \Lambda} M_\lambda := \left\{ f : \Lambda \to \bigcup_{\lambda \in \Lambda} M_\lambda \,\Big|\, f(\lambda) \in M_\lambda \right\}$$

with respect to

$$\forall r \in R, f, f' \in \prod_{\lambda \in \Lambda} M_\lambda, \lambda \in \Lambda : \quad \begin{cases} (r \cdot f)(\lambda) = r \cdot f(\lambda) \\ (f + f')(\lambda) = f(\lambda) + f'(\lambda) \end{cases}$$

is a left R-module and

$$\bigoplus_{\lambda \in \Lambda} M_\lambda := \left\{ f \in \prod_{\lambda \in \Lambda} M_\lambda \,\Big|\, f(\lambda) = 0 \text{ for all but finitely many } \lambda \right\}$$

2.1 Structure Theory of Modules   37

is a submodule of $\prod_{\lambda \in \Lambda} M_\lambda$. The R-module $\prod_{\lambda \in \Lambda} M_\lambda$ is called the *direct product* of the M_λ and $\bigoplus_{\lambda \in \Lambda} M_\lambda$ the *direct sum* of the M_λ.

From the definitions, it immediately follows that the projections

$$\pi_{\lambda_0} : \prod_{\lambda \in \Lambda} M_\lambda \longrightarrow M_{\lambda_0}, \quad f \longmapsto f(\lambda_0)$$

and the inclusions

$$\iota_{\lambda_0} : M_{\lambda_0} \longrightarrow \bigoplus_{\lambda \in \Lambda} M_\lambda, \quad m \longmapsto \left(\lambda \mapsto \begin{cases} 0 \in M_\lambda & \lambda \neq \lambda_0 \\ m & \lambda = \lambda_0 \end{cases} \right)$$

for each $\lambda_0 \in \Lambda$ are R-module homomorphisms. □

The following exercise shows that we could also consider inclusions for direct products and projections for direct sums. However, the universal properties for direct sums and products described in Proposition 2.17 then have no correspondence.

Exercise 2.7 (Direct Sums and Products of Modules) Show that the inclusion map $\bigoplus_{\lambda \in \Lambda} M_\lambda \to \prod_{\lambda \in \Lambda} M_\lambda$ is an R-module homomorphism. From this, conclude that the projections

$$\widetilde{\pi}_{\lambda_0} : \bigoplus_{\lambda \in \Lambda} M_\lambda \longrightarrow M_{\lambda_0}, \quad f \longmapsto f(\lambda_0)$$

and the inclusions

$$\widetilde{\iota}_{\lambda_0} : M_{\lambda_0} \longrightarrow \prod_{\lambda \in \Lambda} M_\lambda, \quad m \longmapsto \left(\lambda \mapsto \begin{cases} 0 \in M_\lambda & \lambda \neq \lambda_0 \\ m & \lambda = \lambda_0 \end{cases} \right)$$

for each $\lambda_0 \in \Lambda$ are R-module homomorphisms.

Proposition 2.17 (Homomorphisms for Products and Sums) *Let R be a commutative ring with identity. Further, let M_λ for $\lambda \in \Lambda$ as well as M and N be left R-modules.*

(i) $\varphi : M \to \prod_{\lambda \in \Lambda} M_\lambda$ *is an R-module homomorphism if and only if $\pi_\lambda \circ \varphi : M \to M_\lambda$ is an R-module homomorphism for each $\lambda \in \Lambda$.*

(ii) *Let $\psi_\lambda : M_\lambda \to N$ be R-module homomorphisms. Then there exists exactly one R-module homomorphism $\psi : \bigoplus_{\lambda \in \Lambda} M_\lambda \to N$ with $\psi \circ \iota_\lambda = \psi_\lambda$ for all $\lambda \in \Lambda$.*

Proof For (i) we calculate

$$\varphi(m+m')(\lambda) = (\pi_\lambda \circ \varphi)(m+m') = (\pi_\lambda \circ \varphi)(m) + (\pi_\lambda \circ \varphi)(m')$$
$$= \varphi(m)(\lambda) + \varphi(m')(\lambda) = \big(\varphi(m) + \varphi(m')\big)(\lambda),$$
$$\varphi(rm)(\lambda) = (\pi_\lambda \circ \varphi)(rm) = r \cdot \big((\pi_\lambda \circ \varphi)(m)\big)$$
$$= r\big(\varphi(m)(\lambda)\big) = \big(r \cdot \varphi(m)\big)(\lambda).$$

For (ii) we first show uniqueness: If $\psi \in \mathrm{Hom}_R\Big(\bigoplus_{\lambda \in \Lambda} M_\lambda, N\Big)$ with $\psi \circ \iota_\lambda = \psi_\lambda$ for all $\lambda \in \Lambda$, then

$$\psi(f) = \psi\Big(\sum_{\lambda \in \Lambda} \iota_\lambda\big(f(\lambda)\big)\Big) = \sum_{\lambda \in \Lambda} \psi \circ \iota_\lambda\big(f(\lambda)\big) = \sum_{\lambda \in \Lambda} \psi_\lambda\big(f(\lambda)\big).$$

To prove existence, we set

$$\forall f \in \bigoplus_{\lambda \in \Lambda} M_\lambda : \quad \psi(f) := \sum_{\lambda \in \Lambda} \psi_\lambda\big(f(\lambda)\big).$$

Then

$$\forall m \in M_{\lambda_0} : \quad \psi \circ \iota_{\lambda_0}(m) = \sum_{\lambda \in \Lambda} \psi_\lambda\Big(\big(\iota_{\lambda_0}(m)\big)(\lambda)\Big) = \psi_{\lambda_0}(m)$$

and

$$\psi(rf + r'f') = \sum_{\lambda \in \Lambda} \psi_\lambda\big(r \cdot f(\lambda) + r' \cdot f'(\lambda)\big)$$
$$= \sum_{\lambda \in \Lambda} r\, \psi_\lambda\big(f(\lambda)\big) + r'\psi_\lambda\big(f'(\lambda)\big)$$
$$= r \sum_{\lambda \in \Lambda} \psi_\lambda\big(f(\lambda)\big) + r' \sum_{\lambda \in \Lambda} \psi_\lambda\big(f'(\lambda)\big) = r\,\psi(f) + r'\psi(f').$$

$\square$

Exercise 2.8 (Universal Property of Sum and Product) Let M_λ for $\lambda \in \Lambda$, as well as S and P be left R-modules. Show:

2.1 Structure Theory of Modules

(i) If there are R-module homomorphisms $p_\lambda \colon P \to M_\lambda$ with respect to which P has the following universal property (for all λ)

$$\begin{array}{ccc} & & \overset{\forall \varphi_\lambda}{} \\ N & \longrightarrow & M_\lambda \\ {\scriptstyle \exists ! \varphi} \downarrow & \nearrow {\scriptstyle p_\lambda} & \\ P & & \end{array}$$

then P is isomorphic to the direct product of the M_λ.

(ii) If there are R-module homomorphisms $j_\lambda \colon M_\lambda \to S$ has, with respect to S, the following universal property (for all λ)

$$\begin{array}{ccc} & & \overset{\forall \varphi_\lambda}{} \\ N & \longleftarrow & M_\lambda \\ {\scriptstyle \exists ! \varphi} \uparrow & \swarrow {\scriptstyle j_\lambda} & \\ S & & \end{array}$$

then S is isomorphic to the direct sum of the M_λ.

Proposition 2.17 and Exercise 2.8 provide a first hint of a duality between the constructions *direct sum* and *direct product*: If one describes a property of one construction through a diagram of mappings and reverses all the arrows, one obtains a property of the other construction. We will see a number of such dualities.

Corollary 2.18 (Spaces of Homomorphisms) *Let R be a commutative ring with identity. Further, let M_λ for $\lambda \in \Lambda$ as well as M and N be left R-modules. Then the following left R-modules are isomorphic:*

$$\operatorname{Hom}_R\left(\bigoplus_{\lambda \in \Lambda} M_\lambda, N\right) \cong \prod_{\lambda \in \Lambda} \operatorname{Hom}_R(M_\lambda, N).$$

The respective R-module structures are given by a combination of Construction 2.16 and Exercise 2.5.

Proof We define a mapping

$$\Phi \colon \prod_{\lambda \in \Lambda} \operatorname{Hom}_R(M_\lambda, N) \to \operatorname{Hom}_R\left(\bigoplus_{\lambda \in \Lambda} M_\lambda, N\right)$$

by

$$\forall f \in \prod_{\lambda \in \Lambda} \operatorname{Hom}_R(M_\lambda, N), \; g \in \bigoplus_{\lambda \in \Lambda} M_\lambda \colon \quad (\Phi(f))(g) := \sum_{\lambda \in \Lambda} f_\lambda(g_\lambda),$$

where we set $f_\lambda := f(\lambda) \in \operatorname{Hom}_R(M_\lambda, N)$ and $g_\lambda := g(\lambda) \in M_\lambda$. A longer, but simple calculation shows (exercise!) that Φ is an R-module homomorphism. Next, we define a mapping

$$\Psi : \operatorname{Hom}_R\left(\bigoplus_{\lambda \in \Lambda} M_\lambda, N\right) \to \prod_{\lambda \in \Lambda} \operatorname{Hom}_R(M_\lambda, N)$$

by $\left(\Psi(\psi)\right)_\lambda (m_\lambda) := \psi \circ \iota_\lambda(m_\lambda)$ for $\psi \in \operatorname{Hom}_R\left(\bigoplus_{\lambda \in \Lambda} M_\lambda, N\right)$ as well as $\lambda \in \Lambda$ and $m_\lambda \in M_\lambda$. Again, one checks (exercise!) that Ψ is an R-module homomorphism.

Two more calculations (exercise!) show that Φ and Ψ are inverses of each other, which completes the proof. □

One can also reverse the idea of constructing direct sums of modules and use it to decompose a given module into small pieces.

Remark 2.19 (Internal Direct Sum) Let M be a left R-module and M_λ, $\lambda \in \Lambda$ a family of submodules. Then M is called the *(internal) direct sum* of the M_λ, if the mapping induced by the inclusions $M_\lambda \hookrightarrow M$ via Proposition 2.17

$$\varphi : \bigoplus_{\lambda \in \Lambda} M_\lambda \longrightarrow M$$

is bijective. Note that the image of φ is just

$$\sum_{\lambda \in \Lambda} M_\lambda = \left\{ \sum_{\lambda \in \Lambda} m_\lambda \;\middle|\; \text{finite sums,}\; m_\lambda \in M_\lambda \right\} = \left\langle \bigcup_{\lambda \in \Lambda} M_\lambda \right\rangle$$

The mapping φ is bijective if and only if each element m of M can be expressed in exactly one way as a (finite) sum

$$\sum_{\lambda \in \Lambda} m_\lambda \quad \text{with}\; m_\lambda \in M_\lambda.$$

So M is the (internal) direct sum of the M_λ, if and only if there are R-module homomorphisms $p_\lambda : M \longrightarrow M_\lambda$ for all $\lambda \in \Lambda$ that fulfill the following conditions:

(a) $m = \sum_{\lambda \in \Lambda} p_\lambda(m)$ for all $m \in M$, where all but finitely many summands are 0.
(b) $p_\lambda(m) = m$ for all $m \in M_\lambda$.

The p_λ are called *canonical projections*. □

2.1 Structure Theory of Modules

With that, our excursion about general properties of direct sums and products of modules is finished, and we return to the construction of free modules.

Construction 2.20 (Free Left-R-Modules) Let R be a ring with identity and E a set. Set

$$_R\mathrm{F}(E) := \{f : E \to R \mid f(e) \neq 0 \text{ only for finitely many } e \in E\} = \bigoplus_E R.$$

When the ring R is clear from the context, we also simply write $\mathrm{F}(E)$ instead of $_R\mathrm{F}(E)$. The mapping $E \to {_R\mathrm{F}(E)}$, $e \mapsto f_e$ with

$$f_e(e') := \begin{cases} 1 & e = e' \\ 0 & \text{otherwise} \end{cases}$$

is injective, and we consider E as a subset of $_R\mathrm{F}(E)$. Then $_R\mathrm{F}(E)$ is an R-module with respect to

$$(f_1 + f_2)(e) := f_1(e) + f_2(e) \quad \text{and} \quad (r\,f)(e) := r\,f(e).$$

Now let V be a (left-) R-module and $\varphi : E \to V$ a mapping. Set

$$\overline{\varphi}(f) := \sum_{e \in E} f(e)\varphi(e)$$

(only finitely many summands are different from 0), then we get $\overline{\varphi}(f_e) = \varphi(e)$ and

$$\overline{\varphi}(r_1 f_1 + r_2 f_2) = \sum_{e \in E} \bigl(r_1 f_1 + r_2 f_2\bigr)(e)\varphi(e)$$
$$= \sum_{e \in E} r_1 f_1(e)\varphi(e) + r_2 f_2(e)\varphi(e) = r_1\overline{\varphi}(f_1) + r_2\overline{\varphi}(f_2).$$

The uniqueness of $\overline{\varphi}$ is clear, because each $f \in {_R\mathrm{F}(E)}$ can be written as linear combinations of the f_e. So $_R\mathrm{F}(E)$ is free with basis E according to Theorem 2.14. □

Thus, we have ensured the existence of many free modules. The following proposition shows that—up to isomorphism—every free module is formed in this way and therefore one can speak of *the* free left-R-module over E.

Proposition 2.21 (Uniqueness of the Free Module) *Let M be a free left-R-module with basis E. Then $M \cong {_R\mathrm{F}(E)}$.*

Proof From the three commutative diagrams

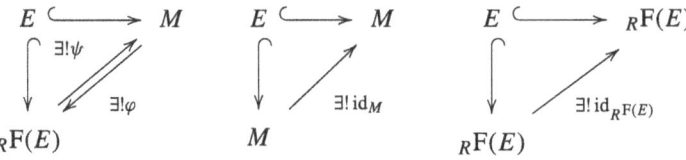

follow the identities $\psi \circ \varphi = \mathrm{id}_M$ and $\varphi \circ \psi = \mathrm{id}_{{}_R\mathrm{F}(E)}$. □

We are now able to prove the earlier claim that every R-module is the quotient module of a free R-module: Let M be a left-R-module and F(M) the free R-module with basis M. Then the universal property (Theorem 2.14) of F(M) provides the commutative diagram

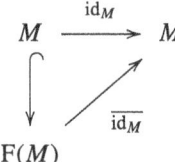

in which $\overline{\mathrm{id}_M}$ is an R-module homomorphism. It is surjective because id_M is surjective. With Construction 2.9(iii), it follows that M is isomorphic to $F(M)/\ker(\overline{\mathrm{id}_M})$.

Exercise 2.9 (Direct Sums of Free Modules) Let R be a commutative ring with identity and $M_1, \ldots, M_k$ free R-modules with disjoint bases $E_1, \ldots, E_k$. Let $E := \bigcup_{j=1}^{k} E_j$. Show that $M = M_1 \oplus \ldots \oplus M_k$ is free with basis E.

We conclude our discussion of free modules with a technical result that allows the decomposition of a module into two summands when it has a free module as a quotient module. We will use this result later together with the Chinese remainder theorem to obtain normal forms for linear mappings and matrices.

Lemma 2.22 (Direct Sums) *Let R be a ring and $\varphi \colon M \to M'$ a surjective R-module homomorphism and $\psi \colon M' \to M$ an R-module homomorphism with $\varphi \circ \psi = \mathrm{id}_{M'}$. Then $M \cong \ker(\varphi) \oplus M'$.*

Proof Since $\mathrm{id}_{M'}$ is injective, ψ and $\varphi|_{\mathrm{im}(\psi)}$ must also be injective. In particular, $\psi \colon M' \to \mathrm{im}(\psi)$ is an isomorphism. It is now sufficient to show that each $m \in M$ can be uniquely written as $m = m_1 + m_2$ with $m_1 \in \ker(\varphi)$ and $m_2 \in \mathrm{im}(\psi)$ (see Remark 2.19). First, we show uniqueness: So $m_1 + m_2 = m_1' + m_2'$ with $m_1, m_1' \in \ker(\varphi)$ and $m_2, m_2' \in \mathrm{im}(\psi)$. Applying φ to this, we get $\varphi(m_2) = \varphi(m_2')$, so $m_2 =$

m'_2, because $\varphi|_{\text{im}(\psi)}$ is injective. To show existence, we write $m = \psi \circ \varphi(m) + l$ for an $l \in M$. Then

$$\varphi(m) = \varphi \circ \psi \circ \varphi(m) + \varphi(l) = \varphi(m) + \varphi(l),$$

so $l \in \ker(\varphi)$. This proves the claim. □

One often encodes the hypotheses of Lemma 2.22 in the form of a *short exact sequence*

$$0 \longrightarrow \ker(\varphi) \longrightarrow M \overset{\varphi}{\longrightarrow} M' \longrightarrow 0,$$

where exact means that at each point the kernel of the outgoing homomorphism is equal to the image of the incoming homomorphism. The zeros represent the 0-module.

Corollary 2.23 *Let $\varphi \colon M \to M'$ be a surjective R-module homomorphism and M' free. Then $M \cong \ker(\varphi) \oplus M'$.*

Proof This follows from a combination of Theorem 2.14 and Lemma 2.22. More precisely, for each element e' of a basis E' of M', one chooses a preimage in $\varphi^{-1}(e') \subseteq M$ and extends the thus constructed mapping $\psi \colon E' \to M$ to a module homomorphism $\psi \colon M' \to M$. □

2.2 Applications to Linear Mappings

In this section, we want to use the structure theory of modules to further investigate linear self-mappings of finite-dimensional vector spaces. The applications are based on the following construction: Let $\mathbb{K}$ be a field and V a finite-dimensional $\mathbb{K}$-vector space. Each $\varphi \in \text{End}_{\mathbb{K}}(V)$ induces a $\mathbb{K}[X]$-module structure on V according to Example 2.2(iii):

$$\left(\sum a_j X^j\right) v := \sum a_j \varphi^j(v).$$

Conversely, from a $\mathbb{K}[X]$-module structure on V, one obtains an endomorphism $\varphi \in \text{End}_{\mathbb{K}}(V)$ via

$$\varphi(v) := Xv.$$

In this way, one finds a bijection between $\text{End}_{\mathbb{K}}(V)$ and the set of $\mathbb{K}[X]$-module structures on V. Therefore, one can assume that the structure theory of $\mathbb{K}[X]$-modules also provides results about endomorphisms.

Decomposition of Finitely Generated Modules over Euclidean Rings

To implement the outlined strategy, we need a decomposition theorem for finitely generated modules, which can be shown using the more specific ring-theoretic results from Ch. 1 for modules over Euclidean rings. We start with a technical lemma that we will use in the proof of the decomposition theorem.

Lemma 2.24 (Annihilators of Prime Elements) *Let R be a Euclidean ring and $p \in R$ prime, $0 \neq r \in R$ and $M := R/rR$. Further let $M_p := \{m \in M \mid pm = 0\}$. Then M_p is a submodule of M and:*

(i) *If $0 \neq r = px \in R$ with $x \in R$, then $M_p = xR/rR \cong R/pR$.*
(ii) *If p is not a divisor of r, then $M_p = \{0\}$.*

Proof

(i) The equality $M_p = xR/rR$ follows from the uniqueness of prime factorization (see Theorem 1.28): $m = s + rR$ is in M_p if and only if $ps \in rR$. Because $r = px$, this is equivalent to $s \in xR$. The isomorphism is obtained from the R-module homomorphism

$$R \to xR/rR, \quad a \mapsto xa + rR,$$

whose kernel is exactly pR, and from the first isomorphism theorem for modules (see Construction 2.9).
(ii) This again follows from the uniqueness of prime factorization: $ps = rr'$ shows that $p|r'$, and thus $s \in rR$.
□

Let R be a Euclidean ring and $p \in R$ prime. According to Propositions 1.25, 1.31, and 1.20, R/pR is a field. Therefore, in Lemma 2.24(i) in particular, M_p is an R/pR-vector space of dimension 1.

Before we can prove the announced decomposition theorem, we need to draw a conclusion from Corollary 2.23 that we will need in the proof.

Lemma 2.25 (Submodules of Free Modules) *Let R be a principal ideal domain and M a submodule of R^n. Then there exists a $d \leq n$ such that M is isomorphic to R^d.*

Proof Let $\pi : R^n \to R$, $(r_1, \ldots, r_n) \mapsto r_n$ the projection onto the last component. Then π is an R-module homomorphism and therefore (see Example 2.7) $I := \pi(M)$ is a submodule of R, i.e., an ideal in R. As R is a principal ideal domain, there exists an $x \in I$ with $I = xR$. Further, $\{x\}$ is a basis for I, i.e., $I \cong R$ is free. According

2.2 Applications to Linear Mappings

to Corollary 2.23, it follows that $M \cong \ker(\pi|_M) \oplus I$. Since $\ker(\pi|_M) \subseteq R^{n-1}$, it follows by induction over n that $\ker(\pi|_M) \cong R^{d'}$ with $d' \leq n-1$. Thus, we have

$$M \cong \ker(\pi|_M) \oplus I \cong R^{d'} \oplus R \cong R^{d'+1}$$

with $d' + 1 \leq n$. □

We now come to the repeatedly announced decomposition theorem for finitely generated modules. Its considerable importance can already be seen from the fact that the special case of $\mathbb{Z}$-modules, i.e., abelian groups, is known under the name *fundamental theorem on finitely generated abelian groups*.

Theorem 2.26 (Decomposition of Finitely Generated Modules I) *Let R be a Euclidean ring and M a finitely generated R-module. Then there are uniquely determined ideals $R \neq I_1 \supseteq I_2 \supseteq \ldots \supseteq I_\ell$ in R such that*

$$M \cong \bigoplus_{j=1}^{\ell} R/I_j.$$

Proof Existence of the ideals: Let $m_1, \ldots, m_n \in M$ be generators of M. Consider the surjective R-module homomorphism

$$\psi: R^n \to M, \quad (r_1, \ldots, r_n) \mapsto \sum_{i=1}^{n} r_i m_i.$$

According to Lemma 2.25, $\ker(\psi) \cong R^d$ for some $d \leq n$. In particular, there is an R-module homomorphism $\varphi: R^d \to R^n$, whose image is exactly $\ker(\psi)$. We then have an exact sequence

$$R^d \xrightarrow{\varphi} R^n \xrightarrow{\psi} M \longrightarrow 0.$$

With the first isomorphism theorem for modules (see Construction 2.9), we have

$$M \cong R^n / \ker(\psi) \cong R^n / \operatorname{im}(\varphi).$$

Let $\{v_1, \ldots, v_d\}$ and $\{w_1, \ldots, w_n\}$ be bases of R^d and R^n (see Example 2.13). Further, let $A \in \operatorname{Mat}(n \times d, R)$ be the *representing matrix* of φ with respect to these bases, i.e., $A = (a_{ij})_{\substack{i=1,\ldots,n \\ j=1,\ldots,d}}$ with

$$\forall j = 1, \ldots, d : \quad \varphi(v_j) = \sum_{i=1}^{n} a_{ij} w_i.$$

Claim: We can choose the bases so that A becomes a "diagonal matrix" ($a_{ij} = 0$ for $i \neq j$) and also $a_{11} \mid a_{22} \mid \ldots \mid a_{dd}$ holds.

With this claim, we then obtain $\operatorname{im}(\varphi) = \{\sum_{i=1}^{d} a_{ii} r_i w_i \mid r_i \in R\}$ and from this

$$M \cong R^n / \operatorname{im} \varphi \cong \left(\bigoplus_{i=1}^{d} R/a_{ii} R\right) \oplus \left(\bigoplus_{i=d+1}^{n} R\right).$$

By now omitting those summands for which $a_{ii} \in \operatorname{Unit}(R)$, i.e., for which $a_{ii} R = R$ (this happens for the first k elements with $k \leq d$, then no more), and appending the null ideal $(n - d)$ times, we find an ideal sequence of the desired kind.

We now prove the claim by induction over n: First, we choose the basis so that $\widetilde{d}(A)$ with

$$\widetilde{d}(A) := \min\{d(a_{ij}) \mid a_{ij} \neq 0; i = 1, \ldots, d; j = 1, \ldots, n\}$$

is minimal, where $d: R \setminus \{0\} \to \mathbb{N}_0$ is the degree function of R. Note that the case $A = 0$ is clear anyway.

The basis changes are effected by elementary row and column transformations. We only use permutations and additions of multiples of a row (column) to another, because only for these can we generally guarantee invertibility. Otherwise, the proof consists of an adaptation of the Gauss algorithm: By row and column swapping we can achieve that $\widetilde{d}(A) = d(a_{11})$. Note that due to the minimality of $\widetilde{d}(A)$ the element a_{11} must divide all coefficients in the first row and the first column of the matrix A, because otherwise by adding suitable multiples of the first row (column) via division with remainder we could generate elements of smaller degree. Now we divide the elements a_{i1} (without) remainder by a_{11} and subtract the corresponding multiple of the first row from the i-th row. This allows us to assume $a_{i1} = 0$ for $i > 1$. In exactly the same way we get $a_{1j} = 0$ for $j > 1$. Thus A has the shape

$$A = \begin{pmatrix} a_{11} & 0 \\ 0 & A' \end{pmatrix}$$

with $A' \in \operatorname{Mat}((d-1) \times (n-1), R)$. If a_{11} divides all coefficients of A', we are done by induction. If not, i.e., if there is an a_{ij} i.e., not divided by a_{11}, then we add the j-th row to the first, divide a_{ij} with remainder r by a_{11} and then subtract the corresponding multiple of the first column from the j-th column. This then creates the entry r in position $1j$, and this contradicts the minimality of $d(a_{11})$. So this case cannot occur at all, and we are done.

Uniqueness of the ideals: We assume

$$M = \bigoplus_{j=1}^{\ell} R/I_j$$

2.2 Applications to Linear Mappings

with the mentioned properties and want to describe the I_j using M. For this, let $I_j = r_j R$ be such that

$$r_1 \mid r_2 \mid \ldots \mid r_\ell.$$

1. Step ("Crossing out"): First we determine how often the zero ideal occurs: For this, set

$$M' := \{m \in M \mid \exists r \in R \setminus \{0\} : rm = 0\}.$$

Then M' is a submodule of M. More precisely, one sees that

$$M' = \bigoplus_{j=1}^{\widetilde{\ell}} R/I_j$$

with $\widetilde{\ell} := \max\{j \mid I_j \neq 0\}$ and $M/M' \cong R^{\ell-\widetilde{\ell}}$. Thus, M determines (see Proposition 2.15), how many summands R occur in M, and we can assume that none of the r_j is equal to 0.

2. Step ("Shortening"): For $p \in R$ prime, now set $M_p := \{m \in M \mid pm = 0\}$. Then M_p is an R/pR-vector space, and by Lemma 2.24 the dimension of M_p counts exactly the number of r_j in which p occurs as a prime factor. Let p now be a prime divisor of r_1 and thus of all other r_j. Then $\dim_{R/pR} M_p = \ell$ holds. If

$$M = \bigoplus_{j=1}^{\ell'} R/s_j R$$

is another sum decomposition of the required kind, then p divides at least ℓ of the elements $s_1, \ldots, s_{\ell'}$. In particular, $\ell \leq \ell'$. From symmetry reasons it follows that $\ell = \ell'$, and p divides all s_j.

Consider the module pM: Again with Lemma 2.24 one sees that

$$pM \cong \bigoplus_{j=1}^{\ell} R/x_j R \cong \bigoplus_{j=1}^{\ell} R/y_j R$$

with $px_j = r_j$ and $py_j = s_j$.

Now we repeat the previous procedure for pM instead of M (no free R-summands will appear anymore), i.e., we cancel a common prime element of all summands. Successively, we find that the r_j and the s_j coincide up to units. □

With the Chinese remainder theorem (see Lemma 1.14) we can further split the sum decomposition from Theorem 2.26:

Theorem 2.27 (Decomposition of Finitely Generated Modules II) *Let R be Euclidean and $r \in R$ have the factor decomposition $r = u p_1^{n_1} \cdots p_k^{n_k}$ with $p_j \in R$ non-associated prime elements and $u \in \mathrm{Unit}(R)$. Then*

$$R/rR \cong \bigoplus_{i=1}^{k} R/p_i^{n_i} R.$$

Proof We perform an induction over k and write $r = r' p_k^{n_k}$. Then r' and $p_k^{n_k}$ are coprime, i.e., we have with Proposition 1.25 $R = r'R + p_k^{n_k} R$. With the Chinese remainder theorem and induction follows

$$R/rR \cong R/r'R \oplus R/p_k^{n_k} R \cong \left(\bigoplus_{i=1}^{k-1} R/p_i^{n_i} R\right) \oplus R/p_k^{n_k} R \cong \bigoplus_{i=1}^{k} R/p_i^{n_i} R.$$

□

Exercise 2.10 (Finitely Generated Modules over Principal Ideal Domains) Generalize (using the exercises at the end of Ch. 1) the Theorems 2.26 and 2.27 for principal ideal domains (instead of Euclidean rings).

With Theorem 2.26 and Theorem 2.27 we have killed two birds with one stone. For $R = \mathbb{K}[X]$ we will use them to find normal forms for linear mappings. For $R = \mathbb{Z}$ they represent prototypes of structural theorems, because they provide for each specimen of the considered structure—finitely generated abelian groups—an isomorphic model, which is assembled in an easy to describe way (direct sum) from building blocks that are fully understood (the cyclic groups $\mathbb{Z}/q\mathbb{Z}$).

More cannot be expected, because there is an unmanageable number of essentially equal models of the same abelian group (see the remark on the concept of isomorphism after Definition 1.8). For each finite or countable set M one can find a bijection φ to one (actually even many different) such group $A := \bigoplus_{j=1}^{n} \mathbb{Z}/q_j\mathbb{Z}$. If one then defines the group operation on such a set by $m * m' := \varphi^{-1}\bigl(\varphi(m) + \varphi(m')\bigr)$, $(M, *)$ is a group isomorphic to $(A, +)$. It is important to note that everything that can be said about structural, i.e., determined by the group multiplication, properties of $(M, *)$ can be read off $(A, +)$ and transferred to $(M, *)$ with φ.

Minimal Polynomial and Rational Normal Form

We now come to the announced applications to normal forms of linear mappings. Let $\mathbb{K}$ be a field and V be a finite-dimensional vector space. Further let $\varphi \in \mathrm{End}_\mathbb{K}(V)$ and $\mathbb{K}[\varphi]$ be the subring of $\mathrm{End}_\mathbb{K}(V)$ generated by $\mathbb{K}\varphi$. The ring homomorphism

$$\mathrm{ev}_\varphi : \mathbb{K}[X] \to \mathbb{K}[\varphi], \quad \sum a_j X^j \mapsto \sum a_i \varphi^j$$

2.2 Applications to Linear Mappings

is called the *evaluation* in φ. The evaluation has a non-trivial kernel because $\mathbb{K}[\varphi]$ is a finite-dimensional $\mathbb{K}$-vector space, but $\mathbb{K}[X]$ is not (and the evaluation is obviously $\mathbb{K}$-linear). The kernel of the evaluation is of the form $q_\varphi \mathbb{K}[X]$, where q_φ is normalized (see Example 1.6) and is thus uniquely determined. We use the fact that $\mathbb{K}[X]$ is Euclidean (see Example 1.23) and therefore a principal ideal domain (see Proposition 1.25) and that the non-zero constant polynomials are precisely the units in $\mathbb{K}[X]$. The polynomial $q_\varphi \in \mathbb{K}[X]$ is called the *minimal polynomial* of φ.

The following proposition shows that the minimal polynomial q_φ plays an important role in the decomposition of V as a $\mathbb{K}[X]$-module.

Proposition 2.28 (Minimal Polynomial) *Let $\varphi \in \mathrm{End}_{\mathbb{K}}(V)$ and*

$$V \cong \bigoplus_{j=1}^{\ell} \mathbb{K}[X]/q_j\mathbb{K}[X]$$

be the sum decomposition from Theorem 2.26, applied to the $\mathbb{K}[X]$-module structure on V induced by φ. If the q_j are normalized, then $q_\ell = q_\varphi$ is the minimal polynomial of φ.

Proof The module structure on V is made so that an $f \in \mathbb{K}[X]$ lies in the kernel of ev_φ precisely when $fV = \{0\}$. But this is the case if and only if f is divisible by all q_j. Because $q_1 \mid q_2 \mid \ldots \mid q_\ell$ it follows $\ker \mathrm{ev}_\varphi = q_\ell \mathbb{K}[X]$, thus the claim. □

Next, we provide natural bases for the $\mathbb{K}$-vector spaces of the form $\mathbb{K}[X]/q\mathbb{K}[X]$. The matrix representation with respect to such bases will yield the rational normal form of φ.

Proposition 2.29 (Cyclic Basis) *Let $0 \neq q \in \mathbb{K}[X]$. Then $\mathbb{K}[X]/q\mathbb{K}[X]$ is a $\mathbb{K}$-vector space of dimension $\deg(q)$. More precisely, the cosets $x^j := X^j + q\mathbb{K}[X]$ with $0 \leq j \leq \deg(q) - 1$ form a basis for this space.*

Proof Verifying that $\mathbb{K}[X]/q\mathbb{K}[X]$ is a $\mathbb{K}$-vector space is routine. The x^j with $0 \leq j \leq \deg(q) - 1$ are linearly independent, as any nontrivial linear relation yields a polynomial i.e., divisible by q. Conversely, division by q with remainder shows that every polynomial modulo $q\mathbb{K}[X]$ is equal to a polynomial of degree less than $\deg(q)$. □

Assume that the decomposition in Proposition 2.28 consists of only one summand. Let $v \in V$ be the vector corresponding to the element $x^0 = 1 + q\mathbb{K}[X]$ in Proposition 2.29. Then $v, \varphi(v) = Xv, \ldots, \varphi^d(v) = X^d v$ with $d = \deg(q) - 1$ form a basis for V. This circumstance motivates the following definition.

Definition 2.30 (φ-**Cyclic Vectors**) Let $\varphi \in \mathrm{End}_{\mathbb{K}}(V)$ be a linear self-mapping of the $\mathbb{K}$–vector space V. We say that $v \in V$ is a φ-*cyclic vector* if V is spanned by the $\varphi^j(v)$ with $j \in \mathbb{N}_0$. This means that

$$\mathbb{K}[X]/q_\varphi \mathbb{K}[X] \to V, \quad f + q_\varphi \mathbb{K}[X] \mapsto fv$$

is a vector space isomorphism (see Construction 2.9).

Assuming the existence of a φ-cyclic vector, we now immediately obtain the rational normal form of φ.

Theorem 2.31 (Rational Normal Form) *Let* $\varphi \in \mathrm{End}_{\mathbb{K}}(V)$ *and* $q_\varphi = X^d + a_{d-1}X^{d-1} + \ldots + a_0$ *be the minimal polynomial of* φ. *If there is a* φ-*cyclic vector* $v_1 \in V$, *then* V *has a basis* $v_1, \ldots, v_d$, *with respect to which the representing matrix of* φ *has the following form:*

$$\begin{pmatrix} 0 & \cdots & & 0 & -a_0 \\ 1 & 0 & & \vdots & -a_1 \\ 0 & 1 & \ddots & \vdots & -a_2 \\ \vdots & 0 & \ddots & 0 & \vdots \\ \vdots & \vdots & \ddots & 1 & 0 & -a_{d-2} \\ 0 & 0 & \cdots & 0 & 1 & -a_{d-1} \end{pmatrix}$$

Proof Set $v_j := \varphi^{j-1}(v_1) = X^{j-1}v_1$ with $j = 2, \ldots, d$. According to Lemma 2.29, the $v_1, \ldots, v_d$ now form a basis for V. The assertion about the representing matrix is now easy to verify. $\square$

In general, there are no vectors in V that are cyclic for all of V. However, Proposition 2.28 shows that V can always be decomposed into a direct sum of φ-invariant subspaces for which there are φ-cyclic vectors.

Exercise 2.11 (Decomposition into Cyclic Spaces) Let $\mathbb{K}$ be a field, V a finite-dimensional $\mathbb{K}$-vector space and $\varphi \in \mathrm{End}_{\mathbb{K}}(V)$. Show that V is the finite direct sum of subspaces V_j, all of which have a φ-cyclic vector.

Exercise 2.12 (Minimal Polynomial) Let $V = \mathbb{R}^3$ and $\varphi \in \mathrm{End}_{\mathbb{R}}(\mathbb{R}^3)$ be given by $\varphi(x) = Ax$ with

$$A := \begin{pmatrix} 1 & 1 & 0 \\ 0 & 1 & 0 \\ 0 & 0 & 2 \end{pmatrix}.$$

Consider V with the $\mathbb{R}[X]$-module structure induced by φ, $p \cdot x := p(\varphi)x$ for $p \in \mathbb{R}[X]$ and $x \in V$. Determine a φ-cyclic vector of V and the minimal polynomial of φ.

Characteristic Polynomial and Jordan Normal Form

The refinement of the decomposition from Theorem 2.26 in Theorem 2.27 leads to the decomposition of the *characteristic polynomial* χ_φ of an endomorphism $\varphi \in \mathrm{End}_\mathbb{K}(V)$ into powers of linear factors and ultimately to the Jordan normal form. We start with the special case in which the decomposition from Theorem 2.27 has only one summand of the form $\mathbb{K}[X]/(X - \lambda)^k \mathbb{K}[X]$ with $\lambda \in \mathbb{K}$. The latter is an additional assumption, because in general, it is not known whether there are eigenvalues of φ in $\mathbb{K}$.

Lemma 2.32 (Jordan Block) *Let $\varphi \in \mathrm{End}_\mathbb{K}(V)$ and $V \cong \mathbb{K}[X]/(X - \lambda)^k \mathbb{K}[X]$ with $\lambda \in \mathbb{K}$. Then there is a basis $\{v_1, \ldots, v_k\}$ of V with respect to which the representative matrix of φ has the form*

$$\begin{pmatrix} \lambda & 1 & 0 & \ldots & & \ldots & 0 \\ 0 & \lambda & 1 & 0 & & & \vdots \\ \vdots & \ddots & \ddots & \ddots & \ddots & \ddots & \vdots \\ \vdots & & & 0 & \lambda & 1 & 0 \\ \vdots & & & & 0 & \lambda & 1 \\ 0 & \ldots & & \ldots & & 0 & \lambda \end{pmatrix}$$

In particular, $(X - \lambda)^k$ is the characteristic polynomial χ_φ of φ.

Proof Let v_j be the image of $(X - \lambda)^{k-j} + (X - \lambda)^k \mathbb{K}[X]$ under the $\mathbb{K}[X]$-module isomorphism

$$\mathbb{K}[X]/(X - \lambda)^k \mathbb{K}[X] \to V.$$

Then, as in the proof of Proposition 2.29, one can see that $v_1, \ldots, v_k$ becomes a basis (exercise). Finally, one calculates

$$Xv_j = (X - \lambda)v_j + \lambda v_j = v_{j-1} + \lambda v_j, \quad j \geq 2$$

and

$$Xv_1 = (X - \lambda)v_1 + \lambda v_1 = \lambda v_1.$$

This proves the assertion. □

For simplicity, we now assume that the field $\mathbb{K}$ is *algebraically closed*, i.e., every non-constant polynomial is of the form

$$f = c(X - \lambda_1) \cdots (X - \lambda_n),$$

where the $\lambda_j \in \mathbb{K}$ are not necessarily different. In particular, the prime elements of $\mathbb{K}[X]$ all have degree 1. This condition is fulfilled for $\mathbb{C}$ according to the fundamental theorem of algebra. It can be shown that every field can be considered as a subfield of an algebraically closed field ([Ke95, Theorem 19.6.1]). This then allows a part of the following results to be transferred to arbitrary fields.

Proposition 2.33 (Characteristic Polynomial) *Let $\varphi \in \mathrm{End}_{\mathbb{K}}(V)$ and $V \cong \bigoplus_{j=1}^{\ell} \mathbb{K}[X]/q_j\mathbb{K}[X]$ be the sum decomposition from Theorem 2.26, applied to the $\mathbb{K}[X]$-module structure on V induced by φ. If the q_j are normalized, then $q_1 \cdots q_{\ell} = \chi_{\varphi}$ is the characteristic polynomial of φ.*

Proof Each direct summand in V is a φ-invariant subspace, and the characteristic polynomials of the restrictions multiply to the characteristic polynomial of φ. So we can assume $\ell = 1$. Then Theorem 2.27 provides another sum decomposition, so we can assume: $V \cong \mathbb{K}[X]/q^k\mathbb{K}[X]$ with $q \in \mathbb{K}[X]$ prime. Since we have assumed $\mathbb{K}$ to be algebraically closed, $q = X - \lambda$ with $\lambda \in \mathbb{K}$. With Lemma 2.32 the assertion follows.
□

With this proposition, we can show that for algebraically closed $\mathbb{K}$, a basis can be found for every $\varphi \in \mathrm{End}_{\mathbb{K}}(V)$ such that the representing matrix is in Jordan normal form, i.e., a block diagonal matrix in which the blocks have the shape from Lemma 2.32.

Theorem 2.34 (Jordan Normal Form) *If $\mathbb{K}$ is algebraically closed, then for every $\varphi \in \mathrm{End}_{\mathbb{K}}(V)$ there is a basis for V, with respect to which the representing matrix of φ is in Jordan normal form.*

Proof Combine Proposition 2.33 with Lemma 2.32.
□

The above considerations yield even more useful results about the nature of linear mappings. We prove a criterion for diagonalizability and a decomposition theorem for matrices.

Corollary 2.35 (Diagonalizability) *Let $\mathbb{K}$ be algebraically closed. An endomorphism $\varphi \in \mathrm{End}_{\mathbb{K}}(V)$ is diagonalizable if and only if the minimal polynomial q_{φ} of φ has no multiple roots.*

Proof φ is diagonalizable if and only if all Jordan blocks are trivial (i.e., 1×1 matrices). According to Proposition 2.33 and Lemma 2.32, this means that each q_j only has simple roots. Now Proposition 2.28 shows that q_{φ} only has simple roots. Conversely, all q_j are divisors of q_{φ}, so they only have simple roots if q_{φ} only has simple roots. This proves the assertion.
□

2.2 Applications to Linear Mappings

Theorem 2.36 (Jordan-Chevalley Decomposition) *Let $\mathbb{K}$ be algebraically closed and $\varphi \in \operatorname{End}_{\mathbb{K}}(V)$. Then there are uniquely determined elements $\varphi_d, \varphi_n \in \operatorname{End}_{\mathbb{K}}(V)$ with the following properties:*

(i) $\varphi = \varphi_d + \varphi_n$.
(ii) φ_d *is diagonalizable.*
(iii) φ_n *is nilpotent.*
(iv) $\varphi_d \circ \varphi_n = \varphi_n \circ \varphi_d$.

There is a polynomial $f_d \in \mathbb{K}[X]$ without a constant term with $f_d(\varphi) = \varphi_d$. In particular, $\varphi_d(U_2) \subseteq U_1$ holds if $U_1 \subseteq U_2$ are subspaces of V with $\varphi(U_2) \subseteq U_1$.

Proof The existence of $\varphi_d, \varphi_n \in \operatorname{End}_{\mathbb{K}}(V)$ with properties (i)–(iv) is obtained from the Jordan normal form (i.e., Theorem 2.34) and the uniqueness from the fact that only the zero mapping is diagonalizable and nilpotent (exercise).

The proof of the existence of f_d provides an independent proof of the existence of φ_d and φ_n: Let $\chi_\varphi = \prod_{i=1}^{k}(X - \lambda_i)^{m_i}$ be the characteristic polynomial of φ. According to the Chinese remainder theorem in the version of Exercise 1.3, we find a polynomial $f_d \in \mathbb{K}[X]$ with

$$\forall i = 1, \ldots, k: \quad f_d \in \lambda_i + (X - \lambda_i)^{m_i} \mathbb{K}[X]$$

and $f_d \in X\mathbb{K}[X]$ (if all $\lambda_i \neq 0$).

Consider the φ-invariant subspaces $V_i := \ker(\varphi - \lambda_i \operatorname{id})^{m_i}$. It follows that $f_d(\varphi)|_{V_i} = \lambda_i \operatorname{id}|_{V_i}$ and

$$\left((\varphi - f_d(\varphi))|_{V_i}\right)^{m_i} = 0.$$

Since $V = \bigoplus_{i=1}^{k} V_i$ is a φ-invariant direct sum decomposition, it follows that $f_d(\varphi)$ is diagonalizable and $\varphi - f_d(\varphi)$ is nilpotent. This proves the claim. □

Literature: The material presented in this chapter can be found e.g., in [Ke95] and in [La93].

Multilinear Algebra 3

Multilinear algebra is an extension of linear algebra, in which the study of inner products and other bilinear mappings as well as determinants is systematically embedded. The starting point is the definition of a multilinear mapping, as known from higher derivatives of differentiable functions in several variables and the determinant as a function of column vectors. The crucial idea is then the introduction of the tensor product of modules. This is a module that transforms multilinear mappings into linear mappings, thus allowing their study using methods of linear algebra. Tensor products appear in almost all areas of mathematics and play an important role especially in differential geometry, algebraic geometry, algebraic topology, and functional analysis. They are also the starting point for the derivation of various universal structures with multiplications. Prominent examples are the exterior algebra, which is a far-reaching generalization of differential forms, the symmetric algebra, and the universal enveloping algebra of a Lie algebra.

In this chapter, we deal with tensor products and the tensor algebra derived from them as a prototype of structural constructions, but also briefly touch on how the abstract constructions are related to the tensor calculus of differential geometry (as applied, e.g., in general relativity theory). We follow the principle established in Section 2.1 in the construction of free modules, introducing new objects via their desired (universal) properties, which are each to be formulated in such a way that there is at most one such object up to isomorphism, and then proving the existence of such an object.

3.1 Tensor Products

We begin with a formal definition of multilinear mappings in the context of modules.

Definition 3.1 (Multilinear Mappings) Let R be a ring and $(M_\lambda)_{\lambda \in \Lambda}$ a family of R-modules and P an R-module. A mapping

$$\varphi : \prod_{\lambda \in \Lambda} M_\lambda \longrightarrow P$$

is called *R-multilinear*, if for each $\lambda_0 \in \Lambda$, all $r, s \in R$, and all $f, g, h \in \prod_{\lambda \in \Lambda} M_\lambda$
with

(a) $f(\lambda) = g(\lambda) = h(\lambda)$ for all $\lambda \neq \lambda_0$,
(b) $f(\lambda_0) = r\, g(\lambda_0) + s\, h(\lambda_0)$

it holds that

$$\varphi(f) = r\, \varphi(g) + s\, \varphi(h).$$

If $P = R$, then φ is called *multilinear form*. The set of R-multilinear mappings is denoted by $L_R(M_\lambda; P)$. If $\Lambda = \{1, \ldots, n\}$, then we also write $L_R(M_1, \ldots, M_n; P)$.

Multilinear mappings are generalizations of linear mappings. Bilinear mappings often appear as multiplications. Higher multilinearity is found in determinants, but also in the description of geometric quantities such as curvatures (see Section 9.2).

Example 3.2 (Multilinear Mappings)

(i) If $|\Lambda| = 1$, then the multilinear mappings are just the module homomorphisms.
(ii) $\varphi : M_1 \times M_2 \to P$ is bilinear if for all $m_1, m_1' \in M_1, m_2, m_2' \in M_2$ and $r, r' \in R$ it holds:

$$\varphi(r\, m_1 + r'm_1', m_2) = r\varphi(m_1, m_2) + r'\varphi(m_1', m_2),$$
$$\varphi(m_1, r\, m_2 + r'm_2') = r\varphi(m_1, m_2) + r'\varphi(m_1, m_2').$$

(iii) If R is a commutative ring, then the matrix multiplication

$$\text{Mat}(m \times n, R) \times \text{Mat}(n \times l, R) \longrightarrow \text{Mat}(m \times l, R), \quad (A, B) \mapsto AB$$

is bilinear.

□

Tensor Products of Two Modules over R

We begin our discussion of tensor products with the conversion of bilinear mappings into linear mappings. In this context, the role of the ring R in the construction is easier to understand. To make the subtlety of the construction transparent, we also

3.1 Tensor Products

carry it out here for not necessarily commutative rings. It should be noted that every R-module M, whether right or left, carries a $\mathbb{Z}$-module structure compatible with the R-module structure as an abelian group (see Example 2.2). By compatible we mean that

$$\forall r \in R, m \in M, z \in \mathbb{Z}: \quad r \cdot (z \cdot m) = z \cdot (r \cdot m) \quad \text{or} \quad (z \cdot m) \cdot r = z \cdot (m \cdot r). \quad (3.1)$$

This follows immediately (exercise!) from the properties of the R-module structure and the construction of the $\mathbb{Z}$-module structure.

Definition 3.3 (Tensor Product over R) Let R be a ring, M a right R-module and N a left R-module. A pair (T, π) is called a *tensor product* of M and N if:

(a) T is an abelian group.
(b) $\pi: M \times N \to T$ is $\mathbb{Z}$-bilinear and satisfies

$$\forall m \in M, n \in N, r \in R: \quad \pi(m \cdot r, n) = \pi(m, r \cdot n). \quad (3.2)$$

(c) (T, π) satisfies the following *universal property*: For every abelian group C and every $\mathbb{Z}$-bilinear mapping $\varrho: M \times N \to C$ with (3.2) there exists exactly one group homomorphism $\overline{\varrho}: T \to C$ with $\overline{\varrho} \circ \pi = \varrho$, i.e., we have the following commutative diagram:

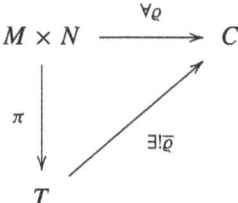

Note that in Definition 3.3 no left R-module structure on T is required. For this we would need an additional left R-module structure on M. We will analyze this situation in a later step. However, it is clear here that the situation simplifies if R is commutative. Then the right R-module structure on M provides a left R-module structure by $r \cdot m := m \cdot r$ and it would be natural to require an R-module structure on T in (a) and to assume in (b) that π is an R-bilinear mapping.

Construction 3.4 (Tensor Product over R) Under the assumptions of Definition 3.3, let ${}_{\mathbb{Z}}F(M \times N)$ be the free $\mathbb{Z}$-module over $M \times N$. Further, let $Y \subseteq {}_{\mathbb{Z}}F(M \times N)$ be the subset of elements in ${}_{\mathbb{Z}}F(M \times N)$ of the following type:

$$(m + m', n) - (m, n) - (m', n),$$
$$(m, n + n') - (m, n) - (m, n'),$$
$$(m \cdot r, n) - (m, r \cdot n),$$

where $m, m' \in M, n, n' \in N$ and $r \in R$. Note here that $M \times N$ is considered as a subset of $_{\mathbb{Z}}F(M \times N)$ (see Theorem 2.14 and Construction 2.20). We denote the quotient $\mathbb{Z}$-module $_{\mathbb{Z}}F(M \times N)/\langle Y \rangle$ as $M \otimes_R N$ and define $\pi : M \times N \to M \otimes_R N$ by $\pi(m, n) = (m, n) + \langle Y \rangle$ (see Definition 2.10 and Construction 2.20). We denote $\pi(m, n)$ as $m \otimes n$. If $R = \mathbb{Z}$, then we simply write $M \otimes N$ instead of $M \otimes_{\mathbb{Z}} N$. □

With this, we have a candidate for the tensor product over R, but we still need to prove that $M \otimes_R N$ is indeed a tensor product over R.

Proposition 3.5 (Existence of Tensor Products over R) *Under the conditions of Definition 3.3, $(M \otimes_R N, \pi)$ is a tensor product over R of M and N.*

Proof It holds that $M \otimes_R N = \langle \{m \otimes n \mid m \in M, n \in N\} \rangle$. The mapping π is $\mathbb{Z}$-bilinear and satisfies Eq. (3.2) according to the definition of Y. Now let $\varrho : M \times N \to C$ be as in the definition of the tensor product. Then, from Construction 2.20, it follows that ϱ can be extended on $_{\mathbb{Z}}F(M \times N)$ to a $\mathbb{Z}$-module homomorphism $\widetilde{\varrho} : {_{\mathbb{Z}}F(M \times N)} \to C$. Since ϱ was assumed to be $\mathbb{Z}$-bilinear, $\widetilde{\varrho}|_{\langle Y \rangle} \equiv 0$, and (3.2) provides the existence of a mapping $\overline{\varrho} : M \otimes_R N = {_{\mathbb{Z}}F(M \times N)}/\langle Y \rangle \longrightarrow C$ with

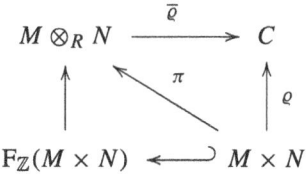

From the diagram, we read off $\overline{\varrho} \circ \pi = \varrho$. The uniqueness statement is clear because $\overline{\varrho}$ is determined by its values on $\{m \otimes n \mid m \in M, n \in N\}$. □

We motivated our discussion of tensor products at the beginning with the possibility of making linear mappings from bilinear mappings. The universal property of the tensor product over R is constructed in such a way that this transformation is guaranteed. Conversely, if C is an abelian group, i.e., a $\mathbb{Z}$-module, and $\varphi\, M \otimes_R N \to C$ is a $\mathbb{Z}$-module homomorphism, then the mapping

$$\varrho = \varphi \circ \pi : M \times N \to C, \quad (m, n) \mapsto \varphi(m \otimes n)$$

is $\mathbb{Z}$-bilinear and satisfies (3.2). Thus,

$$\{\varrho \in L_{\mathbb{Z}}(M, N; C) \mid \varrho \text{ satisfies } ((3.2))\} \longrightarrow \mathrm{Hom}_{\mathbb{Z}}\left(M \otimes_R N, C\right), \quad \varrho \mapsto \overline{\varrho} \tag{3.3}$$

is a bijection. In this case, bilinear mappings (of a certain type) are even equivalent to linear mappings. Statements of this form also result for other tensor products.

3.1 Tensor Products

In the formation of tensor products of modules, surprising effects can occur compared to tensor products of vector spaces, comparable to the new phenomena in the multiplication on rings compared to the multiplication in fields. In particular, there is the analogue of zero divisors.

Example 3.6 (Vanishing Tensor Products of Modules) Let $R = \mathbb{Z}$, $M = \mathbb{Z}/3\mathbb{Z}$ and $N = \mathbb{Z}/2\mathbb{Z}$. Then

$$F_{\mathbb{Z}}(M \times N) = \left\{ f : \mathbb{Z}/3\mathbb{Z} \times \mathbb{Z}/2\mathbb{Z} \to \mathbb{Z} \right\} \cong \mathrm{Mat}(3 \times 2, \mathbb{Z})$$

and $(m, 3 \cdot n) - (m \cdot 3, n) = (m, n) - (0, n) \in \langle Y \rangle$. From

$$((0, n) + \langle Y \rangle) + ((0, n) + \langle Y \rangle) = (0 + 0, n) + \langle Y \rangle = (0, n) + \langle Y \rangle$$

it follows that $(0, n) \in \langle Y \rangle$, so $(m, n) \in \langle Y \rangle$ for all $m \in M$ and all $n \in N$, which shows $\langle Y \rangle = F_{\mathbb{Z}}(M \times N)$. But then, according to Construction 3.4, $\mathbb{Z}/3\mathbb{Z} \otimes \mathbb{Z}/2\mathbb{Z} = \{0\}$. □

One question that we did not systematically address in Ch. 2 when examining direct sums and products is the *naturality* of these constructions. By this we mean that we have not explained how these constructions behave with respect to module homomorphisms. For example, if M and N are two R-modules and $\varphi \colon M \to M'$ and $\psi : N \to N'$ are two R-module homomorphisms, is there a connection between $M \oplus N$ and $M' \oplus N'$? According to Construction 2.16, the elements of $M \oplus N$ can simply be written as pairs (m, n) with $m \in M$ and $n \in N$. Then one immediately verifies that

$$\varphi \oplus \psi : M \oplus N \to M' \oplus N', \quad (m, n) \mapsto \big(\varphi(n), \psi(m)\big) \tag{3.4}$$

is an R-module homomorphism (exercise). In a similar way, we will always ask when constructing new objects from initial objects whether homomorphisms between two sets of initial objects also provide homomorphisms between the corresponding new objects. This will then be our criterion for the naturality of a construction. As soon as we have the language of category theory available, we will also speak of the *functoriality* of the construction.

The following proposition shows that the construction of the tensor product over R is a natural construction in the sense just explained.

Proposition 3.7 (Tensor Product of Homomorphisms) *Let R be a ring, M and M' right-R-modules and N and N' left-R-modules. Further let $\varphi \in \mathrm{Hom}_R(M, M')$ and $\psi \in \mathrm{Hom}_R(N, N')$. Then there is exactly one $\mathbb{Z}$-module homomorphism $\tau : M \otimes_R N \to M' \otimes_R N'$ with $\tau(m \otimes n) = \varphi(m) \otimes \psi(n)$. It is denoted by $\varphi \otimes \psi$.*

Proof The mapping $\varrho : M \times N \to M' \otimes_R N'$, $(m, n) \mapsto \varphi(m) \otimes \psi(n)$ is $\mathbb{Z}$-bilinear, and it holds $\varrho(m \cdot r, n) = \varphi(m \cdot r) \otimes \psi(n) = \varphi(m) \otimes \psi(r \cdot n) = \varrho(m, r \cdot n)$. So according to Proposition 3.5 there is exactly one homomorphism $\overline{\varrho} : M \otimes_R N \longrightarrow M' \otimes_R N'$ with $\overline{\varrho} \circ \pi = \varrho$, i.e., $\overline{\varrho}(m \otimes n) = \varphi(m) \otimes \psi(n)$. □

Similar to Example 3.6, one finds for tensor products of module homomorphisms zero divisor phenomena: e.g., in general it is not true that $\overline{\varphi} : N \otimes M' \to M \otimes M'$ for $N \subseteq M$ is injective.

Example 3.8 (Vanishing Tensor Products of Homomorphisms) We consider $R = \mathbb{Z}$, $M = \mathbb{Z}$, $N = 2\mathbb{Z}$, $M' = \mathbb{Z}/2\mathbb{Z}$. Since $M \cong N$ we have

$$N \otimes M' \cong M \otimes M' \cong M' \quad (R \otimes_R M' \cong M').$$

With the embedding $j : N \hookrightarrow M$ and $n = 2k \in N$, it follows that

$$j(n) \otimes m' = 2k \otimes m' = k \otimes 2m' = k \otimes 0 = 0.$$

This shows that $j \otimes \mathrm{id}_{M'} \ N \otimes M' \to M \otimes M'$ is the zero mapping. □

We now return to the question of how to endow a tensor product $M \otimes_R N$ over R with a left-R-module structure. It turns out that this is possible as soon as M carries a left-R-module structure that is compatible with the right-R-module structure. It even works for compatible left-S-right-R-module structures on M. This leads to the concept of an S-R-bimodule: An abelian group M that carries a right-R-module structure and a left-S-module structure that are *compatible* in the sense of

$$\forall s \in S, m \in M, r \in R : \quad s \cdot (m \cdot r) = (s \cdot m) \cdot r \tag{3.5}$$

is called a (S, R)-*bimodule*. Equation (3.5) allows us to use the notation $s \cdot m \cdot r$ for the products $s \cdot (m \cdot r)$ and $(s \cdot m) \cdot r$.

Note that due to (3.1) every left-R-module is an $(R, \mathbb{Z})$-bimodule and every right-R-module is a $(\mathbb{Z}, R)$-bimodule.

Proposition 3.9 (Module Structures on Tensor Products over R) *Let A, R and B be rings, N an (R, B)-bimodule and M an (A, R)-bimodule. Then:*

(i) $M \otimes_R N$ *is an (A, B)-bimodule with*

$$a \cdot \left(\sum_j m_j \otimes n_j \right) \cdot b = \sum_j (a \cdot m_j) \otimes (n_j \cdot b).$$

3.1 Tensor Products

(ii) *Let C be an (A, B)-bimodule and $\varrho : M \times N \to C$ a $\mathbb{Z}$-bilinear mapping with*

$$\varrho(m \cdot r, n) = \varrho(m, r \cdot n),$$
$$\varrho(a \cdot m, n) = a \cdot \varrho(m, n),$$
$$\varrho(m, n \cdot b) = \varrho(m, n) \cdot b$$

for $m \in M, n \in N, a \in A, b \in B$ and $r \in R$. Then the induced $\mathbb{Z}$-module homomorphism $\overline{\varrho}\ M \otimes_R N \to C$ is an (A, B)-bimodule homomorphism, i.e.,

$$\forall a \in A, t \in M \otimes_R N, b \in B : \quad \overline{\varrho}(a \cdot t \cdot b) = a \cdot \overline{\varrho}(t) \cdot b.$$

Proof

(i) For $a \in A$, define the mapping $L_a : M \times N \to M \times N$, $(m, n) \mapsto (a \cdot m, n)$ to be a $\mathbb{Z}$-module homomorphism $\overline{L}_a : F_{\mathbb{Z}}(M \times N) \to F_{\mathbb{Z}}(M \times N)$:

$$\begin{array}{ccc} F_{\mathbb{Z}}(M \times N) & \xrightarrow{\overline{L}_a} & F_{\mathbb{Z}}(M \times N) \\ \uparrow & & \uparrow \\ M \times N & \xrightarrow{L_a} & M \times N \end{array}$$

Because

$$a \cdot (m + m') = (a \cdot m) + (a \cdot m') \quad \text{and} \quad (a \cdot m) \cdot r = a \cdot (m \cdot r)$$

the mapping $\overline{L}_a$ maps the $\mathbb{Z}$-module $\langle Y \rangle$ (in the notation of Construction 3.4) into itself and factors to a $\mathbb{Z}$-module homomorphism

$$\begin{array}{ccc} M \otimes_R N & \xrightarrow{\widetilde{L}_a} & M \otimes_R N \\ \pi \uparrow & & \uparrow \pi \\ F_{\mathbb{Z}}(M \times N) & \xrightarrow{\overline{L}_a} & F_{\mathbb{Z}}(M \times N) \end{array}$$

Set $a \cdot t := \widetilde{L}_a(t)$ for all $t \in M \otimes_R N$. Then we have

$$a \cdot \left(\sum_j m_j \otimes n_j \right) = \widetilde{L}_a \circ \pi \left(\sum_j (m_j, n_j) \right) = \pi \left(\sum_j L_a(m_j, n_j) \right)$$
$$= \pi \left(\sum_j (a \cdot m_j, n_j) \right) = \sum_j \left((a \cdot m_j) \otimes n_j \right).$$

Similarly, for $b \in B$, consider the mapping $R_b : M \times N \to M \times N$, $(m, n) \mapsto (m, n \cdot b)$ and extend it to a $\mathbb{Z}$-module homomorphism $\overline{R}_b : F_{\mathbb{Z}}(M \times N) \to F_{\mathbb{Z}}(M \times N)$: Because

$$(n + n') \cdot b = (n \cdot b) + (n' \cdot b) \quad \text{and} \quad r \cdot (n \cdot b) = (r \cdot n) \cdot b$$

$\overline{R}_b$ also maps the $\mathbb{Z}$-module $\langle Y \rangle$ into itself and factors to a $\mathbb{Z}$-module homomorphism:

$$\begin{array}{ccc} M \otimes_R N & \xrightarrow{\widetilde{R}_b} & M \otimes_R N \\ \pi \uparrow & & \uparrow \pi \\ F_{\mathbb{Z}}(M \times N) & \xrightarrow{\overline{R}_b} & F_{\mathbb{Z}}(M \times N) \end{array}$$

With $t \cdot b := \widetilde{R}_b(t)$, the equation $\left(\sum_j m_j \otimes n_j\right) \cdot b = \sum_j \left(m_j \otimes (n_j \cdot b)\right)$ holds for all $t \in M \otimes_R N$, and we get $(a \cdot t) \cdot b = a \cdot (t \cdot b)$. The remaining proof of (i) consists of routine calculations (exercise).

(ii) It suffices to show that $\overline{\varrho}(a \cdot (m \otimes n) \cdot b) = a \cdot \overline{\varrho}(m \otimes n) \cdot b$. For this, we calculate $\overline{\varrho}(a \cdot (m \otimes n) \cdot b) = \overline{\varrho}((a \cdot m) \otimes (n \cdot b)) = \varrho(a \cdot m, n \cdot b) = a \cdot \varrho(m, n) \cdot b = a \cdot \overline{\varrho}(m \otimes n) \cdot b$.

$\square$

Proposition 3.9 allows us to consider iterated tensor products starting from suitable bimodules. Analogous to the investigation of iterated products of numbers, the question arises whether the order of multiplication matters, i.e., whether there is an associativity law for tensor products. The following proposition shows that this is the case if we do not distinguish between isomorphic modules.

Proposition 3.10 (Associativity of Tensor Products) *Let A, B, R and S be rings, M an (A, R)-bimodule, N an (R, S)-bimodule and L an (S, B)-module. Then $M \otimes_R N$ is an (A, S)-bimodule, $N \otimes_S L$ is an (R, B)-bimodule, and there is exactly one (A, B)-bimodule isomorphism*

$$\Phi : (M \otimes_R N) \otimes_S L \longrightarrow M \otimes_R (N \otimes_S L)$$

with $\Phi((m \otimes n) \otimes l) = m \otimes (n \otimes l)$ for $m \in M, n \in N, l \in L$.

Proof The statements about the existence of the bimodule structures follow directly from Proposition 3.9, and the uniqueness of Φ is clear, because the elements $(m \otimes n) \otimes l$ generate the $\mathbb{Z}$-module $(M \otimes_R N) \otimes_S L$.

3.1 Tensor Products

To show existence, we consider for each $l \in L$ the R-module homomorphism (exercise: verify homomorphism!) $R_l : N \to N \otimes_S L$, $n \mapsto n \otimes l$. According to Proposition 3.7,

$$\psi_l := \mathrm{id}_M \otimes R_l : M \otimes_R N \to M \otimes_R (N \otimes_S L)$$

a $\mathbb{Z}$-module homomorphism. Now consider the $\mathbb{Z}$-bilinear mapping

$$\varphi : (M \otimes_R N) \times L \to M \otimes_R (N \otimes_S L), \quad (t, l) \mapsto \psi_l(t).$$

Claim: $\psi_{s \cdot l}(t) = \psi_l(t \cdot s)$, i.e., $\varphi(t \cdot s, l) = \varphi(t, s \cdot l)$.
This follows from the identities

$$\psi_{s \cdot l}(m \otimes n) = m \otimes \big(n \otimes (s \cdot l)\big),$$
$$\psi_l\big((m \otimes n) \cdot s\big) = \psi_l\big(m \otimes (n \cdot s)\big) = m \otimes \big((n \cdot s) \otimes l\big).$$

So there exists a $\mathbb{Z}$-module homomorphism $\Phi : (M \otimes_R N) \otimes_S L \longrightarrow M \otimes_R (N \otimes_S L)$ with

$$\Phi\big((m \otimes n) \otimes l\big) = \varphi\big((m \otimes n), l\big) = \psi_l(m \otimes n) = m \otimes (n \otimes l).$$

Similarly, one finds $\Psi : M \otimes_R (N \otimes_S L) \to (M \otimes_R N) \otimes_S L$ with $\Psi\big(m \otimes (n \otimes l)\big) = (m \otimes n) \otimes l$, and it follows that $\Psi = \Phi^{-1}$.

All that remains is to show that Φ and Ψ are even (A, B)-bimodule homomorphisms. To do this, it suffices to verify the equations

$$a \cdot \Phi\big((m \otimes n) \otimes l\big) \cdot b = \Phi\big(a \cdot \big((m \otimes n) \otimes l\big) \cdot b\big)$$

and

$$a \cdot \Psi\big(m \otimes (n \otimes l)\big) \cdot b = \Psi\big(a \cdot \big(m \otimes (n \otimes l)\big) \cdot b\big)$$

for $a \in A, b \in B, m \in M, n \in N$ and $l \in L$. These follow from the identities

$$a \cdot \big((m \otimes n) \otimes l\big) \cdot b = a \cdot \big((m \otimes n) \otimes (l \cdot b)\big) = \big((a \cdot m) \otimes n\big) \otimes (l \cdot b)$$

and

$$a \cdot \big(m \otimes (n \otimes l)\big) \cdot b = \big((a \cdot m) \otimes (n \otimes l)\big) \cdot b = (a \cdot m) \otimes \big(n \otimes (l \cdot b)\big).$$

$\square$

Tensor Products of Modules over Commutative Rings

As hinted at earlier, the theory of tensor products is simpler when restricted to commutative rings. We now consider the special case where not only the rings are commutative, but all are equal to a fixed ring R. This amounts to considering all modules as (R, R)-bimodules. The theorems proven above then show that we can form tensor products of finitely many modules, with the number of modules being arbitrary. The stronger assumptions not only allow for shorter proofs, but also better results. In particular, we do not have to limit ourselves to finite products. Note also that in the following definition, the tensor products automatically carry an R-module structure.

Definition 3.11 (Tensor Product of R-Modules) Let R be a commutative ring and M_λ, $\lambda \in \Lambda$ a family of R-modules. A pair (T, π) is called a *tensor product* of the M_λ, if:

(a) T is an R-module.
(b) $\pi : \prod_{\lambda \in \Lambda} M_\lambda \to T$ is R-multilinear.
(c) (T, π) satisfies the following universal property: For every R-module C and every R-multilinear mapping $\varrho : \prod_{\lambda \in \Lambda} M_\lambda \to C$ there exists exactly one R-module homomorphism $\overline{\varrho} : T \to C$ with $\overline{\varrho} \circ \pi = \varrho$, i.e., we have the following commutative diagram:

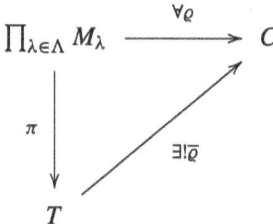

The construction of such tensor products follows the same idea as Construction 3.4 of a tensor product over R, but due to the multitude of factors, it is technically a bit complicated to write down.

Construction 3.12 (Tensor Product of R-Modules) Consider the free $\mathbb{Z}$-module A generated by $\prod_{\lambda \in \Lambda} M_\lambda$. Let Y be the set of elements in A of the type

$$f - g - h; \quad \begin{cases} f(\lambda) = g(\lambda) = h(\lambda) & \lambda \neq \lambda_0 \\ f(\lambda_0) = g(\lambda_0) + h(\lambda_0) \end{cases}$$

3.1 Tensor Products

$$f - g; \quad \begin{cases} f(\lambda) = g(\lambda) & \lambda_0 \neq \lambda \neq \lambda_1 \\ r \cdot f(\lambda_0) = g(\lambda_0) \\ f(\lambda_1) = r \cdot g(\lambda_1) \end{cases}$$

where $\lambda_0, \lambda_1 \in \Lambda$ are each fixed. If Λ is countable and we write the functions $f \in \prod_{\lambda \in \Lambda} M_\lambda$ as sequences $(\ldots, m_{\lambda_0}, \ldots)$, this means we consider elements of the form

$$(\ldots, m + m', \ldots) - (\ldots, m, \ldots) - (\ldots, m', \ldots)$$
$$\uparrow \qquad\qquad \uparrow \qquad\qquad \uparrow$$
$$\lambda_0 \qquad\qquad \lambda_0 \qquad\qquad \lambda_0$$

and

$$(\ldots, r \cdot m, \ldots, n, \ldots) - (\ldots, m, \ldots, r \cdot n, \ldots) \, .$$
$$\uparrow \quad\;\; \uparrow \qquad\qquad \uparrow \quad\;\;\;\;\; \uparrow$$
$$\lambda_0 \quad \lambda_1 \qquad\qquad \lambda_0 \quad\;\; \lambda_1$$

Denote $A/\langle Y \rangle$ with $\bigotimes_{\lambda \in \Lambda} M_\lambda$ and the image of $f = \bigl(f(\lambda)\bigr)_{\lambda \in \Lambda}$ under $\pi : \prod_{\lambda \in \Lambda} M_\lambda \to \bigotimes_{\lambda \in \Lambda} M_\lambda$ with $\bigotimes_{\lambda \in \Lambda} f(\lambda)$. □

In analogy to Proposition 3.5, we can now show the existence of arbitrary tensor products of R-modules.

Theorem 3.13 (Existence of Tensor Products) *Let R be a commutative ring and M_λ, $\lambda \in \Lambda$ a family of R-modules, then $\bigl(\bigotimes_{\lambda \in \Lambda} M_\lambda, \pi\bigr)$ is a tensor product of the M_λ.*

Proof Choose $\lambda_0 \in \Lambda$ arbitrarily and set for $r \in R$

$$L_r : \prod_{\lambda \in \Lambda} M_\lambda \to \prod_{\lambda \in \Lambda} M_\lambda, \quad f \mapsto L_r f$$

with

$$L_r f(\lambda) = \begin{cases} r \cdot f(\lambda) & \lambda = \lambda_0 \\ f(\lambda) & \lambda \neq \lambda_0 \end{cases}$$

to a $\mathbb{Z}$-module homomorphism $\overline{L}_r : A \to A$, where A as in Construction 3.12 is the free $\mathbb{Z}$-module generated by $\prod_{\lambda \in \Lambda} M_\lambda$. The mapping $\overline{L}_r$ preserves Y and factors into a $\mathbb{Z}$-module homomorphism $\widetilde{L}_r : \bigotimes_{\lambda \in \Lambda} M_\lambda \to \bigotimes_{\lambda \in \Lambda} M_\lambda$. Set $r \cdot f := \widetilde{L}_r(f)$ for all $f \in \bigotimes_{\lambda \in \Lambda} M_\lambda$. Thus, we have

$$r \cdot (\ldots \otimes m \otimes \ldots) = (\ldots \otimes r \cdot m \otimes \ldots),$$
$$\underset{\lambda_0}{\uparrow} \qquad \qquad \underset{\lambda_0}{\uparrow}$$

and it is easy to verify that $\bigotimes_{\lambda \in \Lambda} M_\lambda$ becomes an R-module (exercise; see Proposition 3.9).

Now let $\varrho \colon \prod_{\lambda \in \Lambda} M_\lambda \to C$ be as in Definition 3.11. Extend ϱ to a $\mathbb{Z}$-module homomorphism $\widetilde{\varrho} : A \to C$. Then $\widetilde{\varrho}$ vanishes on Y, because ϱ is R-multilinear. Thus, $\widetilde{\varrho}$ factors into a mapping $\overline{\varrho} : \bigotimes_{\lambda \in \Lambda} M_\lambda \to C$, which is initially only a $\mathbb{Z}$-module homomorphism (exercise; see Proposition 3.5).

As in Proposition 3.9(ii), we now show that $\overline{\varrho}$ is automatically an R-module homomorphism. □

As in Proposition 3.7, we can now see that Construction 3.12 is also natural (exercise!).

A number of further conclusions can be drawn from Theorem 3.13. For example, we can specify and at the same time sharpen the initial claim that tensor products of R-modules transform multilinear mappings into linear mappings. Analogous to the situation described in (3.3) for tensor products of two modules, we even obtain an identification of multilinear mappings on products with linear mappings on tensor products.

Remark 3.14 (Multilinear versus Linear) Let R be a commutative ring and M_λ, $\lambda \in \Lambda$ a family of R-modules. Then the assignment $\varrho \mapsto \overline{\varrho}$ provides a bijection

$$\Phi : L_R(M_\lambda; C) \longrightarrow \mathrm{Hom}_R \left(\bigotimes_{\lambda \in \Lambda} M_\lambda, C \right),$$

where C is another R-module (see Definition 3.1). With the pointwise R-module structures (see Exercise 2.5), Φ is even an R-module isomorphism (exercise). □

Since we have now obtained n-fold tensor products not as iterations of tensor products, the question arises again whether the result differs from an iteratively obtained tensor product. Using the universal property of tensor products, one sees that this is not the case, considering isomorphic modules as equal again.

3.1 Tensor Products

Corollary 3.15 (Associativity of the Tensor Product) *Let R be a commutative ring and $M_1, \ldots, M_k$ R-modules. Then for each successive tensor product $\big((M_1 \otimes_R \ldots) \otimes_R (\ldots \otimes_R M_k)\big)$ there is one (and only one) R-module isomorphism:*

$$\mu : \big((M_1 \otimes_R \ldots) \otimes_R (\ldots \otimes_R M_k)\big) \to M_1 \otimes \ldots \otimes M_k := \bigotimes_{j=1}^{k} M_j$$

with $\mu\big((m_1 \otimes \ldots) \otimes (\ldots \otimes m_k)\big) = m_1 \otimes \ldots \otimes m_k = \otimes_{j=1}^{k} m_j.$

Proof With Proposition 3.10 and induction, one sees (exercise) that

$$M_1 \times \ldots \times M_k \longrightarrow (M_1 \otimes_R \ldots) \otimes_R (\ldots \otimes_R M_k),$$
$$(m_1, \ldots, m_k) \longmapsto (m_1 \otimes \ldots) \otimes (\ldots \otimes m_k)$$

is a tensor product of the M_j, $j = 1, \ldots, k$. On the other hand, it follows from Theorem 3.13 that $\prod_{j=1}^{k} M_j \to \bigotimes_{j=1}^{k} M_j$ with $(m_1, \ldots, m_k) \mapsto \otimes_{j=1}^{k} m_j$ also defines a tensor product of the M_j, $j = 1 \ldots k$. Now one argues with the uniqueness of the tensor product (in the universal property):

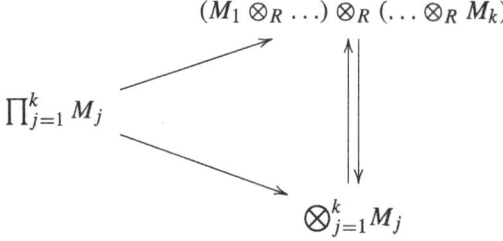

$\square$

We have already seen various analogies between tensor products and products of natural numbers. We have not addressed in Section 2.1 during the discussion of direct sums that there are analogies to the addition of natural numbers, e.g., an associative law. The analogies go even further: Forming tensor products of direct sums, one even gets a distributive law.

Theorem 3.16 (Distributive Law for Tensor Products) *Let R be a commutative ring and $M_1, \ldots, M_k$ R-modules. Furthermore, let each M_j be of the form $M_j = \bigoplus_{\lambda \in \Lambda_j} N_{j,\lambda}$. Set $\Gamma = \Lambda_1 \times \ldots \times \Lambda_k$ and*

$$N := \bigoplus_{\gamma \in \Gamma} \big(N_{1,\gamma_1} \otimes \ldots \otimes N_{k,\gamma_k}\big).$$

Then
$$M_1 \otimes \ldots \otimes M_k \cong N.$$

Proof Consider the $N_{j,\lambda} \leq M_j$ as submodules. Each of the mappings

$$\varphi_\gamma : N_{1,\gamma_1} \times \ldots \times N_{k,\gamma_k} \longrightarrow M_1 \otimes \ldots \otimes M_k,$$
$$(n_{1,\gamma_1}, \ldots, n_{k,\gamma_k}) \longmapsto n_{1,\gamma_1} \otimes \ldots \otimes n_{k,\gamma_k}$$

induces an R-module homomorphism

$$\overline{\varphi_\gamma} : N_{1,\gamma_1} \otimes \ldots \otimes N_{k,\gamma_1} \longrightarrow M_1 \otimes \ldots \otimes M_k,$$

and according to Proposition 2.17(ii) the $\overline{\varphi_\gamma}$ can be combined to a module homomorphism

$$\overline{\varphi} : N \longrightarrow M_1 \otimes \ldots \otimes M_k$$

which fulfills

$$\overline{\varphi}\Big(\sum_\gamma \underbrace{n_{1,\gamma_1} \otimes \ldots \otimes n_{k,\gamma_k}}_{\in N_{1,\gamma_1} \otimes \ldots \otimes N_{k,\gamma_k}}\Big) = \sum_\gamma \underbrace{n_{1,\gamma_1} \otimes \ldots \otimes n_{k,\gamma_k}}_{\in M_1 \otimes \ldots \otimes M_k}.$$

Set

$$\varrho : M_1 \times \ldots \times M_k \longrightarrow N,$$
$$(m_1, \ldots, m_k) \longmapsto \sum_\gamma \big(\pi_{\gamma_1}(m_1) \otimes \ldots \otimes \pi_{\gamma_k}(m_k)\big),$$

where $\pi_{\gamma_j} : M_j \to N_{j,\gamma_j}$ for $\gamma_j \in \Lambda_j$ is the projection onto the N_{j,γ_j} factor (see Proposition 2.17(i)). Then ϱ induces a module homomorphism $\overline{\varrho} : M_1 \otimes \ldots \otimes M_k \to N$. So it follows

$$\overline{\varphi} \circ \overline{\varrho}\Big(\sum_{\gamma_1} n_{1,\gamma_1} \otimes \ldots \otimes \sum_{\gamma_k} n_{k,\gamma_k}\Big) = \overline{\varphi} \circ \varrho\Big(\sum_{\gamma_1} n_{1,\gamma_1}, \ldots, \sum_{\gamma_k} n_{k,\gamma_k}\Big)$$
$$= \overline{\varphi}\Big(\sum_\gamma (n_{1,\gamma_1} \otimes \ldots \otimes n_{k,\gamma_k})\Big)$$
$$= \sum_\gamma (n_{1,\gamma_1} \otimes \ldots \otimes n_{k,\gamma_k})$$
$$= \Big(\sum_{\gamma_1} n_{1,\gamma_1}\Big) \otimes \ldots \otimes \Big(\sum_{\gamma_k} n_{k,\gamma_k}\Big),$$

3.1 Tensor Products

i.e., $\overline{\varphi} \circ \overline{\varrho}$ and $\mathrm{id}_{M_1 \otimes \ldots \otimes M_k}$ agree on the set $\{m_1 \otimes \ldots \otimes m_k \mid m_j \in M_j\}$, which generates the R-module $M_1 \otimes \ldots \otimes M_k$. Then $\overline{\varphi} \circ \overline{\varrho} = \mathrm{id}_{M_1 \otimes \ldots \otimes M_k}$. Similarly, one sees that $\overline{\varrho} \circ \overline{\varphi} = \mathrm{id}_N$. □

In Example 3.8 we saw that tensor products of embeddings of submodules need not be embeddings. Such a thing cannot happen when considering tensor products of free modules. In this case, the tensor products of basis elements themselves form a basis.

Proposition 3.17 (Tensor Products of Free Modules) *Let R be a commutative ring with one and $M_1, \ldots, M_k$ free R-modules with bases $E_1, \ldots, E_k$. Let $E := \{e_1 \otimes \ldots \otimes e_k \mid e_j \in E_j\}$. Then $M = M_1 \otimes \ldots \otimes M_k$ is free with basis E.*

Proof It is clear that $\langle E \rangle = M$. To show the R-independence of E, we prove the following claim: Let A and B be R-modules and $a_1, \ldots, a_p \in A$, $b_1, \ldots, b_q \in B$ R-independent, then also $\{a_i \otimes b_j \mid (i, j) \in \{1, \ldots, p\} \times \{1, \ldots, q\}\}$ is an R-independent set.
In fact,

$$0 = \sum_{i,j} r_{ij}(a_i \otimes b_j) = \sum_{i,j} r_{ij} a_i \otimes b_j = \sum_j \underbrace{\left(\sum_i r_{ij} a_i\right)}_{=:a_j'} \otimes b_j.$$

For $f \in \mathrm{Hom}_R(A, R)$, $g \in \mathrm{Hom}_R(B, R)$ (see Proposition 3.7) the mapping

$$f \otimes g : A \times B \to R, \quad (a, b) \mapsto f(a)g(b)$$

is R-bilinear. So there is $\overline{\varrho} : A \otimes_R B \to R$ with $\overline{\varrho} \circ \pi = f \otimes g$, i.e.,

$$0 = \overline{\varrho}\left(\sum_j a_j' \otimes b_j\right) = \sum_j f(a_j') \, g(b_j).$$

Now we choose g with $g\left(\sum_{j=1}^q r_j b_j\right) = r_l$ and get $f(a_l') = 0$ for $l = 1, \ldots, q$. If f is chosen so that $f\left(\sum_{i=1}^p r_i a_i\right) = r_n$, it follows that $f\left(\sum_{i=1}^p r_{ij} a_i\right) = r_{nj}$ for $n = 1, \ldots, p$. Together one gets $r_{ij} = 0$ for all for all i, and j and that proves the claim. □

A consequence of Proposition 3.17 is that for vector spaces the dimension formula

$$\dim_\mathbb{K} (V_1 \otimes_\mathbb{K} \ldots \otimes_\mathbb{K} V_k) = (\dim_\mathbb{K} V_1) \cdot \ldots \cdot (\dim_\mathbb{K} V_k)$$

holds. Together with the dimension formula (see Exercise 2.9)

$$\dim_{\mathbb{K}} (V_1 \oplus \ldots \oplus V_k) = \left(\dim_{\mathbb{K}} V_1\right) + \ldots + \left(\dim_{\mathbb{K}} V_k\right)$$

we can see the analogies to products and sums of numbers again.

Remark 3.18 (Finite Tensor Products of Vector Spaces) In books on linear algebra that also discuss tensor products, as well as in texts on differential geometry, a different construction of tensor products is often chosen. It only works for finite-dimensional vector spaces, but can be implemented without the detour via free modules: Let $\mathbb{K}$ be a field and $V_1, \ldots, V_k$ be finite-dimensional $\mathbb{K}$-vector spaces. Furthermore, let $V_1^*, \ldots, V_k^*$ be the corresponding dual spaces. Then the $\mathbb{K}$-vector space $L_{\mathbb{K}}(V_1^*, \ldots, V_k^*; \mathbb{K})$ of $\mathbb{K}$-multilinear mappings is a $\mathbb{K}$-vector space of dimension $\prod_{j=1}^{k} \dim_{\mathbb{K}}(V_j)$. We define for $(v_1, \ldots, v_k) \in V_1 \times \ldots \times V_k$ the $\mathbb{K}$-multilinear mapping $v_1 \otimes \ldots \otimes v_k \in L_{\mathbb{K}}(V_1^*, \ldots, V_k^*; \mathbb{K})$ by

$$(v_1 \otimes \ldots \otimes v_k)(f_1, \ldots, f_k) := \prod_{j=1}^{k} f_j(v_j).$$

Then the mapping

$$\varrho_0 \, V_1 \times \ldots \times V_k \to L_{\mathbb{K}}(V_1^*; \ldots, V_k^*; \mathbb{K}), \quad (v_1, \ldots, v_k) \mapsto v_1 \otimes \ldots \otimes v_k$$

is $\mathbb{K}$-multilinear. Thus, there is a uniquely determined $\mathbb{K}$-linear mapping

$$\overline{\varrho}_0 : V_1 \otimes \ldots \otimes V_k \longrightarrow L_{\mathbb{K}}(V_1^*, \ldots, V_k^*; \mathbb{K})$$

with $\overline{\varrho}_0 \circ \pi = \varrho_0$. Since $L_{\mathbb{K}}(V_1^*, \ldots, V_k^*; \mathbb{K})$ is spanned by the $v_1 \otimes \ldots \otimes v_k$, $\overline{\varrho}_0$ is surjective, and therefore bijective for dimensional reasons. By combining with this isomorphism, it is verified that $\left(L_{\mathbb{K}}(V_1^*, \ldots, V_k^*; \mathbb{K}), \varrho_0\right)$ is a tensor product of the V_j. This of course only works because we already know that $V_1 \otimes \ldots \otimes V_k$ is a tensor product. If we start with $\left(L_{\mathbb{K}}(V_1^*, \ldots, V_k^*; \mathbb{K}), \varrho_0\right)$, we have to show the universal property of the tensor product directly. □

Exercise 3.1 (Calculation Rules for Tensor Products of Vector Spaces) Let $\mathbb{K}$ be a field and V, W, U be finite-dimensional K-vector spaces. Show the following isomorphisms of $\mathbb{K}$-vector spaces:

(i) $(V \otimes W)^* \cong V^* \otimes W^*$.
(ii) $\mathrm{Hom}_{\mathbb{K}}(V, W) \cong V^* \otimes W$.
(iii) $L_{\mathbb{K}}(V, W; U) \cong \mathrm{Hom}_{\mathbb{K}}(V \otimes W, U)$.
(iv) $(V \oplus W) \otimes U \cong (V \otimes U) \oplus (W \otimes U)$.

3.2 Tensor Algebras

We have already pointed out several times the analogies between tensor products and products of numbers as well as between direct sums and sums of numbers. In this section, we go a step further and show that for commutative R on the direct sum of all tensor powers of a given R-module, an R-bilinear product can be introduced that makes this sum a ring. This ring is called the tensor algebra of the module.

Universal Property and Construction

Even though we have already hinted at how tensor algebras are constructed using tensor products, we still follow the strategy outlined in Section 3.1 in our treatment of tensor algebras, characterizing newly introduced objects first by universal properties and then ensuring their existence through construction.

Since the universal property refers to other algebras, we start with the formal definition of an R-algebra.

Definition 3.19 (R-Algebras) Let R be a commutative ring, A an R-module and $\cdot : A \times A \to A$ an R-bilinear multiplication mapping. Then $(A, \cdot)$ is called an *R-algebra*.

It should be emphasized that apart from R-bilinearity, no algebraic conditions are imposed on the multiplication. In particular, it is not required to be associative. In the case of an associative multiplication, the R-algebra is called *associative*. If the multiplication is commutative, the R-algebra is called *commutative*.

Example 3.20 (R-Algebras)

(i) Let R be a ring, then R is a $\mathbb{Z}$-algebra with respect to ring multiplication.
(ii) Let R be a commutative ring. Then R is an R-algebra with respect to ring multiplication.
(iii) Let R be commutative, then $\mathrm{Mat}(n \times n, R)$ is an R-algebra with respect to matrix multiplication.
(iv) Let R be commutative and M an R-module, then $\mathrm{Hom}_R(M, M)$ is an R-algebra with respect to the composition of mappings.
(v) Let $(A, \cdot)$ be an R-algebra, then A is also an R-algebra with respect to

$$(a, b) \mapsto a \cdot b - b \cdot a \quad \text{and} \quad (a, b) \mapsto a \cdot b + b \cdot a$$

□

Since R-algebras are generalizations of rings, it is not surprising that the elementary structural concepts of rings are also transferred to R-algebras (see Definition 1.8 and Definition 1.10).

Definition 3.21 (Subalgebras, Ideals and Homomorphisms) Let R be a commutative ring. If A is an R-algebra, then an R-submodule B of A is called a *subalgebra* of A, if $B \cdot B \subseteq B$ is. If even $B \cdot A$, $A \cdot B \subseteq B$ holds, B is called an *ideal* in A. For two R-algebras A and A', an R-module homomorphism $\varphi : A \to A'$ is called an *R-algebra homomorphism*, if

$$\forall a, b \in A: \quad \varphi(a \cdot b) = \varphi(a) \cdot \varphi(b)$$

and, in the case that both algebras have an identity, $\varphi(1_A) = 1_{A'}$, holds.

Definitions 3.19 and 3.21 allow us to formulate the universal property of the tensor algebra over a module.

Definition 3.22 (Universal Property of Tensor Algebras) Let R be a commutative ring with identity and M an R-module. A pair (T, ϱ) is called a *tensor algebra* over M, if:

(a) T is an associative R-algebra with identity.
(b) $\varrho : M \to T$ is R-linear, i.e., an R-module homomorphism.
(c) (T, ϱ) fulfills the following *universal property*: For every associative R-algebra A with identity and every R-module homomorphism $\varphi : M \to A$, there exists exactly one R-algebra homomorphism $\overline{\varphi} : T \to A$ with $\overline{\varphi} \circ \varrho = \varphi$, i.e., we have the following commutative diagram:

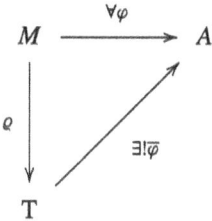

We now carry out the construction of a tensor algebra over M.

Construction 3.23 (Tensor Algebra over M) Let R be a commutative ring with identity and M an R-module. Then for every $n \in \mathbb{N}$, the tensor product $T^n(M) := M^{\otimes n} := M \otimes \ldots \otimes M$ (n factors) is an R-module. Set

$$T(M) := \bigoplus_{n \in \mathbb{N}_0} T^n(M),$$

3.2 Tensor Algebras

where $T^0(M) := R$. Then $T(M)$ is an R-module. Consider the mapping

$$\pi_{p,q}: T^p(M) \times T^q(M) \to T^p(M) \otimes T^q(M), \quad (a,b) \mapsto a \otimes b,$$

which makes $T^p(M) \otimes T^q(M)$ a tensor product of $T^p(M)$ and $T^q(M)$. Then

$$\Theta_{p,q}: T^p(M) \times T^q(M) \to T^{p+q}(M), \quad (a,b) \mapsto a \otimes b$$

according to Theorem 3.13 defines a canonical isomorphism

$$\overline{\Theta}_{p,q}: T^p(M) \otimes T^q(M) \to T^{p+q}(M)$$

with $\overline{\Theta}_{p,q}\big((m_1 \otimes \ldots \otimes m_p) \otimes (m'_1 \otimes \ldots \otimes m'_q)\big) = m_1 \otimes \ldots \otimes m_p \otimes m'_1 \otimes \ldots \otimes m'_q$ and $\Theta_{p,q} = \overline{\Theta}_{p,q} \circ \pi_{p,q}$. Now set

$$\Theta: T(M) \times T(M) \to T(M), \quad \left(\bigoplus_{p \geq 0} a_p, \bigoplus_{q \geq 0} b_q\right) \mapsto \bigoplus_{n \geq 0}\left(\sum_{n=p+q} \Theta_{p,q}(a_p, b_q)\right)$$

and obtain a well-defined R-bilinear mapping (exercise; see Proposition 2.17). To prove that this product makes $T(M)$ an associative R-algebra, it suffices to show

$$\forall a \in T^p(M), b \in T^q(M), c \in T^r(M): \quad \Theta\big(\Theta(a,b), c\big) = \Theta\big(a, \Theta(b,c)\big)$$

For this, we calculate

$$\Theta\big(\Theta(a,b), c\big) = \Theta\big(\Theta_{p,q}(a,b), c\big) = \Theta_{p+q,r}(a \otimes b, c) = a \otimes b \otimes c$$
$$= \Theta_{p,q+r}(a, b \otimes c) = \Theta\big(a, \Theta_{q,r}(b,c)\big) = \Theta\big(a, \Theta(b,c)\big)$$

and note that $1 \in R = T^0(M) \subset T(M)$ is a one for $T(M)$. $\square$

The proof that $(T(M), \varrho)$ is a tensor algebra over M is based on the universal properties of the tensor product and the direct sum.

Theorem 3.24 (Existence of the Tensor Algebra over M) *The pair $(T(M), \varrho)$ from Construction 3.23 is a tensor algebra over M.*

Proof For an R-linear mapping $\varphi M \to A$, consider for $n > 0$ the R-multilinear mapping

$$\tau_n: M \times \ldots \times M \to A, \quad (m_1, \ldots, m_n) \mapsto \varphi(m_1) \cdot \ldots \cdot \varphi(m_n).$$

Then, by the universal property of the tensor product (see Definition 3.11), there is an R-linear mapping $\overline{\tau}_n: M^{\otimes n} \longrightarrow A$ with $\overline{\tau}_n \circ \pi_n = \tau_n$, where $\pi_n: M^n \to M^{\otimes n}$

is the canonical mapping that makes $M^{\otimes n}$ a tensor product of the M factors. Further, we set

$$\overline{\tau}_0 : R \to A, \quad r \mapsto r \cdot 1_A.$$

With Proposition 2.17, which is based on the universal property of the direct sum, we combine the $\overline{\tau}_n$ into an R-module homomorphism

$$\overline{\varphi} : \bigoplus_{n \geq 0} M^{\otimes n} \longrightarrow A$$

with $\overline{\varphi} \circ \iota_n = \overline{\tau}_n$, where $\iota_n : M^{\otimes n} \hookrightarrow \bigoplus_{n \geq 0} M^{\otimes n} = \mathrm{T}(M)$ is the canonical injection. We obtain $\overline{\varphi} \circ \varrho = \overline{\varphi} \circ \iota_1 = \overline{\tau}_1 = \tau_1 = \varphi$ and it remains (for the statement of existence) to show that $\overline{\varphi} : \mathrm{T}(M) \to A$ is an R-algebra homomorphism. The identity $\overline{\varphi}(1) = 1_A$ is clear from the definitions. It is now sufficient to prove

$$\forall a \in \mathrm{T}^p(M),\ b \in \mathrm{T}^q(M): \quad \overline{\varphi}(a \otimes b) = \overline{\varphi}(a)\,\overline{\varphi}(b).$$

If $a = m_1 \otimes \ldots \otimes m_p$, $b = m'_1 \otimes \ldots \otimes m'_q$, then we calculate

$$\overline{\varphi}(a \otimes b) = \overline{\varphi}(m_1 \otimes \ldots \otimes m_p \otimes m'_1 \otimes \ldots \otimes m'_q)$$
$$= \varphi(m_1) \cdot \ldots \cdot \varphi(m_p)\,\varphi(m'_1) \cdot \ldots \cdot \varphi(m'_q)$$
$$= \overline{\varphi}(m_1 \otimes \ldots \otimes m_q)\,\overline{\varphi}(m'_1 \otimes \ldots \otimes m'_q)$$
$$= \overline{\varphi}(a)\,\overline{\varphi}(b).$$

Now the claim follows with R-linearity.

To show the uniqueness statement, we assume that $\overline{\psi} : \mathrm{T}(M) \to A$ with $\overline{\psi} \circ \varrho = \varphi$ is given. Then $\overline{\psi}(m_1 \otimes \ldots \otimes m_n) = \varphi(m_1) \cdots \varphi(m_n)$ and $\overline{\psi}(r) = r \cdot 1_A$, which in turn implies $\overline{\psi} = \overline{\varphi}$, because the set $\{m_1 \otimes \ldots \otimes m_n \mid n > 0,\ m_j \in M\} \cup \{1\}$ generates the algebra $\mathrm{T}(M)$. □

We have already seen that direct sums and tensor products are natural constructions (see the explanation of *naturality* after Example 3.6 and Proposition 3.7). The naturality of tensor algebras is obtained directly from the universal property.

Proposition 3.25 (Extension of Module Homomorphisms) *Let R be a commutative ring with identity, M and N be R-modules and $\varphi \in \mathrm{Hom}_R(M, N)$. Then there exists exactly one R-algebra homomorphism $\mathrm{T}(\varphi) : \mathrm{T}(M) \to \mathrm{T}(N)$ with $\mathrm{T}(\varphi)|_M = \varphi$.*

Proof Because 1 and M generate the algebra $\mathrm{T}(M)$ and by definition $\mathrm{T}(\varphi)(1) = 1$ should hold, the uniqueness is clear. The existence follows from the universal property (see Definition 3.22), applied to $\varrho_N \circ \varphi : M \to \mathrm{T}(N)$, where $\varrho_N : N \hookrightarrow \mathrm{T}(N)$ is the canonical inclusion. □

The extension $\mathrm{T}(\varphi)$ of a module homomorphism φ not only preserves the product, but also the grading of the tensor algebras, i.e., with the notation from Proposition 3.25, it holds that $\mathrm{T}(\varphi)\big(\mathrm{T}^q(M)\big) \subset \mathrm{T}^q(N)$. This follows immediately from the equation

$$\mathrm{T}(\varphi)(m_1 \otimes \ldots \otimes m_q) = \varphi(m_1) \otimes \ldots \otimes \varphi(m_q).$$

Example 3.26 (Tensor Fields) Let $U \subseteq \mathbb{R}^m$ be an open subset and $R := C^\infty(U, \mathbb{R})$. Consider the R-module $M := C^\infty(U, \mathbb{R}^m)$ of vector fields as in Example 2.3. The elements of $\mathrm{T}(M)$ are called *tensor fields* on U. The algebra $\mathrm{T}(M)$ is naturally isomorphic to the algebra $\bigoplus_{n \in \mathbb{N}_0} C^\infty\big(U, \mathrm{T}^n(\mathbb{R}^m)\big)$ with pointwise operations. To be able to write down the R-module isomorphism

$$C^\infty\big(U, \mathrm{T}^n(\mathbb{R}^m)\big) \cong \mathrm{T}^n(M) \tag{3.6}$$

one first notes (exercise) that for a finite-dimensional $\mathbb{R}$-vector space V it holds:

$$C^\infty(U, V) \cong C^\infty(U, \mathbb{R}) \otimes_\mathbb{R} V = R \otimes_\mathbb{R} V$$

as R-modules. With this one finds

$$\bigoplus_{n \in \mathbb{N}_0} C^\infty\big(U, \mathrm{T}^n(\mathbb{R}^m)\big) \cong R \otimes_\mathbb{R} \bigoplus_{n \in \mathbb{N}_0} \mathrm{T}^n(\mathbb{R}^m) = R \otimes_\mathbb{R} \mathrm{T}(\mathbb{R}^m).$$

Let A be an R-algebra. Every R-module homomorphism $R \otimes_\mathbb{R} \mathbb{R}^m \to A$ is uniquely determined by an $\mathbb{R}$-linear mapping $\mathbb{R}^m \to A$, which in turn can be uniquely extended to an $\mathbb{R}$-algebra homomorphism $\mathrm{T}(\mathbb{R}^m) \to A$. Let $\rho : \mathbb{R}^m \to \mathrm{T}(\mathbb{R}^m)$ be the mapping that makes $(\mathrm{T}(\mathbb{R}^m), \rho)$ a tensor algebra over $\mathbb{R}^m$. Then the pair $(R \otimes_\mathbb{R} \mathrm{T}(\mathbb{R}^m), {}_R\rho)$ with

$${}_R\rho : R \otimes_\mathbb{R} \mathbb{R}^m \to R \otimes_\mathbb{R} \mathrm{T}(\mathbb{R}^m), \quad r \otimes v \mapsto r \otimes \rho(v)$$

is a tensor algebra over $R \otimes_\mathbb{R} \mathbb{R}^m \cong C^\infty(U, \mathbb{R}^m)$. Since the universal property uniquely determines the tensor algebra up to isomorphism, (3.6) follows. □

3.3 Symmetric and Exterior Algebras

We motivated the introduction of tensor products by the fact that they allow us to transform multilinear mappings into linear mappings. If additional conditions are imposed on the multilinear mappings, this of course does not change the fact that they can be written as linear mappings on tensor products. But this generates redundancy, meaning it would be possible to find a smaller module on which the special multilinear mapping can be written as a linear mapping. We show how this works in this section for symmetric and antisymmetric multilinear mappings. One could proceed in the same way as we did with the tensor products and construct the so-called *symmetric* and *exterior* products. With the help of tensor algebras, however, we can make the constructions a bit more efficient and construct all symmetric or outer products at once by introducing the *symmetric* and the *exterior algebra*.

Exercise 3.2 (Tensor Products of Hom Spaces) Let R be a commutative ring with identity and $M_1, \ldots, M_k$ as well as A a commutative R-algebra. Show that there is an R-module homomorphism

$$\bigotimes_{j=1}^{k} \mathrm{Hom}_R(M_j, A) \to \mathrm{Hom}_R\left(\bigotimes_{j=1}^{k} M_j, A\right).$$

Hint: First construct an R-module homomorphism

$$\bigotimes_{j=1}^{k} \mathrm{Hom}_R(M_j, A) \to \mathrm{L}_R(M_1, \ldots, M_k; A).$$

The Symmetric Algebra over M

Symmetric algebras can be obtained as subalgebras of tensor algebras, but they can also be characterized by a universal property, just like tensor products and tensor algebras. Following the general philosophy of this chapter, we will take this approach. The construction becomes significantly simpler as it reduces to forming a quotient of the tensor algebra.

Definition 3.27 (Symmetric Algebra over M) Let R be a commutative ring with identity and M an R-module. A pair (S, ψ) is called a *symmetric algebra* over M if:

(a) S is an associative R-algebra with identity.
(b) $\psi : M \to S$ is an R-module homomorphism with

$$\forall m, m' \in M : \quad \psi(m)\psi(m') = \psi(m')\psi(m).$$

(c) (S, ψ) satisfies the following *universal property*: For every associative R-algebra A with identity and every R-module homomorphism $\varphi : M \to A$ with

$$\forall m, m' \in M : \quad \varphi(m)\varphi(m') = \varphi(m')\varphi(m) \tag{3.7}$$

3.3 Symmetric and Exterior Algebras

there exists exactly one R-algebra homomorphism $\overline{\varphi}: S \to A$ with $\varphi = \overline{\varphi} \circ \psi$, i.e., we have the following commutative diagram:

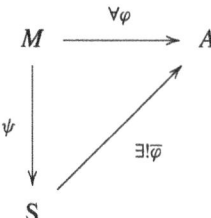

The difference between the universal properties of symmetric and tensor algebras over modules consists only in the additional condition (3.7) on ψ and the mapping φ to be lifted. Thus, one finds an extension of φ to a uniquely determined algebra homomorphism $\overline{\varphi}: T(M) \to A$ with $\varphi = \overline{\varphi} \circ \varrho$ (see Definition 3.22). If this is factored by an R-algebra homomorphism $\pi: T(M) \to S$ with a commutative algebra S, then one can set $\psi := \pi \circ \varrho$. This consideration leads to the following construction.

Construction 3.28 (Symmetric Algebra over M) Let J be the (two-sided) ideal of $T(M)$ generated by $\{m \otimes m' - m' \otimes m \mid m, m' \in M\}$. Then the quotient $T(M)/J$ is both an R-module and a ring. More precisely, $T(M)/J$ is an R-algebra (exercise!). We denote it by $S(M)$. We denote the composition of $M \hookrightarrow T(M)$ and $T(M) \to S(M) = T(M)/J$ as $\psi: M \to S(M)$.

The algebra $S(M)$ is associative with identity by construction. But it is also commutative, because due to $(m + J)(m' + J) = m \otimes m' + J$ it holds

$$(m + J)(m' + J) - (m' + J)(m + J) = (m \otimes m' - m' \otimes m) + J = J.$$

Thus, the condition from Definition 3.27(b) for $S(M)$ is fulfilled. □

Theorem 3.29 (Existence of the Symmetric Algebra over M) $(S(M), \psi)$ *is a symmetric algebra over M.*

Proof Let $\varphi: M \to A$ be as in the definition of the symmetric algebra. The uniqueness of $\overline{\varphi}$ is clear, because $1 + J$ and $\psi(M)$ generate the algebra $S(M)$.

To prove existence, we note that by Definition 3.22 there is exactly one R-algebra homomorphism $\Phi: T(M) \to A$ with $\Phi|_M = \varphi$. With this, one calculates

$$\Phi(m \otimes m' - m' \otimes m) = \Phi(m \otimes m') - \Phi(m' \otimes m) = \Phi(m)\Phi(m') - \Phi(m')\Phi(m)$$
$$= \varphi(m)\varphi(m') - \varphi(m')\varphi(m) = 0$$

and obtains $\Phi|_J \equiv 0$. Thus, Φ factors to an R-algebra homomorphism $\overline{\varphi}: S(M) \to A$, for which then holds:

$$\forall m \in M: \quad \overline{\varphi} \circ \psi(m) = \overline{\varphi}(m + J) = \varphi(m).$$

Since by construction the condition from Definition 3.27(b) for $S(M)$ is also fulfilled, $S(M)$ is a symmetric algebra over M. □

As for all previous constructions, the next step is to show the naturality. This works quite analogously to the case of the tensor algebra solely with the universal property.

Proposition 3.30 (Naturality of the Symmetric Algebra) *Let R be a commutative ring with identity, M and N two R-modules, and $\psi \in \mathrm{Hom}_R(M, N)$. Then there is exactly one R-algebra homomorphism $S(\psi): S(M) \to S(N)$ with $S(\psi) \circ \psi_M = \psi_N \circ \psi$, where $\psi_M : M \to S(M)$ and $\psi_N : N \to S(N)$ are the mappings that make $(S(M), \psi_M)$ and $(S(N), \psi_N)$ symmetric algebras over M and N, respectively. Thus, the following diagram is commutative:*

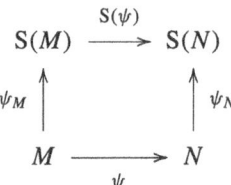

Proof Because $1 + J$ and $\psi_M(M)$ generate the algebra $S(M)$ and by definition $S(\psi_M)(1) = 1$ should hold, the uniqueness is clear. The existence follows from the universal property, applied to $\psi_N \circ \psi : M \to S(N)$. □

Unlike in the case of the tensor algebra, it is not obvious that the mapping $\psi : M \to S(M)$, which makes $(S(M), \psi)$ a symmetric algebra, is injective. In principle, it could be that elements of M are identified with each other during the formation of the quotient. However, this is not the case.

Proposition 3.31 (Embedding of M in $S(M)$) *Let R be a commutative ring with identity and M an R-module. In the notation of Construction 3.28, the mapping $\psi : M \to S(M)$ is injective.*

Proof As an abelian group, J is a direct sum:

$$J = \bigoplus_{q \geq 2} (J \cap T^q(M)),$$

3.3 Symmetric and Exterior Algebras

since J is generated by $\{m \otimes m' - m' \otimes m \mid m, m' \in M\} \subseteq T^2(M)$, and every element of $T(M)$ can be written as a sum $h = \sum_{j \geq 0} h_j$ of homogeneous elements $h_j \in T^j(M)$. It follows that

$$(m \otimes m' - m' \otimes m)h = \sum_{j \geq 0} \underbrace{(m \otimes m' - m' \otimes m)h_j}_{\in T^{j+2}(M)}$$

and a similar formula for multiplication from the left. Thus, every element in J can be written as a sum of elements in $J \cap T^q(M)$ with $q \geq 2$. If now $\psi(m) = 0$ holds, i.e., $m \in M \cap J = T^1(M) \cap J$, then it follows $m = 0$. □

We now return to the original motivation for the introduction of the symmetric algebra, namely the conversion of symmetric multilinear mappings into linear mappings.

Definition 3.32 (Symmetric Mappings) Let M and N be sets and $f: M^q \to N$ a mapping. Then f is called *symmetric*, if for all elements $\sigma \in S_q$ of the symmetric group on q letters, it holds that

$$\forall (m_1, \ldots, m_q) \in M^q: \quad f(m_1, \ldots, m_q) = f(m_{\sigma(1)}, \ldots, m_{\sigma(q)}).$$

The R-module on which a symmetric R-multilinear mapping in q variables becomes an R-linear mapping is the quotient module $S^q(M) := T^q(M)/J \cap T^q(M)$, where J is the ideal from Construction 3.28. This module is called the q-th *symmetric power* of M.

Proposition 3.33 (Symmetric Multilinear Mappings) *Let R be a commutative ring with identity and M, N be R-modules.*

(i) *Let $\gamma \in \mathrm{Hom}_R(S^q(M), N)$. Then*

$$\widetilde{\gamma}: (m_1, \ldots, m_q) \longmapsto \gamma(m_1 \otimes \ldots \otimes m_q + J)$$

defines an element $\widetilde{\gamma} \in L_R(M, \ldots, M; N)$, which is symmetric.

(ii) $\mathrm{Sym}_R(M^q; N) := \{\widetilde{\gamma} \in L_R(M, \ldots, M; N) \mid \widetilde{\gamma} \text{ is symmetric}\}$ *is an R-submodule of $L_R(M, \ldots, M; N)$.*

(iii) *The assignment $\gamma \mapsto \widetilde{\gamma}$ provides an R-module isomorphism*

$$\mathrm{Hom}_R(S^q(M), N) \longrightarrow \mathrm{Sym}_R(M^q; N).$$

Proof Remark 3.14 for $M^{\otimes q}$ provides an R-module isomorphism

$$\Psi : \mathrm{Hom}_R\left(\mathrm{T}^q(M), N\right) \longrightarrow \mathrm{L}_R(M, \ldots, M; N)$$

with $\Psi(\varphi)(m_1, \ldots, m_q) = \varphi(m_1 \otimes \ldots \otimes m_q)$. This gives us the commutative diagram

$$\begin{array}{ccc}
\mathrm{Hom}_R\left(\mathrm{T}^q(M), N\right) & \xrightarrow{\Psi} & \mathrm{L}_R(M, \ldots, M; N) \\
j \uparrow {\scriptstyle \gamma \mapsto \gamma \circ p_q} & & \uparrow \\
\mathrm{Hom}_R\left(\mathrm{S}^q(M), N\right) & \longrightarrow & \mathrm{Sym}_R(M^q; N)
\end{array}$$

where $p_q \colon \mathrm{T}^q(M) \to \mathrm{S}^q(M)$ is the quotient mapping. The assignment $\gamma \mapsto j(\gamma) := \gamma \circ p_q$ is an R-module homomorphism, injective with image

$$B = \{\varphi \in \mathrm{Hom}_R(\mathrm{T}^q(M), N) \mid \varphi|_{\mathrm{J} \cap \mathrm{T}^q(M)} = 0\}.$$

Also, $\Psi \circ j(\gamma)(m_1, \ldots, m_q) = (m_1 \otimes \ldots \otimes m_q + \mathrm{J}) = \tilde{\gamma}(m_1, \ldots, m_q)$. It remains to show that $\Psi(B) = \mathrm{Sym}_R(M^q, N)$. For this, we calculate

$$\Psi \circ j(\gamma)(m_1, \ldots, m_p, m, m', m_{p+3}, \ldots, m_q) - \Psi \circ j(\gamma)(\ldots, m', m, \ldots)$$
$$= j(\gamma)(\ldots \otimes m \otimes m' \otimes \ldots) - j(\gamma)(\ldots \otimes m' \otimes m \otimes \ldots)$$
$$= j(\gamma)(\underbrace{\ldots \otimes (m \otimes m' - m' \otimes m) \otimes \ldots}_{\in \mathrm{J} \cap \mathrm{T}^q(M)}) = 0$$

and conclude that $\Psi \circ j(\gamma)$ is symmetric. But this means $\Psi(B) \subset \mathrm{Sym}_R(M^q; N)$, because S_q is generated by the transpositions.

Conversely, if $\Psi(\varphi)$ is symmetric, then φ vanishes on elements of the form $\ldots \otimes (m \otimes m' - m' \otimes m) \otimes \ldots$ and thus on $\mathrm{J} \cap \mathrm{T}^q(M)$. So φ is of the form $j(\gamma)$. □

At the beginning of this section, it was claimed that the symmetric algebra can be understood as a subalgebra of the tensor algebra, but it was then constructed as a quotient of the tensor algebra. To embed the symmetric algebra into the tensor algebra, one considers *symmetric tensors*.

Definition 3.34 (Symmetric and Antisymmetric Tensors) Let R be a commutative ring with identity and M an R-module. For each permutation $\sigma \in S_q$, the mapping

$$\varphi_\sigma : M^q \to M^{\otimes q} = \mathrm{T}^q(M), \quad (m_1, \ldots, m_q) \mapsto m_{\sigma^{-1}(1)} \otimes \ldots \otimes m_{\sigma^{-1}(q)}$$

3.3 Symmetric and Exterior Algebras

is R-multilinear. Let $\Phi_\sigma : T^q(M) \longrightarrow T^q(M)$ be the associated R-module homomorphism with

$$\Phi_\sigma(m_1 \otimes \ldots \otimes m_q) = m_{\sigma^{-1}(1)} \otimes \ldots \otimes m_{\sigma^{-1}(k)}.$$

We simply write $\sigma \cdot t$ instead of $\Phi_\sigma(t)$.

An element $t \in T^q(M)$ is called a *symmetric tensor*, if

$$\forall \sigma \in S_q : \quad \sigma \cdot t = t,$$

and *antisymmetric tensor*, if

$$\forall \sigma \in S_q : \quad \sigma \cdot t = \text{sign}(\sigma)t,$$

where $\text{sign}(\sigma)$ is the sign of the permutation σ, which can be determined, e.g., as the determinant of the associated permutation matrix. We denote the set of symmetric tensors in $T^q(M)$ with $S_q(M)$ and the set of antisymmetric tensors in $T^q(M)$ with $\Lambda_q(M)$.

We have included the definition of antisymmetric tensors here because we will need them in the discussion of exterior algebras.

Proposition 3.35 (Symmetrization) *Let R be a commutative ring with identity, for which $q!$ is a unit, and M a R-module. Then the mapping*

$$\text{sym} : T^q(M) \to S_q(M), \quad t \mapsto \frac{1}{q!} \sum_{\sigma \in S_q} \sigma \cdot t$$

is a surjective R-module homomorphism with $\text{sym} \circ \text{sym} = \text{sym}$ *(i.e., a projection). The kernel* $\ker(\text{sym}) = \{t \in T^q(M) \mid \text{sym}(t) = 0\}$ *is* $J \cap T^q(M)$ *with the ideal J from Definition 3.28.*

Proof For $\sigma', \sigma \in S_q$ we have $\varphi_{\sigma'\sigma} = \varphi_{\sigma'} \circ \varphi_\sigma$, and the uniqueness statement in the universal property of $S(M)$ yields that $\Phi_{\sigma'\sigma} = \Phi_{\sigma'} \circ \Phi_\sigma$ also holds. With this we calculate

$$\sigma' \cdot \text{sym}(t) = \sigma' \cdot \frac{1}{q!} \sum_{\sigma \in S_q} \sigma \cdot t = \frac{1}{q!} \sum_{\sigma \in S_q} (\sigma \cdot t) = \frac{1}{q!} \sum_{\sigma \in S_q} (\sigma'\sigma) \cdot t$$

$$= \frac{1}{q!} \sum_{\sigma'' \in S_q} \sigma'' \cdot t = \text{sym}(t),$$

i.e., we find sym: $T^q(M) \to S_q(M)$. If $t \in S_q(M)$, then

$$\text{sym}(t) = \frac{1}{q!} \sum_{\sigma \in S_q} \sigma \cdot t = \frac{1}{q!} q! t = t,$$

so sym $\circ$ sym = sym. Because $\Phi_\sigma : T^q(M) \to T^q(M)$ is an R-module homomorphism for each σ, it is easy to verify that sym is also an R-module homomorphism.

By the proof of Proposition 3.31, $J_q := J \cap T^q(M)$ is generated as an R-module by elements of the form

$$\left(\ldots \otimes (m \otimes m') \otimes \ldots \right) - \left(\ldots \otimes (m' \otimes m) \otimes \ldots \right),$$

i.e., by elements $t - \tau_{j,j+1} \cdot t$ with

$$\tau_{j,j+1} := \begin{pmatrix} 1 & j & j+1 & \ldots & q \\ 1 & j+1 & j & \ldots & q \end{pmatrix} \in S_q.$$

If $\sigma \in S_q$ with $\sigma = \sigma_1 \sigma_2$, then $t - \sigma \cdot t = (t - \sigma_2 \cdot t) + (\sigma_2 \cdot t - \sigma_1 \sigma_2 \cdot t)$. So we find

$$J_q = \langle t - \sigma \cdot t \mid \sigma \in S_q, t \in T^q(M) \rangle_{R\text{-module}}$$

(the R-module generated by the $t - \sigma \cdot t$ with $\sigma \in S_q$ and $t \in T^q(M)$), because S_q is generated by the $\tau_{j,j+1}$. Due to

$$\text{sym}(t - \sigma \cdot t) = \frac{1}{q!} \sum_{\sigma' \in S_q} \sigma' \cdot t - \frac{1}{q!} \sum_{\sigma' \in S_q} \sigma'\sigma \cdot t = 0$$

we have $J_q \subseteq \ker(\text{sym})$. Conversely, if $t \in \ker(\text{sym})$, then $t = t - \text{sym}(t) \in J_q$, because

$$q!(t - \text{sym}(t)) = q!t - \sum_{\sigma \in S_q} \sigma \cdot t = \sum_{\sigma \in S_q} (t - \sigma \cdot t) \in J_q.$$

□

With the symmetrization mapping, we can now at least for rings that are algebras over a field of characteristic zero, embed the symmetric algebra as an R, submodule into the tensor algebra.

Corollary 3.36 (Symmetric Tensors and Symmetric Powers) *Let $q!$ be a unit in R. Then:*

(i) $T^q(M) = S_q(M) \oplus J_q$ *as an R-module.*

(ii) *The mapping $p_q|_{S_q(M)} \colon S_q(M) \to S^q(M)$ is an R-module isomorphism, where $p_q \colon T^q(M) \to S^q(M)$ is the canonical projection.*
(iii) *Let $\varphi \colon \bigoplus_{q \in \mathbb{N}_0} S_q(M) \to S(M) = \bigoplus_{q \in \mathbb{N}_0} S^q(M)$ be the isomorphism of R-modules composed from the $p_q|_{S_q(M)}$. Then the algebra multiplication transported by φ^{-1} from $S(M)$ to $\bigoplus_{q \in \mathbb{N}_0} S_q(M)$ is given by*

$$S_q(M) \times S_{q'}(M) \to S_{q+q'}(M), \quad (a, a') \mapsto \mathrm{sym}(a \otimes a').$$

Proof

(i) For $t \in T^q(M)$, $\mathrm{sym}(t) \in S_q(M)$ and $t - \mathrm{sym}(t) \in J_q$, so we have $T^q(M) = S_q(M) + J_q$. To prove the directness of the sum, we note that $t \in S_q(M) \cap \ker(\mathrm{sym}) = J_q$ implies the identity $t = \mathrm{sym}(t) = 0$.
(ii) Because $J_q = \ker(p_q)$, (i) provides both $p_q(S_q(M)) = p_q(T^q(M)) = S^q(M)$ and $\ker(p_q) \cap S_q(M) = \{0\}$. According to the first isomorphism theorem (see Construction 2.9), this implies the assertion.
(iii) For $(a, a') \in S_q(M) \times S_{q'}(M)$, it holds

$$\varphi(a)\varphi(a') = (a + J)(a' + J) = (a \otimes a') + J = \mathrm{sym}(a \otimes a') + J.$$

Since $\mathrm{sym} \colon T^{q+q'}(M) \to S_{q+q'}(M)$ is just the projection with kernel $J_{q+q'}$, it follows that $\varphi^{-1}(\varphi(a)\varphi(a')) = \mathrm{sym}(a \otimes a')$. □

The Exterior Algebra over M

For the exterior algebras, mutatis mutandis, the same motivations, theorems, and proofs apply as for the symmetric algebras. The difference is that now one wants to consider not symmetric multilinear mappings, but *antisymmetric* ones, which are also called *skew-symmetric* or *alternating*. In the case of tensors, this leads to the skew-symmetric tensors from Definition 3.34, and as an algebra, one obtains the *exterior* algebra.

Definition 3.37 (Exterior Algebra over M) Let R be a commutative ring with identity and M an R-module. A pair (Λ, η) is called an *exterior algebra* over M, if the following holds:

(a) Λ is an associative R-algebra with identity.
(b) $\eta \colon M \to \Lambda$ is an R-module homomorphism with

$$\forall m \in M: \quad \eta(m)^2 = 0.$$

(c) (Λ, η) satisfies the following *universal property*: For every associative R-algebra A with identity and every R-module homomorphism $\varphi : M \to A$ with

$$\forall m \in M : \quad \varphi(m)^2 = 0 \tag{3.8}$$

there exists exactly one R-algebra homomorphism $\overline{\varphi} : \Lambda \to A$ with $\varphi = \overline{\varphi} \circ \eta$, i.e., we have the following commutative diagram:

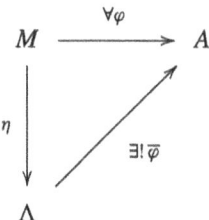

Remark 3.38 (Finitely Generated Modules) Let R be a commutative ring with identity, M an R-module and (Λ, η) an exterior algebra over M. If M is generated by the elements $e_1, \ldots, e_k \in M$, then for every $k + 1$-tuple $(m_1, \ldots, m_{k+1})$ of elements in M, $\eta(m_1) \wedge \ldots \wedge \eta(m_{k+1}) = 0$, where $\wedge$ denotes the multiplication in Λ. To see this, write each of the m_j as an R-linear combination of the e_i and then multiply out. The result is a linear combination of expressions of the form $\eta(e_{i_1}) \wedge \ldots \wedge \eta(e_{i_{k+1}})$. Note that at least one e_i must appear twice in this expression. Because

$$0 = \eta(e_i + e_j) \wedge \eta(e_i + e_j)$$
$$= \eta(e_i) \wedge \eta(e_i) + \eta(e_i) \wedge \eta(e_j) + \eta(e_j) \wedge \eta(e_i) + \eta(e_j) \wedge \eta(e_j)$$
$$= \eta(e_i) \wedge \eta(e_j) + \eta(e_j) \wedge \eta(e_i)$$

we have $\eta(e_i) \wedge \eta(e_j) = -\eta(e_j) \wedge \eta(e_i)$ for $i, j \in \{1, \ldots, k\}$. Therefore, the order of the e_{i_m} in the above expression can only be changed at the cost of a sign. This way, the two identical e_i's are brought next to each other and we find that $\eta(e_{i_1}) \wedge \ldots \wedge \eta(e_{i_{k+1}}) = 0$. □

Construction 3.39 (Exterior Algebra over M) Let I be the (two-sided) ideal of $T(M)$ generated by $\{m \otimes m \mid m \in M\}$. Then the quotient $\Lambda(M) := T(M)/I$ is an R-algebra (exercise). We denote the product on $\Lambda(M)$ by $\wedge : \Lambda(M) \times \Lambda(M) \to \Lambda(M)$ and the composition of $M \hookrightarrow T(M)$ and $T(M) \to \Lambda(M)$ by $\eta \colon T(M) \to \Lambda(M)$. □

Theorem 3.40 (Existence of the Exterior Algebra over M) $\bigl(\Lambda(M), \eta\bigr)$ *is an exterior algebra over M.*

3.3 Symmetric and Exterior Algebras

Proof $\Lambda(M)$ is associative as a quotient of an associative algebra, and η is an R-module homomorphism. Now let $\varphi : M \to A$ be as in the definition of the exterior algebra. The uniqueness of $\overline{\varphi}$ is clear, because $1 + I$ and $\{m + I \mid m \in M\}$ generate the algebra $\Lambda(M)$. One finds an R-algebra homomorphism $\Phi : T(M) \to A$ with $\varphi = \Phi|_M$. Because $\varphi(m)^2 = 0$ for all $m \in M$, we have

$$\Phi(m \otimes m) = \Phi(m) \Phi(m) = \varphi(m)^2 = 0.$$

This yields $\Phi|_I \equiv 0$, so Φ factorizes to an R-algebra homomorphism $\overline{\varphi} : T(M)/I \to A$ with $\overline{\varphi} \circ \eta(m) = \overline{\varphi}(m + I) = \Phi(m) = \varphi(m)$.

It remains to show that $\eta(m) \wedge \eta(m) = 0$ for all $m \in M$. But this is clear with the calculation

$$\eta(m) \wedge \eta(m) = (m + I)(m + I) = m \otimes m + I = I = 0 \in T(M)/I.$$

□

Proposition 3.41 (Naturality of the Exterior Algebra) *Let R be a commutative ring with identity, M and N two R-modules and $\psi \in \mathrm{Hom}_R(M, N)$. Then there exists exactly one R-algebra homomorphism $\Lambda(\psi) \colon \Lambda(M) \to \Lambda(N)$ with*

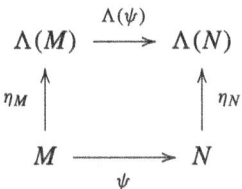

where $\eta_M \colon M \to \Lambda(M)$ and $\eta_N \colon N \to \Lambda(N)$ are the mappings that make $(\Lambda(M), \psi_M)$ and $(\Lambda(N), \psi_N)$ exterior algebras over M and N, respectively.

Proof Because $1 + I$ and $\eta_M(M)$ generate the algebra $\Lambda(M)$ and by definition $\Lambda(\psi)(1) = 1$ should hold, the uniqueness is clear. The existence follows from the universal property of the exterior algebra (see Definition 3.37), applied to the mapping $\eta_N \circ \psi : M \to \Lambda(N)$. □

Example 3.42 (Differential Forms) Let $U \subseteq \mathbb{R}^m$ be an open subset and $R := C^\infty(U, \mathbb{R})$. Consider the R-module $M := C^\infty(U, \mathbb{R}^m)$ of vector fields as in Example 2.3. The elements of the dual R-module $M^\vee := \mathrm{Hom}_R(M, R) \cong C^\infty(U, (\mathbb{R}^m)^*)$ are called 1-*forms* on U. The elements of $\Lambda(M^\vee)$ are called *differential forms* on U. As for the tensor fields in Example 3.26, one shows that the algebra $\Lambda(M^\vee)$ is naturally isomorphic to the algebra $C^\infty(U, \Lambda((\mathbb{R}^m)^*)) = R \otimes_\mathbb{R} \Lambda((\mathbb{R}^m)^*)$. The proof is even somewhat simplified by the fact that $\Lambda((\mathbb{R}^m)^*)$, unlike $T((\mathbb{R}^m)^*)$, is finite-dimensional (see Remark 3.38, which shows that for $q > m$, $T^q((\mathbb{R}^m)^*) \subseteq I$). □

Proposition 3.43 (Embedding of M in $\Lambda(M)$) *Let R be a commutative ring with identity and M an R-module. In the notation of Construction 3.39, $\eta : M \to \Lambda(M)$ is injective.*

Proof As in the proof of Proposition 3.31, one shows (exercise) that

$$I = \bigoplus_{q \geq 2} (I \cap T^q(M)).$$

If $\eta(m) = 0$, then $m \in M \cap I = T^1(M) \cap I$ and therefore $m = 0$. □

In contrast to the symmetry of mappings (see Definition 3.32), antisymmetry can only be formulated if the elements of the range can be multiplied by ± 1. Here, we are content to define antisymmetric multilinear mappings.

Definition 3.44 (Alternating Multilinear Mappings) Let R be a commutative ring with identity and M, N be R-modules. An R-multilinear mapping $f : M^q \to N$ is called *alternating*, *antisymmetric* or *skew-symmetric*, if for all permutations $\sigma \in S_q$ it holds that

$$\forall (m_1, \ldots, m_q) \in M^q : \quad f(m_1, \ldots, m_q) = \text{sign}(\sigma) f(m_{\sigma(1)}, \ldots, m_{\sigma(q)}).$$

The most prominent example of an alternating mapping is the determinant of square matrices with entries in a commutative ring, when viewed as a function $\det : (R^n)^n \to R$ of the column vectors.

The R-module on which an alternating R-multilinear mapping in q variables becomes an R-linear mapping is the quotient module $\Lambda^q(M) := T^q(M)/I \cap T^q(M)$, where I is the ideal from Construction 3.39. This module is called the q-th *exterior power* of M.

Proposition 3.45 (Alternating Multilinear Mappings) *Let R be a commutative ring with identity and M, N be R-modules.*

(i) *Let $\gamma \in \text{Hom}_R(\Lambda^q(M), N)$. Then*

$$\tilde{\gamma} : (x_1, \ldots, x_q) \mapsto \gamma(x_1 \otimes \ldots \otimes x_q + I_q) = \gamma((x_1 + I) \wedge \ldots \wedge (x_q + I))$$

defines an alternating mapping $M^q \to N$.

(ii) $\text{Alt}_R(M^q, N) := \{\tilde{\gamma} \in L_R(M, \ldots, M; N) \mid \tilde{\gamma} \text{ alternating}\}$ *is an R-submodule of $L_R(M, \ldots, M; N)$.*

3.3 Symmetric and Exterior Algebras

(iii) *The assignment* $\gamma \mapsto \tilde{\gamma}$ *provides an R-module isomorphism*

$$\mathrm{Hom}_R\left(\Lambda^q(M), N\right) \longrightarrow \mathrm{Alt}_R(M^q, N).$$

Proof The proof is analogous to that of Proposition 3.33. We only sketch it here and leave the details to the reader as an exercise.

$$\begin{array}{ccc} \mathrm{Hom}_R\left(\mathrm{T}^q(M), N\right) & \xrightarrow{\Psi} & \mathrm{L}_R(M, \ldots, M; N) \\ \uparrow & & \uparrow \\ \mathrm{Hom}_R\left(\Lambda^q(M), N\right) & \longrightarrow & \mathrm{Alt}_R(M^q; N) \end{array}$$

A direct calculation shows

$$\mathrm{Alt}_R(M^q, N) = \left\{ \Psi(\varphi) \mid \varphi \in \mathrm{Hom}_R(\mathrm{T}^q(M), N),\ \varphi|_{I_q} \equiv 0 \right\}.$$

Conversely, if $\Psi(\varphi)$ is alternating, then φ vanishes on a generating system of I_q thus on all of I_q. □

Since we have already introduced antisymmetric tensors in Definition 3.34, we can now immediately introduce the *antisymmetrization* or *alternation*.

Proposition 3.46 (Alternation) *Let R be a commutative ring with identity, for which* $q!$ *is a unit and M an R-module. Then the mapping*

$$\mathrm{alt} : \mathrm{T}^q(M) \to \Lambda_q(M), \quad t \mapsto \frac{1}{q!} \sum_{\sigma \in S_q} \mathrm{sign}(\sigma) \sigma \cdot t$$

is a surjective R-module homomorphism with $\mathrm{alt} \circ \mathrm{alt} = \mathrm{alt}$ *and* $\ker(\mathrm{alt}) = I_q :=$ $I \cap \mathrm{T}^q(M)$ *with the ideal I from Construction 3.39.*

Proof The proof is analogous to the proof of Proposition 3.35:

$$\sigma' \cdot \mathrm{alt}(t) = \sigma' \cdot \frac{1}{q!} \sum_{\sigma \in S_q} \mathrm{sign}(\sigma) \sigma \cdot t = \frac{1}{q!} \sum_{\sigma \in S_q} \mathrm{sign}(\sigma) \cdot (\sigma' \sigma \cdot t)$$

$$= \mathrm{sign}(\sigma') \frac{1}{q!} \sum_{\sigma \in S_q} \mathrm{sign}(\sigma' \sigma) \cdot (\sigma' \sigma \cdot t)$$

$$= \mathrm{sign}(\sigma') \frac{1}{q!} \sum_{\sigma'' \in S_q} \mathrm{sign}(\sigma'')(\sigma'' \cdot t)$$

$$= \mathrm{sign}(\sigma')\, \mathrm{alt}(t).$$

So we have alt : $T^q(M) \to \Lambda_q(M)$, and for $t \in \Lambda_q(M)$ we calculate

$$\text{alt}(t) = \frac{1}{q!} \sum_{\sigma \in S_q} \text{sign}(\sigma) \sigma \cdot t = \frac{1}{q!} \sum_{\sigma \in S_q} \text{sign}(\sigma)^2 t = \frac{1}{q!} \sum_{\sigma \in S_q} t = t.$$

This shows alt $\circ$ alt $=$ alt.

Because $\Phi_\sigma : T^q(M) \to T^q(M)$ is an R-module homomorphism for every $\sigma \in S_q$, alt is also an R-module homomorphism. Since $q!$ is a unit in R, for $q \geq 2$ also 2 is a unit in R. We have

$$x \otimes x = \frac{1}{2}(x \otimes x + x \otimes x)$$

and

$$x \otimes x' + x' \otimes x = (x + x') \otimes (x + x') - (x \otimes x) - (x' \otimes x'),$$

so I_q is generated by the elements of the form

$$(\ldots \otimes x \otimes x' \otimes \ldots) + (\ldots \otimes x' \otimes x \otimes \ldots),$$

i.e., by the elements of the form $t + \tau_{j,j+1} \cdot t$. If $\sigma \in S_q$ with $\sigma = \sigma_1 \sigma_2$, then

$$t - \text{sign}(\sigma)\sigma \cdot t = (t - \text{sign}(\sigma_1)\sigma_1 \cdot t) + \text{sign}(\sigma_1)(\sigma_1 \cdot t - \text{sign}(\sigma_2)\sigma_2(\sigma_1 \cdot t))$$

and thus

$$I_q = \langle t - \text{sign}(\sigma)\sigma \cdot t \mid \sigma \in S_q, \ t \in T^q(M)\rangle_{R\text{-Module}}.$$

One finds $t - \text{alt}(t) \in I_q$ for $t \in T^q(M)$ and because

$$\text{alt}(t - \text{sign}(\sigma')\sigma' \cdot t) = \frac{1}{q!} \sum_{\sigma \in S_q} \text{sign}(\sigma) \sigma \cdot t - \frac{1}{2} \sum_{\sigma \in S_q} \text{sign}(\sigma \sigma') \sigma \sigma' \cdot t = 0,$$

finally $I_q \subseteq \ker(\text{alt})$. Conversely $\text{alt}(t) = 0$ implies $t = t - \text{alt}(t) \in I_q$. $\square$

With alt: $T^q(M) \to \Lambda_q(M)$ instead of sym: $T^q(M) \to S_q(M)$, copying the proof we now obtain the following analogue of Corollary 3.36.

Corollary 3.47 (Skew-Symmetric Tensors and Exterior Powers) *Let $q!$ be a unit in R. Then the following holds:*

(i) $T^q(M) = \Lambda_q(M) \oplus I_q$ *as an R-module.*
(ii) $\Lambda_q(M) \longrightarrow \Lambda^q(M)$ *is an R-module isomorphism.*

3.3 Symmetric and Exterior Algebras

(iii) Let $\varphi \colon \bigoplus_{q \in \mathbb{N}_0} \Lambda_q(M) \to \Lambda(M) = \bigoplus_{q \in \mathbb{N}_0} \Lambda^q(M)$ *be the isomorphism of R-modules composed from the isomorphisms given in (ii). Then the algebra multiplication transported by* φ^{-1} *from* $\Lambda(M)$ *to* $\bigoplus_{q \in \mathbb{N}_0} \Lambda_q(M)$ *is given by*

$$\Lambda_q(M) \times \Lambda_{q'}(M) \to \Lambda_{q+q'}(M), \quad (a, a') \mapsto \mathrm{alt}(a \otimes a)$$

Literature: The concepts and results presented here can be found mostly in the sources already mentioned in Ch. 2, [Ke95] and [La93]. A truly comprehensive source is [Bo70].

Pattern Recognition 4

This chapter aims to uncover certain patterns in the considerations of Chs. 2 and 3 and provide evidence for the claim from the introduction to Part I that the presented structure theory of modules would be exemplary for algebraic structures. In particular, we address the parallels mentioned in Section 2.1 between the constructions of sub- and quotient structures for vector spaces, rings, and modules, as well as the repeatedly mentioned naturality of the constructions presented in Ch. 3. The former leads us to basic concepts of *universal algebra*, the latter to elementary concepts of *category theory*.

4.1 Universal Algebra

Universal algebra formalizes the concept of an algebraic structure as a set together with any number of operations on this set. The concept of operation is broader than in groups or rings. Operations on a set M are allowed that combine more than two elements, but also those that only have one element as input or even work without any input. This has the advantage that certain elements can be distinguished, such as the neutral element in a group or the identity in a ring, by describing them as the image of a mapping $M^0 := \{\emptyset\} \to M$. Operations with only one element as input, i.e., mappings of the form $M^1 := M \to M$, allow modeling operations with other sets, as they occur in modules, where for each element r of the ring R there is a mapping $M \to M, m \mapsto r \cdot m$. This leads to the concept of the n-ary operation on M, where $n \in \mathbb{N}_0 = \mathbb{N} \cup \{0\}$.

Definition 4.1 (n-**Ary Operation**) Let M be a set and $n \in \mathbb{N}_0$. An *n-ary operation* is a mapping $M^n \to M$.

An algebraic structure is supposed to be a set with operations, but in concrete examples like groups or rings, there are also laws such as associativity or distributivity. To precisely capture what we want to understand by an algebraic structure, we need to specify what kind of laws these should be. In universal algebra, there is the concept of a *equationally defined class*, which we want to lean on here. The key idea is to allow as laws only equations of expressions that are valid, no matter which elements of the set are used. To formulate this idea, we first introduce the *term algebra* for a set of operations, whose elements are the expressions that can be equated.

Definition 4.2 (Term Algebras) Let Φ be a set with a mapping $s\colon \Phi \to \mathbb{N}_0$. We set $\Phi_n := s^{-1}(n)$ and fix a set X of variables. The *term algebra* $T_\Phi(X)$ to Φ is defined inductively by the following settings:

(a) $\forall x \in X: \quad x \in T_\Phi(X)$.
(b) $\forall n \in \mathbb{N}_0, \varphi \in \Phi_n, t_1, \ldots t_n \in T_\Phi(X): \quad \varphi(t_1, \ldots, t_n) \in T_\Phi(X)$.

Condition (b) means that one can "substitute" n terms into a $\varphi \in \Phi_n$ and thereby form a new term. The elements of Φ_n thus provide mappings $T_\Phi(X)^n \to T_\Phi(X)$. We consider the elements as n-ary operations on $T_\Phi(X)$. The term algebra of Φ therefore consists of terms that can be formed by nesting the operations, where variables from a given set of variables are used. In the multiplicative associative law of a ring, e.g., the terms $x_1 \cdot (x_2 \cdot x_3)$ and $(x_1 \cdot x_2) \cdot x_3$ occur, which would be written as $\varphi(x_1, \varphi(x_2, x_3))$ and $\varphi(\varphi(x_1, x_2), x_3)$, respectively, if multiplication is denoted as $\varphi\colon R \times R \to R$.

Now let M be a set and Φ a set of operations on M and $s\colon \Phi \to \mathbb{N}_0$ the mapping that assigns to each operation its arity. Since only finitely many variables occur in each term $t \in T_\Phi(X)$, for each *evaluation mapping* $\mathrm{ev}\colon X \to M$ the terms of the term algebra can also be evaluated, resulting in an evaluation mapping $\mathrm{ev}\colon T_\Phi(X) \to M$. With this, we can formally grasp the concept of an algebraic structure in the sense of universal algebra.

Definition 4.3 (Algebraic Structure) An *algebraic structure* consists of a set M, a set Φ of operations on M and a set $\Gamma \subseteq T_\Phi(X)^2$ of pairs of terms, for which the following holds:

(a) Only finitely many variables occur in Γ.
(b) $\forall (t_1, t_2) \in \Gamma, \mathrm{ev}\colon X \to M: \quad \mathrm{ev}(t_1) = \mathrm{ev}(t_2)$.

The following example of group structure illustrates in particular the role of the 0- and 1-ary operations.

4.1 Universal Algebra

Example 4.4 (Group Structure) Let G be a set with three operations:

$$1 : G^0 \to G, \; \emptyset \mapsto 1,$$
$$\iota : G^1 \to G, \; g \mapsto g^{-1},$$
$$\mu : G^2 \to G, \; (g, h) \mapsto g * h.$$

The set Γ consists of the pairs of terms

$$\mu\bigl(x_1, \mu(x_2, x_3)\bigr) \, , \, \mu\bigl(\mu(x_1, x_2), x_3\bigr),$$
$$\mu(1, x_1) \, , \, x_1,$$
$$x_1 \, , \, \mu(x_1, 1),$$
$$\mu\bigl(\iota(x_1), x_1\bigr) \, , \, 1,$$
$$1 \, , \, \mu\bigl(x_1, \iota(x_1)\bigr).$$

That is, the operations are subject to the following laws:

(a) μ is associative.
(b) $\forall g \in G : \; 1 * g = g = g * 1$.
(c) $\forall g \in G : \; g^{-1} * g = 1 = g * g^{-1}$.

Then $(1, \iota, \mu)$ is called a *group structure* on G and $(G, 1, \iota, \mu)$ a *group*. □

The already considered algebraic structures, such as (abelian) groups, rings, vector spaces, R-modules or R-algebras, can all be understood as algebraic structures in the sense of Definition 4.3 (exercise). In contrast, fields or integral domains are not algebraic structures in the sense of Definition 4.3, but only rings with special properties. This is because these structures are not defined by universally valid equations, but certain properties are required for certain elements (e.g., those different from zero), such as being invertible.

Beyond the examples already mentioned, there are many more algebraic structures in mathematics, including those with more than 2-ary operations.

Example 4.5 (Algebraic Structures) Let R be a commutative ring with identity and V an R-module.

(i) V is a *Lie algebra* over R, if V carries an R-algebra structure $[\cdot, \cdot] : V \times V \to V$ in the sense of Definition 3.19 that has the following properties:
   (a) $\forall x \in V : \quad [x, x] = 0$ (*antisymmetry*).
   (c) $\forall x, y, z \in V : \quad [[x, y], z] + [[y, z], x] + [[z, x], y] = 0$ (*Jacobi identity*).
(ii) Let $V = V_0 \oplus V_1$ be a direct sum of R-modules. Then V is a $\mathbb{Z}_2$-*graded Lie algebra* over R, if V carries an R-algebra structure $[\cdot, \cdot] : V \times V \to V$ in the sense of Definition 3.19 that has the following properties:

(a) $\forall x \in V_i, y \in V_j: \quad [x, y] = -(-1)^{ij}[y, x] \in V_{ij}$
(*graded antisymmetry*).
(b) $\forall x \in V_i, y \in V_j, z \in V_k$:

$$(-1)^{ik}[x, [y, z]] + (-1)^{ji}[y, [z, x]] + (-1)^{kj}[z, [x, y]] = 0$$

(*graded Jacobi identity*).
(iii) V is a *Jordan algebra* over R, if V carries an R-algebra structure $V \times V \to V, (x, y) \mapsto xy$ in the sense of Definition 3.19 that has the following properties:
  (a) $\forall x, y \in V: \quad xy = yx$ (*commutativity*).
  (b) $\forall x, y \in V: \quad x(x^2 y) = x^2(xy)$ (*Jordan identity*).
(iv) V is a *Lie triple system* over R, if V carries an R-multilinear 3-ary operation $[\cdot, \cdot, \cdot]: V^3 \to V$ that has the following properties:
  (a) $\forall x, y \in V: \quad [x, x, y] = 0$.
  (b) $\forall x, y, z \in V: \quad [x, y, z] + [y, z, x] + [z, x, y] = 0$.
  (c) $\forall x, y, z, a, b \in V$:

$$[x, y, [z, a, b]] = [z, [x, y, a], b] + [[x, y, z], a, b] + [z, a, [x, y, b]].$$

(v) V is a *Jordan triple system* over R, if V carries an R-multilinear 3-ary operation $\{\cdot, \cdot, \cdot\}: V^3 \to V$, which has the following properties:
  (a) $\forall x, y, z \in V: \quad \{x, y, z\} = \{x, z, y\}$.
  (b) $\forall x, y, z, a, b \in V$:

$$\{x, y, \{z, a, b\}\} = \{z, \{x, y, a\}, b\} - \{\{y, x, z\}, a, b\} + \{z, a, \{x, y, b\}\}.$$

□

With Definition 4.3, a generalization of the terms "subvector space", "subring", "submodule" etc. is immediate.

Definition 4.6 (Algebraic Substructure) Let (M, Φ, Γ) be an algebraic structure. Let $N \subseteq M$ be a subset and $\Phi|_N$ the family of mappings $\varphi_N : N^{s(\varphi)} \to M$, where $s(\varphi)$ is the arity of $\varphi \in \Phi$. If

$$\forall \varphi \in \Phi: \quad \varphi(N^{s(\varphi)}) \subseteq N$$

holds, i.e., if N is *closed* under Φ, then $(N, \Phi|_N, \Gamma|_N)$ or simply N is called a *substructure* of (M, Φ, Γ). Here, $\Gamma|_N$ consists of the pairs of terms in $T_{\Phi|_N}(X)$, which are obtained when each operation that occurs in terms of Γ is restricted to N.

We test this definition on groups: Let $(G, 1, \iota, *)$ be a group and $H \subseteq G$ a substructure in the sense of Definition 4.6. Then $1 \in H$ and $h_1 * h_2 \in H$ as well as $h^{-1} \in H$ for all $h, h_1, h_2 \in H$. Thus, H is a subgroup of G in the usual sense.

4.1 Universal Algebra

Conversely, for every subgroup H of G, $1 \in H$ and also H is closed under inversion and multiplication. Thus, Definition 4.6 actually provides exactly the concept of a subgroup for group structures.

Exercise 4.1 (Algebraic Structures) Formulate the operations and laws for rings, vector spaces, modules, and algebras as algebraic structures in the sense of Definition 4.3 and show that Definition 4.6 provides the correct concepts of substructures.

While the formulation of a substructure was quite obvious, one has to introduce an additional concept for the definition of quotient structures of algebraic structures: the *congruence relation*.

Definition 4.7 (Congruence Relation) Let (M, Φ, Γ) be an algebraic structure and $\sim$ an equivalence relation on M. If for all $\varphi \in \Phi$ with $s := s(\varphi) \geq 1$

$$\forall a_1, b_1, \ldots, a_s, b_s \in M \text{ with } a_j \sim b_j : \quad \varphi(a_1, \ldots, a_s) \sim \varphi(b_1, \ldots, b_s), \tag{4.1}$$

holds, then $\sim$ is called a *congruence relation* for (M, Φ, Γ).

The set of equivalence classes with respect to a congruence relation is the sought-after common generalization of the quotient structures from Ch. 2.

Construction 4.8 (Algebraic Quotient Structures) Let (M, Φ, Γ) be an algebraic structure and $\sim$ a congruence relation for (M, Φ, Γ). We denote the set of equivalence classes $[m]$ of elements $m \in M$ with $M_\sim$. Due to (4.1), for each s-ary operation $\Phi \ni \varphi : M^s \to M$, an s-ary operation $\varphi_\sim : M_\sim^s \to M_\sim$ is defined by

$$\forall m \in M : \quad \varphi_\sim([m_1], \ldots, [m_s]) := [\varphi(m_1, \ldots, m_s)]$$

With $\Phi_\sim := \{\varphi_\sim\}$, $(M_\sim, \Phi_\sim, \Gamma_\sim)$ then becomes an algebraic structure, which we call the *quotient structure* of (M, Φ, Γ) with respect to $\sim$. Analogous to the case of algebraic substructures, the pairs of terms in the term algebra $T_{\Phi_\sim}(X)$ are obtained from the terms of Γ, in which each $\varphi \in \Phi$ is replaced by the corresponding $\varphi_\sim \in \Phi_\sim$. □

To explain the connection between Construction 4.8 and the already considered quotient structures, we consider the special case of an algebraic structure with a (commutative) group addition. This includes rings, modules, and vector spaces. We set $I := [0]$. If $x \in I$, then (4.1) yields that $\varphi(x, m_1, \ldots) \sim \varphi(0, m_1, \ldots)$ for each $\varphi \in \Phi$. For the addition, it follows for $x, y \in I$ that $x + y \sim 0 + 0$, i.e., $x + y \in I$. The unary operation $m \to -m$ yields $-x \sim 0$ for $x \in I$. Thus, I is a subgroup. If now (M, Φ, Γ) is an R-module, then the unary operations $m \mapsto r \cdot m$ yield

$$\forall x \in I : \quad r \cdot x \sim r \cdot 0 = 0,$$

i.e., $r \cdot I \subseteq I$. In other words, I is a submodule. Two elements $m, m' \in M$ are congruent, i.e., $m \sim m'$, if and only if $m - m' \sim 0$. This means that the equivalence class $[m]$ of $\sim$ is precisely the coset $m + I$. It is clear that the quotient structure from Construction 4.8 for modules is nothing other than the quotient module from Construction 2.9. For rings and vector spaces, the proof looks practically the same (exercise!).

As the next example, we consider the connection of congruence relations and *normal subgroups* in a not necessarily commutative group G. In this example, the equivalence class $N = [1]$ of the unit is the relevant object.

Example 4.9 (Normal Subgroups and Quotient Groups) Let $(G, 1, \iota, \mu)$ be a group and $\sim$ a congruence relation for it. With the operations μ and ι from Example 4.4, it is shown, as above for the abelian subgroups of a module, that $N := [1]$ is a subgroup. Now let $n \in N$ and $g \in G$. Then one calculates

$$gng^{-1} = \mu(gn, g^{-1}) = \mu(\mu(g, n), g^{-1}) \sim \mu(\mu(g, 1), g^{-1}) = \mu(g, g^{-1}) = 1,$$

because with $n \sim 1$ it follows that $\mu(g, n) \sim \mu(g, 1) = g$. Thus,

$$\forall g \in G, n \in N : \quad gng^{-1} \in N,$$

and this is exactly the condition for N to be a normal subgroup. As before, it is seen that $g \sim h$ is equivalent to $g \in hN$. Thus, the quotient structure from Construction 4.8 is precisely the space $G/N := \{gN \mid g \in G\}$ of *cosets* of N. The induced operations on G/N are then:

$$[1] : (G/N)^0 \to G/N, \; \emptyset \mapsto [1] = 1N = N,$$
$$\iota_N : (G/N)^1 \to G/N, \; gN \mapsto g^{-1}N,$$
$$\mu_N : (G/N)^2 \to G/N, \; (gN, hN) \mapsto ghN.$$

If, conversely, $N \subseteq G$ is a normal subgroup, then

$$\forall g, h \in G : \quad g \sim h \;:\Leftrightarrow\; g \in hN$$

defines a congruence relation (exercise). □

An example of an algebraic structure that only has unary operations, but is still rich, is the *G-set* for a group G.

Example 4.10 (G-Set) Let M be a set and G a group. For each $g \in G$, a mapping $\varphi_g : M \to M$ is given. With $\Phi = \{\varphi_g \mid g \in G\}$, (M, Φ) is called a *G-set*, if each φ_g fulfills the following conditions, where we write $g \cdot m := \varphi_g(m)$:

(a) $\forall m \in M: \quad 1 \cdot m = m$.
(b) $\forall g, h \in G, m \in M: \quad g \cdot (h \cdot m) = (gh) \cdot m$.

Instead of a G-set M, we also speak of a G-action on M.

From (a) and (b) it follows in particular that each φ_g is bijective.

A congruence relation $\sim$ for (M, Φ) is an equivalence relation with

$$\forall g \in G, m, m' \in M: \quad m \sim m' \Rightarrow g \cdot m \sim g \cdot m'.$$

An example of this is $\sim_G$, which is defined by

$$\forall m, m' \in M: \quad m \sim_G m' :\Leftrightarrow \exists g \in G \text{ with } m' = g \cdot m$$

The equivalence classes with respect to $\sim_G$ are precisely the *orbits*

$$G \cdot m := [m] = \{g \cdot m \in M \mid g \in G\},$$

and the associated quotient structure is the G-set $G \backslash M$ of all orbits with the *trivial action*, i.e.,

$$\forall g \in G, [m] \in G \backslash M: \quad g \cdot [m] = [m].$$

$\square$

Exercise 4.2 (Congruence Relations) Describe the congruence relations for Lie algebras and Lie triple systems as well as for Jordan algebras and Jordan triple systems.

4.2 Naive Category Theory

Category theory was founded in the mid-twentieth century as a formalization of certain considerations of algebraic topology and plays a significant role in the mathematics of the early twenty-first century. In this chapter, we limit ourselves to the introduction of some basic concepts from category theory, which we then, similar to universal algebra in Section 4.1, apply as a kind of formal pattern recognition to the constructions of the preceding chapters. This not only clarifies already observed similarities, but also sharpens the view for possible constructions in other areas. We illustrate the potential of categorical ideas by also applying them to topological concepts, thus building a bridge to Part II of this book.

Categories

A category consists of objects and a kind of mappings, called *morphisms*, between the objects. Already with the very first definition of the field, one has to decide how to deal with set theory: Should the objects of a category form a set or not? The question is important, since the basic example for a category is supposed to be the category of sets with the mappings as morphisms. But if every set is an object of this category, then the set of objects is the *set of all sets*, and this is known to be a problematic assumption.

Since axiomatic set theory is not listed in the prerequisites for reading this book printed in the preface, we will ignore this problem in this chapter and, in analogy to naive set theory, which one deals with at the beginning of mathematics studies, we will conduct a naive category theory. We do this by talking about the *class* of objects of a category, but we do not go into detail about the meaning of the word "class".

Definition 4.11 (Category) A *category* **C** has the following components:

(a) A class ob(**C**) of *objects*.
(b) For any two objects X and Y, we write $X, Y \in$ ob(**C**), there is a set $\mathrm{Hom}_{\mathbf{C}}(X, Y)$, also written as $\mathbf{C}(X, Y)$, of *morphisms*.
(c) For any three objects $X, Y, Z \in$ ob(**C**) there is a mapping

$$\mathbf{C}(X, Y) \times \mathbf{C}(Y, Z) \to \mathbf{C}(X, Z), \quad (f, g) \mapsto g \circ f,$$

called *composition*, which has the following properties:
  (i) For compatible, i.e., composable, triples f, g, h of morphisms, the associative law applies

  $$(f \circ g) \circ h = f \circ (g \circ h).$$

  (ii) For each object $X \in$ ob(**C**) there is a morphism $1_X \in \mathbf{C}(X, X)$, called *identity*, with

  $$\forall f \in \mathbf{C}(Y, X), g \in \mathbf{C}(X, Y): \quad 1_X \circ f = f \quad \text{and} \quad g \circ 1_X = g.$$

We also write $f: X \to Y$ or $X \xrightarrow{f} Y$ for $f \in \mathbf{C}(X, Y)$. Accordingly, morphisms are often also called *arrows*. In addition, there are the notations $\mathrm{End}_{\mathbf{C}}(X)$ and $\mathbf{C}(X)$ for $\mathbf{C}(X, X) = \mathrm{Hom}_{\mathbf{C}}(X, X)$.

When you review the mathematical concepts that you learn in elementary mathematics courses, you realize that you have already seen quite a large number of examples of categories. It starts with the category **Set**, whose objects are the sets and whose morphisms are the mappings between sets. For two sets X and Y, **Set**(X, Y) is then the set of mappings from X to Y. The composition in this example is nothing other than the composition of mappings. Thus, the associativity of ∘ is clear. The identity 1_X in this example is simply the identity mapping $x \mapsto x$, provided $X \neq \emptyset$ applies. The set **Set**($\emptyset$) has only one element anyway, and that is $1_\emptyset$.

To speak of *the* identity is not only justified in the example **Set**. For two identities $1_X, 1'_X \in \mathbf{C}(X)$ always $1_X = 1_X \circ 1'_X = 1'_X$ applies.

In order not to have to constantly prove the associativity of the composition and the neutrality of the identity in the following, we introduce the concept of a *subcategory* here and then take advantage of the fact that associativity and neutrality are inherited by subcategories.

Definition 4.12 (Subcategory) Let **C** be a category. A *subcategory* **S** of **C** consists of a subclass ob(**S**) of ob(**C**) as objects and, for each pair $X, Y \in$ ob(**S**), subsets $\mathbf{S}(X, Y) \subseteq \mathbf{C}(X, Y)$ with:

(i) $\forall f \in \mathbf{S}(X, Y), g \in \mathbf{S}(Y, Z): g \circ f \in \mathbf{S}(X, Z)$.
(ii) $\forall X \in$ ob(**S**) $: 1_X \in \mathbf{S}(X, X)$.

The subcategory **S** is called *full*, if $\mathbf{S}(X, Y) = \mathbf{C}(X, Y)$ for all $X, Y \in$ ob(**S**).

With this definition, it is clear that a subcategory is itself a category with respect to the composition and identities of the original category.

We continue our list of elementary examples of categories: The category **Grp**, whose objects are all groups and whose morphisms between two groups G, H are the group homomorphisms $\varphi: G \to H$, is a subcategory of **Set**. To show this, one must verify that the composition of two group homomorphisms is itself a group homomorphism. So let $(G_1, *_1), (G_2, *_2), (G_3, *_3)$ be groups and $\varphi_1: G_1 \to G_2$ and $\varphi_2: G_2 \to G_3$ be group homomorphisms. Then for $g, h \in G_1$, it holds that

$$\varphi_2 \circ \varphi_1(g *_1 h) = \varphi_2\big(\varphi_1(g) *_2 \varphi_1(h)\big) = \varphi_2\big(\varphi_1(g)\big) *_3 \varphi_2\big(\varphi_1(h)\big)$$
$$= \big(\varphi_2 \circ \varphi_1(g)\big) *_3 \big(\varphi_2 \circ \varphi_1(h)\big),$$

i.e., $\varphi_2 \circ \varphi_1$ is indeed a group homomorphism. Additionally, one must verify that the identity mapping on a group is a group homomorphism, which is trivial.

A full subcategory **Ab** of **Grp** is obtained when only considering abelian groups as objects. A subcategory of **Ab**, which is not full, is **Ring**, whose objects are the rings. The elements of **Ring**(R, S) are the ring homomorphisms $\varphi: R \to S$. Here again, one must verify that the composition of ring homomorphisms is itself a ring homomorphism. The corresponding calculation is practically identical to the above calculation for the group homomorphisms. The proof that the identity mapping is a ring homomorphism is again trivial. The rings with identity form a subcategory **Ring**$_1$ of **Ring**. This subcategory is also not full, as one requires from a ring homomorphism that it preserves the identity to count it as a morphism of the category (see Example 1.9). In **Ring**$_1$, one finds as a full subcategory the category **CRing**$_1$ of commutative rings with identity. Even though fields and integral

Fig. 4.1 Categories of algebraic structures

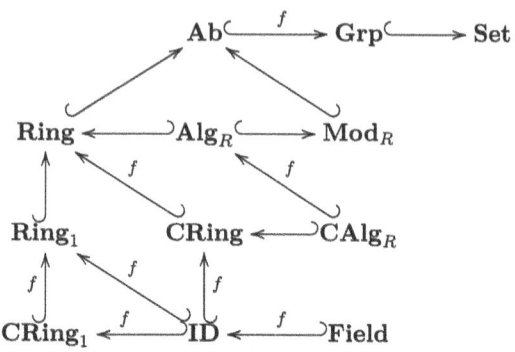

domains are not algebraic structures in the sense of Definition 4.3, both the fields and the integral domains form full subcategories of **CRing**$_1$. We denote them with **Field** and **ID**. Another family of subcategories of **Ab** are the categories $_R$**Mod** and **Mod**$_R$ of left and right R-modules for a fixed ring R, where the morphisms between two R-modules are the R-module homomorphisms (exercise). For commutative R, left-R-modules and right-R-modules are one and the same thing: one simply sets $r \cdot m = m \cdot r$ and then does not distinguish between $_R$**Mod** and **Mod**$_R$. In this case, we also find the category **Alg**$_R$ of associative R-algebras as a subcategory of **Mod**$_R$. The morphisms of this subcategory are the R-algebra homomorphisms, which one again has to show are closed under composition (exercise). Again, one has to make the trivial remark that the identity mapping on an R-algebra is an R-algebra homomorphism. In **Alg**$_R$ one finds the full subcategory **CAlg**$_R$ of commutative R-algebras.

We summarize the relationships between the categories of algebraic structures just described in Figure 4.1 in a diagram. Here, **S** $\hookrightarrow$ **C**, means that **S** is a subcategory of **C**. If **S** is even a full subcategory of **C**, we write **S** $\stackrel{f}{\hookrightarrow}$ **C**.

Exercise 4.3 (Categories of Algebraic Structures) Let X be a finite set of variables and Φ a set of operations. Further, let $\Gamma \subseteq T_\Phi(X)^2$ be a set of pairs of terms (see Definitions 4.2 and 4.3).

(i) Define a category $\mathbf{C}_{\Phi,\Gamma}$ of algebraic structures of type (Φ, Γ) such that as special cases one obtains the categories of groups, rings, and R-modules.
(ii) Describe in detail objects and morphisms for the categories of Lie algebras, Lie triple systems, Jordan algebras, and Jordan triple systems.

Apart from the algebraic structures discussed in Chs. 2 and 3, one finds in elementary mathematics courses a number of other subcategories of **Set** that are of a completely different nature. The most important of these subcategories is **Top**, whose objects are the topological spaces. The morphisms of this subcategory are the continuous mappings. To show that one obtains such a subcategory, one must ascertain that compositions of continuous mappings are continuous and also that the identity on a topological space is always continuous.

4.2 Naive Category Theory

One can combine the algebraic categories from Figure 4.1 with **Top** by requiring that the sets not only carry the corresponding algebraic structure, but also have a topology. In addition, one requires that all structure mappings such as additions, multiplications, inversions etc. are continuous. As morphisms, one then takes the homomorphisms of the algebraic structure, which are in addition continuous. For example, one obtains the category **Grp**$_{\text{top}}$ of *topological groups*.

Definition 4.13 (Topological Group) A *topological group* is a topological space G, together with a continuous group multiplication $G \times G \to G$, $(g, h) \mapsto gh$, for which the inversion $G \to G$, $g \mapsto g^{-1}$ is also a continuous mapping. It is assumed that $G \times G$ carries the product topology (see Example 4.27).

In functional analysis, the relevant category is that of topological vector spaces over $\mathbb{C}$. We refrain from a detailed description here and also from listing further examples from this series of categories of *topological algebra*. Instead, we turn to the material of calculus courses and consider differentiable mappings. Such mappings are initially defined on open subsets of finite-dimensional real vector spaces, and one learns that the composition of two differentiable mappings is itself differentiable. Since the identity mapping on an open subset of a finite-dimensional real vector space is also differentiable, the following definition of a category $\mathbf{C}^{\text{diff}}_{\mathbb{R}-\text{Vect}}$ is suggested: The objects of this category should be the open subsets of finite-dimensional real vector spaces. As morphisms between two such open sets $X, Y \in \text{ob}(\mathbf{C}^{\text{diff}}_{\mathbb{R}-\text{Vect}})$, one chooses the differentiable mappings $\varphi \colon X \to Y$. In practice, one often considers better differentiability properties, e.g., k-times continuous differentiability. This leads to subcategories $\mathbf{C}^{k}_{\mathbb{R}-\text{Vect}}$ of $\mathbf{C}^{\text{diff}}_{\mathbb{R}-\text{Vect}}$, which have the same objects, but whose morphism sets consist only of the mappings of the corresponding differentiability class. Here, one can choose $k \in \mathbb{N}$, but also set $k = \infty$ if one wants to talk about infinitely often differentiable mappings. In any case, one knows that these differentiability classes are preserved under composition and the identity mapping of an open set belongs to each of these differentiability classes. In the example of differentiable functions, one can replace the *real* vector spaces with *complex* vector spaces and require the functions to be complex differentiable. This leads to analogous categories $\mathbf{C}^{\text{diff}}_{\mathbb{C}-\text{Vect}}$ and $\mathbf{C}^{k}_{\mathbb{C}-\text{Vect}}$. If one has dealt more closely with complex differentiability, one knows that the latter categories are all equal, because complex differentiable mappings are automatically infinitely often complex differentiable. It is even true that such functions are $\mathbb{C}$-*analytic*, meaning, all component functions (with respect to any basis) can be locally expanded into complex power series. Since the composition of analytic mappings (real or complex) is again analytic, two more categories are obtained, denoted by $\mathbf{C}^{\omega}_{\mathbb{K}-\text{Vect}}$, where $\mathbb{K}$ is either $\mathbb{R}$ or $\mathbb{C}$.

In the discussion of the $\mathbb{K}$-differentiable functions, we silently used that there are canonical topologies on finite-dimensional $\mathbb{K}$-vector spaces, which are obtained from norms. This uses the fact that all norms are equivalent on such spaces. Since it is also known that linear mappings between finite-dimensional $\mathbb{K}$-vector spaces are automatically continuous with respect to the canonical topologies, one finds

Fig. 4.2 Subcategories of **Top**

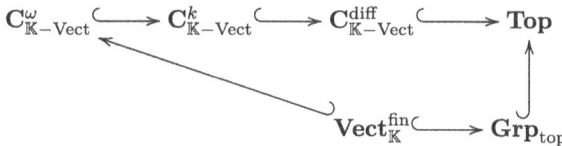

the category $\mathbf{Vect}_{\mathbb{K}}^{\text{fin}} = \mathbf{Mod}_{\mathbb{K}}^{\text{fg}}$ of finite-dimensional $\mathbb{K}$-vector spaces, i.e., the finitely generated $\mathbb{K}$-modules, as a subcategory of $\mathbf{Grp}_{\text{top}}$. Linear mappings are automatically $\mathbb{K}$-analytic, so $\mathbf{Vect}_{\mathbb{K}}^{\text{fin}}$ is even a subcategory of $\mathbf{C}_{\mathbb{K}-\text{Vect}}^{\omega}$. The various inclusions are compiled in Figure 4.2.

It is by no means mandatory that the sets of morphisms of a category consist of mappings. Consider, e.g., the following construction of the *opposite category* $\mathbf{C}^{\text{op}}$ to a category $\mathbf{C}$, in which one simply reverses all arrows. When applied to **Set**, one obtains a category in which only very few morphisms $\varphi \in \mathbf{Set}^{\text{op}}(X, Y)$ are mappings, although the objects $X, Y \in \text{ob}(\mathbf{Set}^{\text{op}})$ are sets.

Example 4.14 (Opposite Category) Let $\mathbf{C}$ be a category. Then one forms the *opposite category* $\mathbf{C}^{\text{op}}$ of $\mathbf{C}$ by

$\text{ob}(\mathbf{C}^{\text{op}}) := \text{ob}(\mathbf{C})$,
$\mathbf{C}^{\text{op}}(X, Y) := \mathbf{C}(Y, X)$ (with the appropriate composition). □

If Example 4.14 seems contrived, one can refer to the following example of a category whose morphisms are not mappings.

Example 4.15 (Ordered Sets) Let M be a set and $\leq$ a *partial order* on M, i.e., a reflexive, transitive, and antisymmetric relation. We consider the elements of M as the objects of a category $\mathbf{M}$. For $m, m' \in \text{ob}(\mathbf{M}) = M$ we set

$$\mathbf{M}(m, m') := \begin{cases} \{m \to m'\} & \text{if } m \leq m', \\ \emptyset & \text{otherwise.} \end{cases}$$

As composition, we choose $(m' \to m'') \circ (m \to m') := m \to m''$. If $m = m'$, then $m \to m'$ is the identity 1_m on m. The transitivity and antisymmetry of the relation $\leq$ imply that the composition $\circ$ is associative. Reflexivity guarantees the existence of the identities 1_m.

If one reverses the order in this construction, one obtains the opposite category. □

Categories, for which the objects form a set as in Example 4.15, are called *small categories*.

Similar to sets or modules, one can also construct new categories from existing categories. Here is, e.g., the product of two categories.

Example 4.16 (Product of Two Categories) Let **A** and **B** be categories. Then one forms the *product category* $\mathbf{A} \times \mathbf{B}$ by

$\text{Ob}(\mathbf{A} \times \mathbf{B}) := \text{Ob}(\mathbf{A}) \times \text{Ob}(\mathbf{B})$ (pairs of objects),
$(\mathbf{A} \times \mathbf{B})((X, X'), (Y, Y')) := \mathbf{A}(X, Y) \times \mathbf{B}(X', Y')$. □

We have mentioned several times in Chs. 2 and 3 that one should consider isomorphic algebraic objects as equal, as long as one is only interested in the algebraic properties of these objects. The concept of isomorphism is easily transferred to general categories.

Definition 4.17 (Isomorphism in Categories) A morphism $f \in \mathbf{C}(X, Y)$ in a category is called an *isomorphism*, if there is a morphism $g \in \mathbf{C}(Y, X)$ with $f \circ g = 1_Y$ and $g \circ f = 1_X$ (*automorphism*, if $X = Y$). Two objects X and Y of a category are called *isomorphic*, if there is an isomorphism $f \in \mathbf{C}(X, Y)$. We then write $X \cong Y$ or $X \cong_\mathbf{C} Y$, if one wants to emphasize the category **C**.

For the algebraic categories from Figure 4.1, this definition provides exactly the concept of isomorphism that we introduced in the respective examples, provided it was not already assumed to be known.

Exercise 4.4 (Monomorphisms and Epimorphisms) Let **C** be a category. A morphism $f \in \mathbf{C}(A, B)$ is called a *monomorphism*, if

$$\forall C \in \text{Ob}(\mathbf{C}), g, h \in \mathbf{C}(C, A): \quad f \circ g = f \circ h \Rightarrow g = h,$$

and a *epimorphism*, when

$$\forall C \in \text{Ob}(\mathbf{C}), g, h \in \mathbf{C}(B, C): \quad g \circ f = h \circ f \Rightarrow g = h.$$

(i) Show that every isomorphism $f \in \mathbf{C}(A, B)$ is both a monomorphism and an epimorphism.
(ii) Show that the embedding $\iota : \mathbb{Z} \to \mathbb{Q}$ in the category **Ring** is both a monomorphism and an epimorphism, but not an isomorphism.

Functors

In Ch. 3, we called some of the constructions we presented *natural*. The concepts of category theory allow us to specify why each was an expression of the same notion of naturality, namely the *functoriality* of the respective construction. Functors between categories are an analogue of (structure-preserving) mappings between sets.

Definition 4.18 (Functor between Categories) Let **A** and **B** be categories. A *functor* $F : \mathbf{A} \to \mathbf{B}$ consists of a prescription

$$\operatorname{ob}(\mathbf{A}) \to \operatorname{ob}(\mathbf{B}), \quad X \mapsto FX$$

and a family of mappings

$$\forall X, Y \in \operatorname{ob}(\mathbf{A}) : \quad \mathbf{A}(X, Y) \to \mathbf{B}(FX, FY), \quad \varphi \mapsto F\varphi$$

with the following properties:

(a) For composable morphisms φ and ψ in **A**: $F(\varphi \circ \psi) = (F\varphi) \circ (F\psi)$.
(b) $\forall X \in \operatorname{ob}(\mathbf{A}) : F(1_X) = 1_{FX}$.

An obvious example of a functor is the *identity* $\operatorname{id}_\mathbf{C}$, which exists for every category **C**. It is defined by the prescriptions $X \mapsto X$ on the objects and the mappings $\varphi \mapsto \varphi$ on the sets of morphisms.

A still abstract and very easy to understand class of examples for functors are the *forgetful functors* from a subcategory to a category. Here, one considers objects of the subcategory as objects of the larger category and morphisms between two objects of the subcategory as morphisms in the larger category between these two objects (considered as objects of the larger category). In the case of the subcategory **Grp** of **Set**, this means that one simply "forgets" the group structure of a group and considers group homomorphisms only as mappings.

The forgetful functors will play an important role in the context of adjoint functors (see Section 4.4), but first we want to discuss a series of functors that we have already seen in the previous chapters, but have not yet recognized as functors. For example, the considerations that led to the consideration of the R-module homomorphism (3.4) showed that forming the direct sum of two modules is a functor.

Example 4.19 (Functoriality of the Direct Sum of Modules) Let R be a ring. We set (see Example 4.16)

$$\operatorname{ob}(\mathbf{Mod}_R \times \mathbf{Mod}_R) \to \operatorname{ob}(\mathbf{Mod}_R), \quad (X, Y) \mapsto X \oplus Y$$

and, for $(X, Y), (X', Y') \in \operatorname{ob}(\mathbf{Mod}_R \times \mathbf{Mod}_R)$ as well as a morphism $(\varphi, \psi) \in \operatorname{Hom}_{\mathbf{Mod}_R \times \mathbf{Mod}_R}\big((X, Y), (X', Y')\big)$:

$$\oplus (\varphi, \psi) := \varphi \oplus \psi : X \oplus Y \to X' \oplus Y', \quad (x, y) \mapsto \big(\varphi(x), \psi(y)\big).$$

One easily verifies (exercise) that in this way one has defined a functor $\oplus : \mathbf{Mod}_R \times \mathbf{Mod}_R \to \mathbf{Mod}_R$. □

Proposition 3.7 shows that the construction of the tensor product of modules is also a functor.

Example 4.20 (Functoriality of the Tensor Product) The functor $\otimes_R : \mathbf{Mod}_R \times_R \mathbf{Mod} \to \mathbf{Ab}$ is given by

$$\mathrm{ob}(\mathbf{Mod}_R \times_R \mathbf{Mod}) \to \mathrm{ob}(\mathbf{Ab}), \quad (M, N) \mapsto M \otimes_R N$$

and $\otimes_R(\varphi, \psi) := \varphi \otimes \psi$ from Proposition 3.7. □

Similarly, Proposition 3.25 shows that the construction of the tensor algebra is also functorial.

Example 4.21 (Functoriality of the Tensor Algebra) Let R be a commutative ring with identity. The functor $\mathrm{T} \colon \mathbf{Mod}_R \to \mathbf{Alg}_R$ is given by

$$\mathrm{ob}(\mathbf{Mod}_R) \to \mathrm{ob}(\mathbf{Alg}_R), \quad M \mapsto \mathrm{T}(M)$$

and $\mathrm{T}(\varphi)$ from Proposition 3.25. □

The functoriality of the symmetric algebra is obtained from Proposition 3.30.

Example 4.22 (Functoriality of the Symmetric Algebra) Let R be a commutative ring with identity. The functor $\mathrm{S} \colon \mathbf{Mod}_R \to \mathbf{CAlg}_R$ is given by

$$\mathrm{ob}(\mathbf{Mod}_R) \to \mathrm{ob}(\mathbf{CAlg}_R), \quad M \mapsto \mathrm{S}(M)$$

and $\mathrm{S}(\varphi)$ from Proposition 3.30. □

With Proposition 3.41, we see that the construction of the exterior algebra is also a functor.

Example 4.23 (Functoriality of the Exterior Algebra) Let R be a commutative ring with identity. The functor $\Lambda \colon \mathbf{Mod}_R \to \mathbf{Alg}_R$ is given by

$$\mathrm{ob}(\mathbf{Mod}_R) \to \mathrm{ob}(\mathbf{Alg}_R), \quad M \mapsto \Lambda(M)$$

and $\Lambda(\varphi)$ from Proposition 3.41. □

Our last example in this series is obtained from Theorem 2.14. It shows that the construction of a free module is also a functor.

Example 4.24 (Functoriality of the Free Module) Let R be a ring with identity. The functor $_R\mathrm{F}\colon \mathbf{Set} \to {}_R\mathbf{Mod}$ is given by

$$\mathrm{ob}(\mathbf{Set}) \to \mathrm{ob}(_R\mathbf{Mod}), \quad E \mapsto {}_R\mathrm{F}(E)$$

and $_R\mathrm{F}(\varphi) := \overline{\varphi}$ from Theorem 2.14. □

Exercise 4.5 (Free right-R-modules) Define a suitable concept of free right-R-modules, provide a construction $\mathbf{Set} \to \mathbf{Mod}_R$, $E \mapsto F_R(E)$ and show that it is functorial.

The following example describes, similar to the forgetful functors, a very general class of functors that are also easy to understand.

Example 4.25 (Represented Functors) Let $\mathbf{C}$ be a category and $X \in \mathrm{ob}(\mathbf{C})$.

(i) For $Y, Y' \in \mathrm{ob}(\mathbf{C})$ and $\varphi \in \mathbf{C}(Y, Y')$ consider the mapping

$$\varphi_* : \mathbf{C}(X, Y) \to \mathbf{C}(X, Y'), \quad f \mapsto \varphi \circ f.$$

Then

$$\mathrm{ob}(\mathbf{C}) \to \mathrm{ob}(\mathbf{Set}), \quad Y \mapsto \mathrm{H}^X(Y) := \mathbf{C}(X, Y),$$

$$\mathbf{C}(Y, Y') \to \mathbf{Set}\big(\mathbf{C}(X, Y), \mathbf{C}(X, Y')\big), \quad \varphi \mapsto \mathrm{H}^X(\varphi) := \varphi_*$$

defines a functor $\mathrm{H}^X : \mathbf{C} \to \mathbf{Set}$.

(ii) For $Y, Y' \in \mathrm{ob}(\mathbf{C}^{\mathrm{op}})$ and $\psi \in \mathbf{C}^{\mathrm{op}}(Y, Y') = \mathbf{C}(Y', Y)$ consider the mapping

$$\psi^* : \mathbf{C}(Y, X) \to \mathbf{C}(Y', X), \quad h \mapsto h \circ \psi.$$

Then

$$\mathrm{ob}(\mathbf{C}^{\mathrm{op}}) \to \mathrm{ob}(\mathbf{Set}), \quad Y \mapsto \mathrm{H}_X(Y) := \mathbf{C}(Y, X),$$

$$\mathbf{C}^{\mathrm{op}}(Y, Y') \to \mathbf{Set}\big(\mathbf{C}(Y, X), \mathbf{C}(Y', X)\big), \quad \psi \mapsto \mathrm{H}_X(\psi) := \psi^*$$

defines a functor $\mathrm{H}_X : \mathbf{C}^{\mathrm{op}} \to \mathbf{Set}$. □

Functors that can be *represented* by objects as in Example 4.25 play a role in modern mathematics that is comparable to the role of integers within the rational numbers. Just as linear equations with integer coefficients can be solved in $\mathbb{Q}$, but not in $\mathbb{Z}$, some constructions cannot be carried out a priori in the class of objects of a category, but can be with a functor as a result. And just as in special cases one can prove that the solution of an equation was indeed an integer, in special cases one can also prove that the resulting functor could be represented by an object.

4.2 Naive Category Theory

A functor $F\colon \mathbf{A}^{\mathrm{op}} \longrightarrow \mathbf{B}$, as it occurs in Example 4.25(ii), is also called a *contravariant* functor or *cofunctor* from $\mathbf{A}$ to $\mathbf{B}$. If one wants to emphasize the contrast, one speaks of an ordinary functor also as a *covariant* functor.

Exercise 4.6 (Dual Modules) Let R be a commutative ring with identity and $M \in \mathrm{ob}(\mathbf{Mod}_R)$. We endow $M^\vee := \mathrm{Hom}_R(M, R) = \mathbf{Mod}_R(M, R)$ with the R-module structure from Exercise 2.5 and call $M^\vee$ the *dual R-module*.

For $M, N \in \mathrm{ob}(\mathbf{Mod}_R)$ and $\varphi \in \mathrm{Hom}_R(M, N)$ set $\varphi^\vee : N^\vee \to M^\vee$, $f \mapsto f \circ \varphi$. Show that through

$$\mathrm{ob}(\mathbf{Mod}_R) \to \mathrm{ob}(\mathbf{Mod}_R)\,,\ M \mapsto M^\vee,$$

$$\mathbf{Mod}_R(M, N) \to \mathbf{Mod}_R(N^\vee, M^\vee)\,,\ \varphi \mapsto \varphi^\vee$$

a contravariant functor $^\vee : \mathbf{Mod}_R \to \mathbf{Mod}_R$ is defined. □

Exercise 4.7 (Hom-Functor) Let $\mathbf{A}$ be a category and $\mathbf{A}^{\mathrm{op}} \times \mathbf{A}$ the product category. Show that through

$$\mathrm{ob}(\mathbf{A}^{\mathrm{op}} \times \mathbf{A}) \to \mathrm{ob}(\mathbf{Set}), (A, A') \mapsto \mathbf{A}(A, A'),$$

$$(\mathbf{A}^{\mathrm{op}} \times \mathbf{A})\big((A, A'), (X, X')\big) \to \mathbf{Set}\big(\mathbf{A}(A, A'), \mathbf{A}(X, X')\big),$$

$$(\varphi, \psi) \mapsto (\alpha \mapsto \psi \circ \alpha \circ \varphi)$$

a functor $\mathrm{Hom}_\mathbf{A} : \mathbf{A}^{\mathrm{op}} \times \mathbf{A} \to \mathbf{Set}$ is defined.

Exercise 4.8 (Product Functor) Let $F : \mathbf{A} \to \mathbf{B}$ and $F' : \mathbf{A}' \to \mathbf{B}'$ be functors. Show that by

$$\mathrm{ob}(\mathbf{A} \times \mathbf{A}') \to \mathrm{ob}(\mathbf{B} \times \mathbf{B}')\,,\ (A, A') \mapsto \big(F(A), F'(A')\big)$$

and

$$(\mathbf{A} \times \mathbf{A}')\big((A, A'), (X, X')\big) \to (\mathbf{B} \times \mathbf{B}')\big((F(A), F'(A')), (F(X), F'(X'))\big),$$

$$(\varphi, \varphi') \mapsto \big(F(\varphi), F'(\varphi')\big)$$

a functor $F \times F' : \mathbf{A} \times \mathbf{A}' \to \mathbf{B} \times \mathbf{B}'$ is defined.

Exercise 4.9 (Composition of Functors) Let $F : \mathbf{A} \to \mathbf{B}$ and $G : \mathbf{B} \to \mathbf{C}$ be functors. Show that by

$$\mathrm{ob}(\mathbf{A}) \to \mathrm{ob}(\mathbf{C})\,,\ A \mapsto G\big(F(A)\big),$$

$$\mathbf{A}(A, A') \to \mathbf{C}\big(G(F(A)), G(F(A'))\big)\,,\ \varphi \mapsto G\big(F(\varphi)\big)$$

a functor $G \circ F : \mathbf{A} \to \mathbf{C}$ is defined.

Exercise 4.10 (Topological Spaces) Let $(X, \mathfrak{T})$ be a topological space, i.e., $\mathfrak{T}$ is the set of open subsets of X. Consider on $\mathfrak{T}$ the partial order given by inclusion. Then Example 4.15 provides a category whose objects are the elements of $\mathfrak{T}$. We denote this category by $\mathbf{T}$ and note that there is exactly one morphism $V \to U$ between two open subsets U and V of X if $V \subseteq U$.

Now let $(X', \mathfrak{T}')$ be another topological space and $f: X \to X'$ a continuous mapping. Show that by $U' \mapsto f^{-1}(U')$ both a functor $\mathbf{T}' \to \mathbf{T}$ and a functor $\mathbf{T}'^{\mathrm{op}} \to \mathbf{T}^{\mathrm{op}}$ are defined.

4.3 Categorical Constructions: Limits

The starting point for our considerations in this section are the constructions of products and direct sums of modules (see Construction 2.16). Unlike various other functorial constructions in Chs. 2 and 3, they make objects of the category $\mathbf{Mod}_R$ into such objects again. It turns out that here again construction patterns can be recognized that can be transferred to other situations.

Products and Sums

The crucial idea in the definition of products and sums of a family of objects in a category is not to imitate Construction 2.16 for modules, but to promote the universal properties from Exercise 2.8 to a principle.

Definition 4.26 (Categorical Product) Let $\mathbf{C}$ be a category and $(X_i)_{i \in I}$ a family of objects in $\mathbf{C}$. A *product* of the X_i is an object P in $\mathbf{C}$ together with morphisms

$$p_i : P \longrightarrow X_i,$$

called *projections*, with the following property: If Y is an object in $\mathbf{C}$ and $q_i : Y \longrightarrow X_i$ are morphisms for $i \in I$, then there exists exactly one morphism $q : Y \to P$ in $\mathbf{C}$, for which $p_i \circ q = q_i$ holds for all $i \in I$. Thus, we have the following commutative diagram:

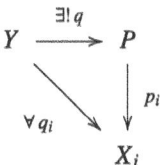

q is denoted by $(q_i)_{i \in I}$ or, in the case of finite subsets, by $(q_1, \cdots, q_n)$.

Products are unique up to isomorphisms (exercise). Therefore, we write $\prod_{i \in I} X_i$ or $X_1 \times \cdots \times X_n$ for *the* product of the X_i.

Example 4.27 (Topological Product) In the category **Set**, the usual Cartesian product $\prod_{i \in I} X_i$ is a product in the categorical sense (exercise). If the X_i are all endowed with topologies, one can consider on $\prod_{i \in I} X_i$ the smallest topology, which contains all pre-images of open sets under the projections $p_k : \prod_{i \in I} X_i \to X_k$. That is, one considers the smallest family of subsets of $\prod_{i \in I} X_i$, which is stable under finite intersections and arbitrary unions and contains both these pre-images as well

4.3 Categorical Constructions: Limits

as the empty set and the full space. The topology thus obtained is called the *product topology*, and it makes $\prod_{i \in I} X_i$ a product of the X_i in the category **Top** (exercise). □

The categories of algebraic structures from Exercise 4.3 all allow products. One simply considers the Cartesian product of the objects and defines the operations componentwise (exercise). However, it is by no means the case that the product of even two objects must exist in every category. Consider, e.g., the category **C**, which consists of two objects X and Y and has no morphisms other than the identities. In **C** there can be no product simply because no object of **C** has morphisms to both X and Y.

Example 4.28 (Products of Algebraic Structures) Let $\mathbf{C} = \mathbf{C}_{\Phi,\Gamma}$ be a category of algebraic structures. Then there are categorical products in **C**. They are given by the set-theoretical products with the componentwise operations. □

Already in the context of Exercise 2.8 we pointed out that there is a simple duality between the universal properties of products and sums of modules, which is based on reversing arrows. For categories, this means going to the opposite category. This leads us to the definition of the *categorical sum*, which is also called *coproduct*.

Definition 4.29 (Categorical Sum or Coproduct) Let **C** be a category and $(X_i)_{i \in I}$ a family of objects in **C**. A *sum* of the X_i is an object S in **C** together with morphisms

$$\iota_i : X_i \longrightarrow S,$$

called *inclusions*, with the following property: If Y is an object in **C** and $q_i : X_i \longrightarrow Y$ are morphisms for $i \in I$, then there exists exactly one morphism $q : S \to Y$ in **C**, for which $q \circ \iota_i = q_i$ holds for all $i \in I$. Thus, we have the following commutative diagram:

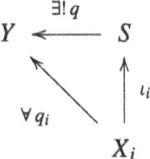

Sums, like products, are unique up to isomorphisms (exercise). We write $\coprod_{i \in I} X_i$ for *the* sum of the X_i.

Example 4.30 (Set-Theoretic and Topological Sums) In the category **Set**, the disjoint union $\coprod_{i \in I} X_i$ of the X_i with the embeddings $\iota_k : X_k \to \coprod_{i \in I} X_i$ is the sum in the categorical sense (exercise). If the X_i are all equipped with topologies, the union of all these topologies itself is a topology, with respect to which each X_k is an open and closed subset of $\coprod_{i \in I} X_i$. With this topology, $\coprod_{i \in I} X_i$ is called the *topological sum* of the X_i. It is indeed a sum of the X_i in **Top** (exercise). □

The same example as for products also shows that the sum itself of only two objects does not have to exist in every category: Consider again the category **C**, which consists of two objects X and Y and has no morphisms other than the identities. In **C**, there can be no sum simply because there are no morphisms from both X and Y into any object of **C**.

Example 4.31 (Tensor Products of Algebras) Let R be a commutative ring with identity and $\mathbf{Alg}_{R,1}$ the subcategory of $\mathbf{Alg}_R$ of R-algebras with identity (with morphisms that preserve the identity). For $A, B \in \mathrm{ob}(\mathbf{Alg}_{R,1})$, the tensor product $A \otimes_R B$ is an R-module, but it also carries an R-bilinear multiplication: The multiplication $\mu_A : A \times A \to A$ can be considered as an R-module homomorphism $\overline{\mu}_A : A \otimes_R A \to A$ according to Remark 3.14, and similarly one has $\overline{\mu}_B : B \otimes_R B \to B$. This gives us the R-module homomorphism

$$\overline{\mu}_A \otimes \overline{\mu}_B : (A \otimes_R A) \otimes_R (B \otimes_R B) \to A \otimes_R B,$$

which, combined with the isomorphism

$$(A \otimes_R B) \otimes_R (A \otimes_R B) \to (A \otimes_R A) \otimes_R (B \otimes_R B),$$

induced by the R-multilinear mapping

$$A \times B \times A \times B \to (A \otimes_R A) \otimes_R (B \otimes_R B), \quad (a, b, a', b') \mapsto (a \otimes a') \otimes (b \otimes b')$$

provides an R-bilinear mapping

$$\mu_{A \otimes_R B} : (A \otimes_R B) \times (A \otimes_R B) \to (A \otimes_R B)$$

Then we have $\mu_{A \otimes_R B}(a \otimes b, a' \otimes b') = aa' \otimes bb'$, and because the elements of the form $a \otimes b$ span the R-module $A \otimes_R B$, the associativity of the product $\mu_{A \otimes_R B}$ follows immediately from the associativity of the products μ_A and μ_B. Moreover, it is clear that $1_A \otimes 1_B$ is the identity of $A \otimes_R B$ if 1_A and 1_B are the identities of A and B, respectively.

Note that $\iota_A: A \to A \otimes_R B$, $a \mapsto a \otimes 1_B$ and $\iota_B: B \to A \otimes_R B$, $b \mapsto 1_A \otimes b$ are morphisms of $\mathbf{Alg}_{R,1}$. Now let $C \in \mathrm{ob}(\mathbf{Alg}_{R,1})$ and $\varphi_A \in \mathbf{Alg}_{R,1}(A, C)$ as well as $\varphi_B \in \mathbf{Alg}_{R,1}(B, C)$. If $\overline{\mu}_C: C \otimes_R C \to C$ is the R-module homomorphism induced by the multiplication μ_C on C, then

$$\varphi := \overline{\mu}_C \circ (\varphi_A \otimes \varphi_B) : A \otimes_R B \to C$$

is an R-module homomorphism that makes the diagrams

commutative. If one could now show that φ is a morphism of R-algebras, one would have proven that $A \otimes_R B$ is the categorical sum of A and B. However, this proof encounters difficulties when C is not commutative:

$$\varphi\big((a \otimes b)(a' \otimes b')\big) = \varphi_A(a)\varphi_A(a')\varphi_B(b)\varphi_B(b') \in C,$$
$$\varphi(a \otimes b)\varphi(a' \otimes b') = \varphi_A(a)\varphi_B(b)\varphi_A(a')\varphi_B(b') \in C.$$

We therefore conclude that the tensor product of R-algebras is the categorical sum in the full subcategory $\mathbf{CAlg}_{R,1}$ of commutative algebras in $\mathbf{Alg}_{R,1}$. □

Example 4.32 (Partially Ordered Sets) We revisit Example 4.15 of a partially ordered set $(M, \leq)$ and consider the category $\mathbf{M}$ constructed there. Then we have

$$\forall m, m' \in M: \quad m \leq m' \Leftrightarrow \mathbf{M}(m, m') \neq \emptyset.$$

An element $i \in M$ is a product of m and m' if there are morphisms $i \to m$ and $i \to m'$ such that for every element $a \in M$ with morphisms $a \to m$ and $a \to m'$ there exists a morphism $a \to i$ that makes the diagram

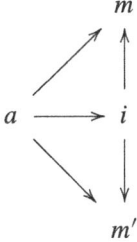

commutative. Expressed through the order, this means nothing else than: i is the *greatest lower bound* or the *infimum* of m and m'.

Since the opposite category $\mathbf{M}^{op}$ is the same set with the reversed partial order, it is immediately clear that an element $s \in M$ is a categorical sum of m and m' if and only if s is the *least upper bound* or the *supremum* of m and m'.

The category therefore has finite products and sums if and only if for two elements $m, m' \in M$ there always exists both a supremum $m \vee m' \in M$ and an infimum $m \wedge m' \in M$. Such partially ordered sets are called *lattices*. □

As a special case, Example 4.32 provides the lattice of natural numbers with the partial order given by divisibility and the greatest common divisor as infimum and the least common multiple lcm as supremum.

The idea of categorical limits and colimits is to generalize the diagrams from Exercise 2.8.

Definition 4.33 (Diagrams of the Form I in a Category) Let $\mathbf{I}$ be a small category. A *diagram of the form* $\mathbf{I}$ in a category $\mathbf{C}$ is a functor $D : \mathbf{I} \to \mathbf{C}$.

The shape of diagrams in products and sums of n objects is simply the category $\bullet \ldots \bullet$ consisting of n objects and only the trivial morphisms (identities). The diagrams from Exercise 2.8 emerge from this in two further steps.

Definition 4.34 (D-Cone and D-Limit) Let $D : \mathbf{I} \to \mathbf{C}$ be a diagram of the form $\mathbf{I}$ in $\mathbf{C}$.

(i) A D-cone $A \xrightarrow{\varphi} D(\mathbf{I})$ consists of an object $A \in \mathrm{ob}(\mathbf{C})$ and a family $\varphi = (\varphi_i)_{i \in \mathbf{I}}$ of morphisms $\varphi_i \in \mathbf{C}(A, D(i))$, where for each morphism $f_{ij} \in \mathbf{I}(i, j)$ the diagram

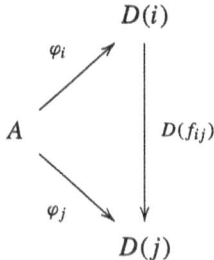

is commutative. The name D-*cone* is explained by the shape of the diagram: A is the tip of the cone and the $D(i)$'s together with the arrows connecting them form the base of the cone.

4.3 Categorical Constructions: Limits

(ii) A *D-limit* is a *D*-cone $L \xrightarrow{\psi} D(\mathbf{I})$, which has the following universal property: For each *D*-cone $A \xrightarrow{\varphi} D(\mathbf{I})$ there exists exactly one $p \in \mathbf{C}(A, L)$ with

$$\forall i \in \mathbf{I}: \quad \psi_i \circ p = \varphi_i \in \mathbf{C}(A, D(i)).$$

Expressed as a diagram, this means

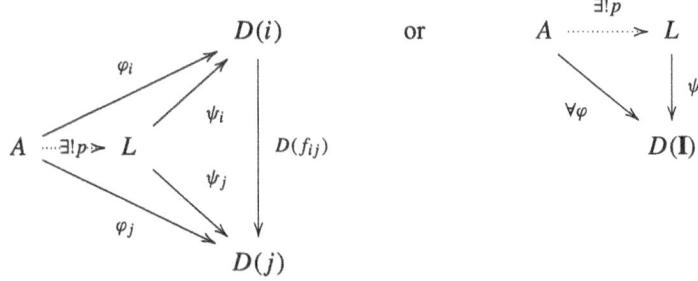

As with objects described by universal properties, it is also clear for *D*-limits that they are unique up to isomorphism, if they exist. It therefore makes sense to introduce the notations $\varprojlim_{\mathbf{I}} D$ or $\varprojlim_{i \in \mathbf{I}} D(i)$ for *the D*-limit.

In the last form of the diagram of a *D*-limit from Definition 4.34, the diagram of a product (see Definition 4.26) is easily recognized. To also be able to deal with sums, we need to introduce the dual concepts. We start with an exercise in which the *opposite functor* is constructed for a functor.

Exercise 4.11 (Opposite Functor) Let $F: \mathbf{C} \to \mathbf{D}$ be a functor. Show that by

$$\forall X \in \mathrm{ob}(C): \quad F^{\mathrm{op}}(X) := F(X)$$

and

$$\forall \varphi \in \mathbf{C}(X, Y): \quad F^{\mathrm{op}}(\varphi) := F(\varphi)^{\mathrm{op}},$$

where $\varphi = \varphi^{\mathrm{op}} \in \mathbf{C}^{\mathrm{op}}(Y, X) = \mathbf{C}(X, Y)$ is, a functor $F^{\mathrm{op}}: \mathbf{C}^{\mathrm{op}} \to \mathbf{D}^{\mathrm{op}}$ is defined. We call this functor the *opposing functor* to F. Further, show that $F^{\mathrm{op}}(\varphi^{\mathrm{op}}) = F(\varphi)^{\mathrm{op}}$ holds.

Definition 4.35 (*D*-Cocone and *D*-Colimit) Let $D : \mathbf{I} \to \mathbf{C}$ be a diagram of shape $\mathbf{I}$ in $\mathbf{C}$. Then $D^{\mathrm{op}}: \mathbf{I}^{\mathrm{op}} \to \mathbf{C}^{\mathrm{op}}$ is a diagram of shape $\mathbf{I}^{\mathrm{op}}$ in $\mathbf{C}^{\mathrm{op}}$. A *D-cocone* is a D^{op}-cone, and a *D-colimit* is a D^{op}-limit. We denote the *D*-colimit with $\varinjlim_{\mathbf{I}} D$ or $\varinjlim_{i \in \mathbf{I}} D(i)$.

The Definition 4.35 of a colimit can also be described without the use of the opposing functor by diagrams, by reversing all arrows. A D-cocone $A \xleftarrow{\varphi} D(\mathbf{I})$ consists of an object $A \in \mathrm{ob}(\mathbf{C})$ and a family $\varphi = (\varphi_i)_{i \in \mathbf{I}}$ of morphisms $\varphi_i \in \mathbf{C}(D(i), A)$, where for each morphism $f_{ij} \in \mathbf{I}(j, i)$ the diagram

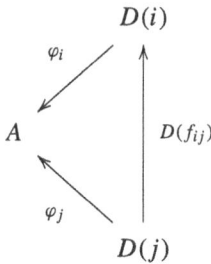

is commutative. A D-colimit is then a D-cocone $K \xleftarrow{\psi} D(\mathbf{I})$, which has the following universal property: For each D-cocone $A \xleftarrow{\varphi} D(\mathbf{I})$ there is exactly $p \in \mathbf{C}(K, A)$ with

$$\forall i \in \mathbf{I}: \quad p \circ \psi_i = \varphi_i \in \mathbf{C}(D(i), A).$$

Expressed as a diagram, this means

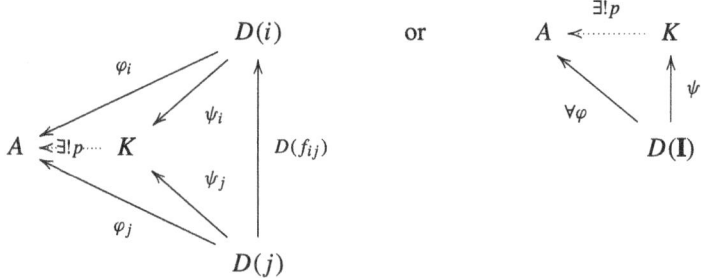

Here, in the last form, the diagram of a sum (see Definition 4.26) is easily recognized.

Example 4.36 (Pullback) We consider diagrams of the shape

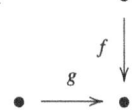

4.3 Categorical Constructions: Limits

This means, the limit diagram has the shape

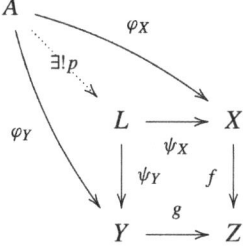

Such a limit is called the *pullback* of

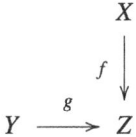

(i) Let **C** = **Set**. Then a limit is given by $L := \{(x, y) \in X \times Y \mid f(x) = g(y)\}$ and $\varphi_X(x, y) = x$ as well as $\varphi_Y(x, y) = y$. Here, we also speak of the *fiber product* $X \times_Z Y$ of X and Y over Z.

(ii) Consider the special case of the fiber product from (i), for which X is a subset of Z and $f: X \to Z$ is the inclusion. Then $L := \{y \in Y \mid g(y) \in X\} = g^{-1}(X) \subseteq Y$ with $\psi_X(y) = g(y)$ and $\psi_Y(y) = y$ is also a pullback.

(iii) A special case of (ii) occurs when Y is also a subset of Z and $g: Y \to Z$ is the inclusion. In this case, $L = X \cap Y$ is simply the intersection of the two sets. □

Fiber products are more interesting when considered in topological categories. We refrain from this here and instead consider the dual concept.

Example 4.37 (Pushout) We consider diagrams of the shape

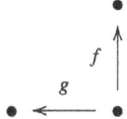

This means, the colimit diagram has the shape

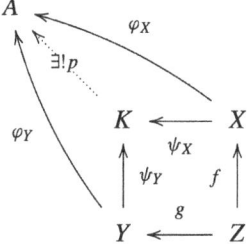

Such a limit is called the *pushout* of

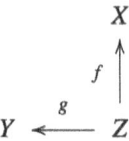

(i) Let **C** = **Set**. Then a colimit is given by the set $K := (X \coprod Y)_\sim$ of equivalence classes of the following defined equivalence relation $\sim$: For $a, b \in X \coprod Y$, $a \sim b$ if and only if $a = b$ or $a = f(z), b = g(z)$ or $a = g(z), b = f(z)$.
(ii) Consider the special case of (i), for which Z is a subset of X and $f : Z \to X$ is the inclusion. Then the points in Z are identified with their images in Y under g. This is also referred to as the *gluing* of X and Y along $g : Z \to Y$.
(iii) A special case of (ii) occurs when $Z = X \cap Y$. In this case, $K = X \cup Y$ with the inclusions $X \cup Y \hookleftarrow X \hookleftarrow X \cap Y$ and $X \cup Y \hookleftarrow Y \hookleftarrow X \cap Y$ is a pushout. □

Gluing also becomes more interesting when performed for topological spaces. We will see a number of such examples.

We conclude our discussion of limits with the actual namesakes of these constructions, the *inductive* and *projective* limits.

Definition 4.38 (Inductive and Projective Limits) Let $(I, \leq)$ be a partially ordered set and **I** the corresponding small category (see Example 4.15). A *projective system* of the shape $(I, \leq)$ in a category **C** is a diagram D of the shape **I** in **C**. The *inverse* or *projective limit* of this system is the D-limit, provided it exists. An *inductive system* of the shape $(I, \leq)$ is a diagram D of the shape $\mathbf{I}^{op}$ in **C**. The *direct* or *inductive limit* of this system is the D-colimit, provided it exists.

Example 4.39 (Inductive and Projective Limits of Sets) With the notations from Definition 4.38 and Definition 4.34, let **C** = **Set**.

(i) The set

$$\varprojlim D := \left\{ (a_i)_{i \in I} \in \prod_{i \in I} D(i) \,\middle|\, \forall i, j \in I : D(f_{ij})a_i = a_j \right\}$$

together with the mappings

$$\psi_j : \varprojlim_\mathbf{I} D = \varprojlim D \to D(j), \quad (a_i)_{i \in I} \mapsto a_j$$

is the projective limit of D.
(ii) Consider the set of equivalence classes

$$\varinjlim D := \left[\coprod_{i \in I} D(i) \right]_\sim$$

in $\coprod_{i\in I} D(i)$ with respect to the equivalence relation defined by

$$\forall i,j \in I: \quad D(i) \ni a_i \sim a_j \in D(j) :\Leftrightarrow \quad \begin{array}{l} \exists \ell \in I, a_\ell \in D(\ell): \\ D(f_{i\ell})a_\ell = a_i, D(f_{j\ell})a_\ell = a_j \end{array}$$

We denote the equivalence class of $a \in \coprod_{i\in I} D(i)$ with respect to $\sim$ by $[a]_\sim$. Then $\varinjlim D$, together with the mappings

$$\psi_j \, D(j) \to \varinjlim D = \varinjlim_I D, \quad a_j \mapsto [a_j]_\sim,$$

is the inductive limit of D. □

The next two examples will be revisited in a generalized form in Ch. 5 and play an important role there.

Example 4.40 (Sheaves of Continuous Functions) Let $(X, \mathfrak{T})$ be a topological space.

(i) Let $x \in X$ and $I := \{U \in \mathfrak{T} \mid x \in U\}$ be the set of all open neighborhoods of x in X. We order I by inclusion and consider for each $U \in I$ the commutative $\mathbb{R}$-algebra $\mathcal{C}(U, \mathbb{R})$ of continuous $\mathbb{R}$-valued functions. Then by

$$D(U) := \mathcal{C}(U, \mathbb{R}), \quad \left(f_{VU}: D(U) \to D(V), \varphi \mapsto \varphi|V\right)$$

a diagram $D: \mathbf{I}^{op} \to \mathbf{CAlg}_\mathbb{R}$ is defined. The inductive limit $\varinjlim_I D$ exists. It consists of the equivalence classes $[\varphi]_\sim$ of functions $\varphi \in \mathcal{C}(U, \mathbb{R})$ with $U \in I$ with respect to the equivalence relation $\sim$, which relates two functions $\varphi \in \mathcal{C}(U, \mathbb{R})$ and $\psi \in \mathcal{C}(V, \mathbb{R})$ if there is a $W \subseteq U \cap V$ in I with $\varphi|_W = \psi|_W$. The mapping $D(U) \to \varinjlim_I D$ is given by $\varphi \mapsto [\varphi]_\sim$.

(ii) Let $M \subseteq X$ be a subset and $I := \{U \in \mathfrak{T} \mid M \subseteq U\}$ be the set of all open neighborhoods of M in X. With this, the construction from (i) can be imitated (details as exercise). □

The following examples are more algebraic in nature, but they can also be equipped with a topological structure i.e., of great importance for further investigations.

Example 4.41 (Inductive and Projective Limits)

(i) For a prime number p, in the diagram

$$\cdots \xrightarrow{f_{n+1}} \mathbb{Z}/p^{n+1}\mathbb{Z} \xrightarrow{f_n} \mathbb{Z}/p^n\mathbb{Z} \xrightarrow{f_{n-1}} \cdots \xrightarrow{f_1} \mathbb{Z}/p\mathbb{Z}$$

in **Ring**$_1$, the mappings $f_n \colon \mathbb{Z}/p^{n+1}\mathbb{Z} \to \mathbb{Z}/p^n\mathbb{Z}$ are given by $z + p^{n+1}\mathbb{Z} \mapsto z + p^n\mathbb{Z}$. An inverse limit is then given by

$$\mathbb{Z}_p := \Big\{(z_n)_{n\in\mathbb{N}} \in \prod_{n\in\mathbb{N}} \mathbb{Z}/p^n\mathbb{Z} \;\Big|\; f_n(z_{n+1}) = z_n\Big\}$$

given. $\mathbb{Z}_p$ is also called the ring of *p-adic integers*. It can be shown that $\mathbb{Z}_p$ is an integral domain, whose field of quotients is isomorphic to the field $\mathbb{Q}_p$ of *p-adic numbers*, i.e., the completion of $\mathbb{Q}$ with respect to the *p*-adic distance.

(ii) For a commutative ring R with identity, in the diagram

$$\cdots \xleftarrow{f_{n+1}} \mathrm{GL}_{n+1}(R) \xleftarrow{f_n} \mathrm{GL}_n(R) \xleftarrow{f_{n+1}} \cdots \xleftarrow{f_1} \mathrm{GL}_1(R)$$

in **Grp** the mappings $f_n \colon \mathrm{GL}_n(R) \to \mathrm{GL}_{n+1}(R)$ are given by

$$g \mapsto \begin{pmatrix} g & 0 \\ 0 & 1 \end{pmatrix}$$

A direct limit is then given by the group of (infinitely large) matrices $A := (a_{ij})_{i,j\in\mathbb{N}}$, whose entries lie in R and satisfy the following conditions:
(a) Only finitely many diagonal elements a_{ii} are not equal to 1.
(b) Only finitely many off-diagonal elements $a_{ij}, i \neq j$ are not equal to 0.
(c) If

$$A = \begin{pmatrix} A_N & 0 & \cdots \\ 0 & 1 & \\ \vdots & & \ddots \end{pmatrix}$$

with $N \in \mathbb{N}$ and $A_N \in \mathrm{Mat}_{N\times N}(R)$, then A_N is invertible in $\mathrm{Mat}_{N\times N}(R)$, i.e., $\det(A_N)$ is a unit in R.

On such matrices, the usual formula for matrix multiplication can be used, i.e., we set $AB := C$ with

$$c_{ij} := \sum_{k\in\mathbb{N}} a_{ik} b_{kj}.$$

The group thus obtained is denoted by $\mathrm{GL}_\infty(R)$. □

4.3 Categorical Constructions: Limits

There are general theorems that provide criteria for categories under which the limits and/or colimits of arbitrary diagrams or only finite diagrams exist. We do not go into this (important!) question in this book, but refer to [HS73], [Le14] and [ML98].

Exercise 4.12 (Inductive and Projective Limits of R-Modules) Show that in the category $\mathbf{Mod}_R$ both injective and projective limits exist.

Exercise 4.13 (Tensor Products Commute with Colimits) Let $D_1, D_2 \colon \mathbf{I} \to \mathbf{Mod}_R$ be diagrams, whose colimits $\varinjlim_\mathbf{I} D_1$ and $\varinjlim_\mathbf{I} D_2$ exist. Show that

$$D_1 \otimes D_2 \colon \mathbf{I} \to \mathbf{Mod}_R, \ i \mapsto D_1(i) \otimes_R D_2(i), \ f_{ij} \mapsto D_1(f_{ij}) \otimes D_2(f_{ij})$$

is a diagram, whose colimit exists and is given by $(\varinjlim_\mathbf{I} D_1) \otimes_R (\varinjlim_\mathbf{I} D_2)$. Hint: The following diagram may be helpful.

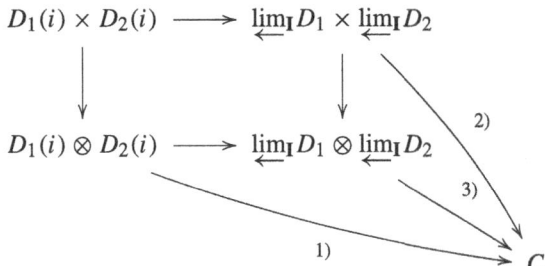

Exercise 4.14 (Homomorphisms Commute with Limits) Let $D \colon \mathbf{I} \to \mathbf{Mod}_R$ be a diagram for which the limit $\varprojlim_\mathbf{I} D$ exists, and $M \in \mathrm{ob}(\mathbf{Mod}_R)$. Show that by

$$i \mapsto \mathrm{Hom}_R(M, D(i)) \quad \text{and} \quad f_{ij} \mapsto \bigl(\varphi(j) \mapsto D(f_{ij}) \circ \varphi(j)\bigr)$$

a diagram $\mathrm{Hom}_R(M, D) \colon \mathbf{I} \to \mathbf{Mod}_R$ is defined, whose limit exists and is given by $\mathrm{Hom}_R(M, \varprojlim_\mathbf{I} D)$. Hint: The following diagrams may be helpful.

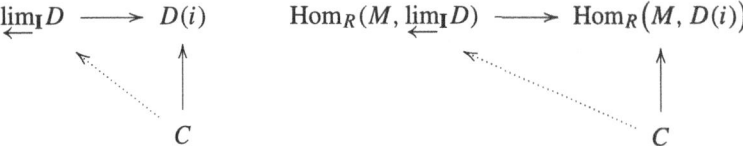

Exercise 4.15 (Homomorphisms and Colimits) Let $D \colon \mathbf{I} \to \mathbf{Mod}_R$ be a diagram for which the limit $\varinjlim_\mathbf{I} D$ exists, and $M \in \mathrm{ob}(\mathbf{Mod}_R)$. Show that by

$$i \mapsto \mathrm{Hom}_R(D(i), M) \quad \text{and} \quad f_{ij} \mapsto \bigl(\varphi(j) \mapsto \varphi(j) \circ D(f_{ij})\bigr)$$

a diagram $\text{Hom}_R(D, M)\colon \mathbf{I} \to \mathbf{Mod}_R$ is defined, whose limit exists and is given by $\text{Hom}_R(\varinjlim_{\mathbf{I}} D, M)$. Hint: The following diagram helps to see that the natural mapping

$$\text{Hom}_R(\varinjlim_{\mathbf{I}} D, M) \to \prod_i \text{Hom}_R(D(i), M)$$

is bijective.

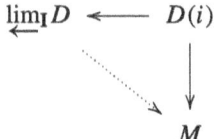

Exercise 4.16 ((Co)limits commute with (Co)limits) Let $D\colon \mathbf{I} \times \mathbf{I}' \to \mathbf{Mod}_R$ be a diagram. For each $i \in \text{ob}(\mathbf{I})$, define a diagram $D(i, \bullet)\colon \mathbf{I}' \to \mathbf{Mod}_R$ by

$$i' \mapsto D(i, i'), \quad f_{i'j'} \mapsto D(f_{(i,i')(i,j')}).$$

Similarly, find a diagram $D(\bullet, i')\colon \mathbf{I}' \to \mathbf{Mod}_R$ for each $i' \in \text{ob}(\mathbf{I}')$. Show:

(i) If the limits $\varprojlim_{\mathbf{I}'} D(i, \bullet)$ and $\varprojlim_{\mathbf{I}} D(\bullet, i')$ exist for all $(i, i') \in \text{ob}(\mathbf{I} \times \mathbf{I}')$, then $\varprojlim_{\mathbf{I} \times \mathbf{I}'} D$ also exists, and it holds

$$\varprojlim_{\mathbf{I} \times \mathbf{I}'} D = \varprojlim_{i \in \mathbf{I}} \bigl(\varprojlim_{i' \in \mathbf{I}'} D(i, i')\bigr) = \varprojlim_{i' \in \mathbf{I}'} \bigl(\varprojlim_{i \in \mathbf{I}} D(i, i')\bigr).$$

(ii) If the colimits $\varinjlim_{\mathbf{I}'} D(i, \bullet)$ and $\varinjlim_{\mathbf{I}} D(\bullet, i')$ exist for all $(i, i') \in \text{ob}(\mathbf{I} \times \mathbf{I}')$, then $\varinjlim_{\mathbf{I} \times \mathbf{I}'} D$ also exists, and it holds

$$\varinjlim_{\mathbf{I} \times \mathbf{I}'} D = \varinjlim_{i \in \mathbf{I}} \bigl(\varinjlim_{i' \in \mathbf{I}'} D(i, i')\bigr) = \varinjlim_{i' \in \mathbf{I}'} \bigl(\varinjlim_{i \in \mathbf{I}} D(i, i')\bigr).$$

Exercise 4.17 (Homomorphisms Commute with (Co)limits) Let $D\colon \mathbf{I} \to \mathbf{Mod}_R$ and $D'\colon \mathbf{I}' \to \mathbf{Mod}_R$ be diagrams for which $\varinjlim_{\mathbf{I}} D$ and $\varprojlim_{\mathbf{I}'} D'$ exist. Show that by

$$(i, i') \mapsto \text{Hom}_R(D(i), D'(i'))$$

and $f_{(i,i')(j,j')} \mapsto \bigl(\varphi(j, j') \mapsto D'(f_{i'j'}) \circ \varphi(j, j') \circ D(f_{ij})\bigr)$

a diagram $\text{Hom}_R(D, D')\colon \mathbf{I} \times \mathbf{I}' \to \mathbf{Mod}_R$ is defined, whose limit exists and is given by

$$\text{Hom}_R(\varinjlim_{\mathbf{I}} D, \varprojlim_{\mathbf{I}'} D') = \varprojlim_{\mathbf{I} \times \mathbf{I}'} \text{Hom}_R(D(i), D'(i'))$$

Hint: The following diagram may be helpful.

$$\begin{array}{ccc}
\text{Hom}_R(D(i), D'(i')) & \longleftarrow & \text{Hom}_R(\varinjlim_{\mathbf{I}} D, D'(i')) \\
\uparrow & & \uparrow \\
\text{Hom}_R(D(i), \varprojlim_{\mathbf{I}'} D') & \longleftarrow & \text{Hom}_R(\varinjlim_{\mathbf{I}} D, \varprojlim_{\mathbf{I}'} D')
\end{array}$$

4.4 Adjoint Functors

The starting point for our considerations in this section is the observation that the central statement of Theorem 2.14, namely the bijectivity of the mapping

$$\mathbf{Set}(E, V) \to {}_R\mathbf{Mod}({}_RF(E), V),$$

where $V \in \mathrm{ob}({}_R\mathbf{Mod})$, appears in a very similar form in the universal property of the tensor algebra (see Definition 3.22) and the symmetric algebra (see Definition 3.27). For the former, for a commutative ring R with identity, we have the bijectivity of the mapping

$$\mathbf{Mod}_R(M, A) \to \mathbf{Alg}_R(\mathrm{T}(M), A),$$

where $A \in \mathrm{ob}(\mathbf{Alg}_R)$, and for the latter the bijectivity of the mapping

$$\mathbf{Mod}_R(M, A) \to \mathbf{Alg}_R(\mathrm{S}(M), A),$$

where $A \in \mathrm{ob}(\mathbf{CAlg}_R)$. In all three cases, we consider objects of a category as objects of a category that encompasses them by "forgetting" the additional properties (see the remarks following Definition 4.18). The similarities of the three statements go even further. All three constructions were functorial, which is reflected in a certain naturality condition for the three bijections: If we denote the forgetful functor in each case by G and write F for ${}_RF$, T and S respectively, then these bijections all have the form

$$\Gamma_{A,B} : \mathbf{A}\bigl(A, G(B)\bigr) \to \mathbf{B}\bigl(F(A), B\bigr)$$

and for $\varphi \in \mathbf{A}(A', A)$ and $\psi \in \mathbf{B}(B, B')$ they satisfy (exercise)

$$\forall \alpha \in \mathbf{A}\bigl(A, G(B)\bigr) : \quad \Gamma_{A',B'}\bigl(G(\psi) \circ \alpha \circ \varphi\bigr) = \psi \circ \Gamma_{A,B}(\alpha) \circ F(\varphi).$$

These properties are summarized in the definition of a pair of *adjoint functors*.

Definition 4.42 (Adjoint Functors) Let $\mathbf{A}$ and $\mathbf{B}$ be two categories and $F : \mathbf{A} \to \mathbf{B}$ and $G : \mathbf{B} \to \mathbf{A}$ be two functors. Then F is called *left adjoint* to G and G is called *right adjoint* to F, if for every pair $(A, B) \in \mathrm{ob}(\mathbf{A} \times \mathbf{B})$ there is a bijection (called *adjunction*)

$$\Gamma_{A,B} : \mathbf{A}\bigl(A, G(B)\bigr) \to \mathbf{B}\bigl(F(A), B\bigr),$$

which satisfies the following naturality condition: For all $\varphi \in \mathbf{A}(A', A)$, $\psi \in \mathbf{B}(B, B')$ and $\alpha \in \mathbf{A}\big(A, G(B)\big)$ it holds

$$\Gamma_{A', B'}\big(G(\psi) \circ \alpha \circ \varphi\big) = \psi \circ \Gamma_{A, B}(\alpha) \circ F(\varphi).$$

In this context, we write $F \dashv G$ or also $\mathbf{A} \;\substack{F \\ \longrightarrow \\ \bot\Gamma \\ \longleftarrow \\ G}\; \mathbf{B}$, when we want to transport all the information in the notation.

It turns out that there are a multitude of examples of pairs of adjoint functors. In particular, there are many examples of adjoint functors to forgetful functors. We will show with Theorem 4.57 that adjoint functors to a given functor are essentially uniquely determined. Therefore, in a certain sense, one inevitably also encounters the more complicated of the described functorial constructions as soon as one has found their simple adjoint constructions.

Example 4.43 (Fields of Fractions) Let **ID** and **Field** be the categories of integral domains and fields as in Section 4.2 and $\mathbf{ID}_m$ the subcategory of **ID**, whose objects are all integral domains, but only the injective *unital* (i.e., preserving the identity) ring homomorphisms are taken as morphisms. Then **Field** is also a subcategory of $\mathbf{ID}_m$. Let $G : \mathbf{Field} \to \mathbf{ID}_m$ be the functor that interprets each field as an integral domain and each field homomorphism (automatically injective) as a unital ring homomorphism. Then the functor $F : \mathbf{ID}_m \to \mathbf{Field}$ defined by

$$\mathrm{ob}(\mathbf{ID}_m) \to \mathrm{ob}(\mathbf{Field}) \,,\ R \mapsto Q(R),$$

$$\mathbf{ID}_m(R, R') \to \mathbf{Field}\big(Q(R), Q(R')\big) \,,\ \varphi \mapsto \left(\frac{r}{s} \mapsto \frac{\varphi(r)}{\varphi(s)}\right)$$

is left adjoint to G (the notations are the same as in Theorem 1.18). The corresponding adjunction

$$\Gamma_{R, \mathbb{K}} : \mathbf{ID}_m\big(R, G(\mathbb{K})\big) \to \mathbf{Field}\big(Q(R), \mathbb{K}\big)$$

for $R \in \mathrm{ob}(\mathbf{ID}_m)$ and $\mathbb{K} \in \mathrm{ob}(\mathbf{Field})$ is given by

$$\Gamma_{R, \mathbb{K}}(\varphi) \mapsto \left(\frac{r}{s} \mapsto \varphi(r)\varphi(s)^{-1}\right)$$

□

Example 4.44 (Abelianization of Groups) Let **Ab** and **Grp** be the categories of abelian groups and groups as in Section 4.2 and $G : \mathbf{Ab} \to \mathbf{Grp}$ the functor that

4.4 Adjoint Functors

simply interprets each abelian group as a group. Then the functor $F : \mathbf{Grp} \to \mathbf{Ab}$ defined by

$$\mathrm{ob}(\mathbf{Grp}) \to \mathrm{ob}(\mathbf{Ab})\,,\; H \mapsto H/[H, H],$$

$$\mathbf{Grp}(H, H') \to \mathbf{Ab}\big(H/[H, H], H'/[H', H']\big)\,,\; \varphi \mapsto \big(g[H, H] \mapsto \varphi(g)[H', H']\big)$$

is left adjoint to G. Here, $[H, H]$ is the subgroup of H generated by the elements of the form $ghg^{-1}h^{-1}$. This is automatically a normal subgroup (see Example 4.9), so that the quotient space $H/[H, H]$ carries a group structure. From the definition, it immediately follows that $H/[H, H]$ is abelian (the notations are the same as in Theorem 1.18). The corresponding adjunction

$$\Gamma_{H,A} : \mathbf{Grp}\big(H, G(A)\big) \to \mathbf{Ab}\big(H/[H, H], A\big)$$

for $H \in \mathrm{ob}(\mathbf{Grp})$ and $A \in \mathrm{ob}(\mathbf{Ab})$ is given by

$$\Gamma_{H,A}(\varphi) \mapsto \big(g[H, H] \mapsto \varphi(g)\big)$$

It should be noted that $\Gamma_{H,A}$ is well-defined because

$$\varphi(ghg^{-1}h^{-1}) = \varphi(g)\varphi(h)\varphi(g^{-1})\varphi(h^{-1}) = 1.$$

$\square$

Example 4.45 (Tensor Products and Hom-Spaces) Let R be a commutative ring with identity and $\mathbf{Mod}_R$ the category of R-modules as in Section 4.2. For $C \in \mathrm{ob}(\mathbf{Mod}_R)$ we consider the functors $G := \mathrm{Hom}_R(C, \cdot) : \mathbf{Mod}_R \to \mathbf{Mod}_R$ and $\cdot \otimes_R C : \mathbf{Mod}_R \to \mathbf{Mod}_R$. Then $F \dashv G$, and the associated adjunction

$$\Gamma_{M,N} : \mathbf{Mod}_R\big(M, \mathrm{Hom}_R(C, N)\big) \to \mathbf{Mod}_R\big(M \otimes_R C, N\big)$$

for $M, N \in \mathrm{ob}(\mathbf{Mod}_R)$ is given by

$$\Gamma_{M,N}(\varphi) \mapsto \big(m \otimes c \mapsto \varphi(m)(c)\big)$$

$\square$

Example 4.46 (Products and Function Spaces) For $B \in \mathrm{ob}(\mathbf{Set})$ we consider the functor $F := (\cdot) \times B : \mathbf{Set} \to \mathbf{Set}$, which is given by

$$\mathrm{ob}(\mathbf{Set}) \to \mathrm{ob}(\mathbf{Set})\,,\; A \mapsto A \times B,$$

$$\mathbf{Set}(A, A') \to \mathbf{Set}\big(A \times B, A' \times B\big)\,,\; \varphi \mapsto \big((a, b) \mapsto (\varphi(a), b)\big)$$

as well as the functor $(\cdot)^B : \mathbf{Set} \to \mathbf{Set}$, which is given by

$$\mathrm{ob}(\mathbf{Set}) \to \mathrm{ob}(\mathbf{Set}) \,,\ A \mapsto A^B := \{f : B \to A\},$$
$$\mathbf{Set}(A, A') \to \mathbf{Set}\big(A^B, (A')^B\big) \,,\ \varphi \mapsto \big(f \mapsto \varphi \circ f\big)$$

Then $F \dashv G$, and the associated adjunction

$$\Gamma_{A,A'} : \mathbf{Set}(A \times B, A') \to \mathbf{Set}\big(A, (A')^B\big)$$

for $A, A' \in \mathrm{ob}(\mathbf{Set})$ is given by

$$\Gamma_{A,A'}(\varphi) \mapsto \big(a \mapsto (b \mapsto \varphi(a,b))\big)$$

□

Example 4.47 (Completion of Metric Spaces) Let **Met** be the category of metric spaces with continuous mappings as morphisms. The complete metric spaces form a full subcategory **CMet**. The functor $G : \mathbf{CMet} \to \mathbf{Met}$ is the natural forgetful functor, which simply regards every complete metric space as a metric space. The completion of a metric space X by embedding it into the set $\overline{X}$ of equivalence classes of Cauchy sequences provides a functor $F : \mathbf{Met} \to \mathbf{CMet}$, because continuous mappings $\varphi : X \to Y$ between metric spaces can be uniquely extended to a continuous mapping $\overline{\varphi} : \overline{X} \to \overline{Y}$ on the completions. Then $F \dashv G$, and the associated adjunction

$$\Gamma_{X,C} : \mathbf{Met}\big(X, G(C)\big) \to \mathbf{CMet}(\overline{X}, C)$$

for $X \in \mathrm{ob}(\mathbf{Met})$ and $C \in \mathrm{ob}(\mathbf{CMet})$ is given by

$$\Gamma_{X,C}(\varphi) \mapsto \overline{\varphi}$$

given, where it must be noted that the embedding $C \to \overline{C}$ is surjective, thus an isomorphism of metric spaces. □

Example 4.48 (Scalar Extension) Let R be a commutative ring with identity and S an R-algebra. Consider the functors $G : {}_S\mathbf{Mod} \to \mathbf{Mod}_R$, which views every left-S-module as an R-module. Then the functor $F := S \otimes_R (\cdot) : \mathbf{Mod}_R \to {}_S\mathbf{Mod}$ is left adjoint to G. The corresponding adjunction

$$\Gamma_{M,N} : \mathbf{Mod}_R(M, N) \to {}_S\mathbf{Mod}(S \otimes_R M, N)$$

for $M \in \mathrm{ob}(\mathbf{Mod}_R)$ and $N \in \mathrm{ob}({}_S\mathbf{Mod})$ is given by

$$\Gamma_{M,N}(\varphi) \mapsto \big(s \otimes m \mapsto s \cdot \varphi(m)\big)$$

□

4.4 Adjoint Functors

Example 4.49 (Frobenius Reciprocity for Groups) Let R be a commutative ring with identity and G a group. An R-module M is called a *G-module*, if there is a G-action $G \times M \to M$ (see Example 4.10) and each of the mappings $m \mapsto g \cdot m$ is an R-module homomorphism. The G-modules together with the R-module homomorphisms $\varphi \in \mathbf{Mod}_R(M, M')$, which satisfy

$$\forall g \in G, m \in M: \quad g \cdot \varphi(m) = \varphi(g \cdot m),$$

form a subcategory $_G\mathbf{Mod}_R$ of $\mathbf{Mod}_R$.

Now let H be a subgroup of G. Then there is a forgetful functor

$$\mathrm{Res}_H^G : {}_G\mathbf{Mod}_R \to {}_H\mathbf{Mod}_R,$$

which arises by simply restricting the G-action to H. This forgetful functor has an adjoint functor $\mathrm{Ind}_H^G : {}_H\mathbf{Mod}_R \to {}_G\mathbf{Mod}_R$, called the *induction* from H to G. It can be constructed as follows: For $M \in \mathrm{ob}({}_H\mathbf{Mod}_R)$, we set

$$\mathrm{Ind}_H^G(M) := \{f : G \to M \mid \forall g \in G, h \in H : f(hg) = h \cdot f(g)\}.$$

Since the H-action is R-linear, $\mathrm{Ind}_H^G(M)$ is an R-submodule of $M^G := \{f : G \to M\}$. With

$$\forall g, g' \in G: \quad (g \cdot f)(g') := f(g'g)$$

we define a G-module structure on $\mathrm{Ind}_H^G(M)$, because for $g, g' \in G, h \in H$ and $f \in \mathrm{Ind}_H^G(M)$ we can calculate as follows:

$$(g \cdot f)(hg') = f(hg'g) = h \cdot f(g'g) = h \cdot (g \cdot f)(g').$$

For $\varphi \in {}_H\mathbf{Mod}_R(M, M')$ we set

$$\mathrm{Ind}_H^G(\varphi) : \mathrm{Ind}_H^G(M) \to \mathrm{Ind}_H^G(M'), \quad f \mapsto \varphi \circ f.$$

Then it is easy to verify (exercise) that $\mathrm{Ind}_H^G : {}_H\mathbf{Mod}_R \to {}_G\mathbf{Mod}_R$ is indeed a functor.

Let $N \in \mathrm{ob}({}_G\mathbf{Mod}_R)$ and $M \in \mathrm{ob}({}_H\mathbf{Mod}_R)$. If $\varphi \in {}_G\mathbf{Mod}_R(N, \mathrm{Ind}_H^G(M))$, then we define

$$\widehat{\varphi} : N \to M, \quad n \mapsto (\varphi(n))(1),$$

where $1 \in G$ is the unit. Then

$$\widehat{\varphi}(h \cdot n) = (\varphi(h \cdot n))(1) = (h \cdot \varphi(n))(1) = \varphi(n)(h) = h \cdot (\varphi(n)(1)) = h \cdot \widehat{\varphi}(n),$$

i.e., $\widehat{\varphi} \in {}_H\mathbf{Mod}_R(\mathrm{Res}_H^G(N), M)$. Conversely, we set

$$\widetilde{\psi} : N \to \mathrm{Ind}_H^G(M), \quad n \mapsto (g' \mapsto \psi(g' \cdot n))$$

for $\psi \in {}_H\mathbf{Mod}_R(\mathrm{Res}_H^G(N), M)$. Then

$$(\widetilde{\psi}(g \cdot n))(g') = \psi(g' \cdot (g \cdot n)) = \psi(g'g \cdot n) = (\widetilde{\psi}(n))(g'g) = (g \cdot (\widetilde{\psi}(n)))(g'),$$

i.e., $\widetilde{\psi}(g \cdot n) = g \cdot (\widetilde{\psi}(n))$ and thus $\widetilde{\psi} \in {}_G\mathbf{Mod}_R(N, \mathrm{Ind}_H^G(M))$. It is now an exercise to verify that $\varphi \mapsto \widehat{\varphi}$ and $\psi \mapsto \widetilde{\psi}$ are inverse mappings to each other. Together, for each object $(N, M) \in \mathrm{ob}({}_G\mathbf{Mod}_R \times {}_H\mathbf{Mod}_R)$ we obtain a bijection

$$\Gamma_{N,M} : {}_G\mathbf{Mod}_R(N, \mathrm{Ind}_H^G(M)) \to {}_H\mathbf{Mod}_R(\mathrm{Res}_H^G(N), M).$$

Another exercise is now to verify that this bijection is natural in (N, M), i.e., to show $\mathrm{Res}_H^G \dashv \mathrm{Ind}_H^G$. □

This concludes our list of elementary examples of pairs of adjoint functors, which is far from exhaustive. We present some more in exercises, but otherwise refer to the tables in [ML98, S. 87] and [HS73, S. 198].

Exercise 4.18 (Frobenius Reciprocity for Algebras) Let R be a commutative ring with identity and B a R-algebra with unit. For a subalgebra A of B, which contains the unit, there is always the natural restriction functor $\mathrm{Res}_A^B : {}_B\mathbf{Mod} \to {}_A\mathbf{Mod}$. Show that the functors

$$\mathrm{Ind}_A^B := \mathrm{Hom}_A(B, \cdot) : {}_A\mathbf{Mod} \to {}_B\mathbf{Mod}$$

and

$$\mathrm{Prod}_A^B := B \otimes_A (B, \cdot) : {}_A\mathbf{Mod} \to {}_B\mathbf{Mod}$$

are right and left adjoint to Res_A^B, respectively, i.e., $\mathrm{Prod}_A^B \dashv \mathrm{Res}_A^B \dashv \mathrm{Ind}_A^B$.

Exercise 4.19 (Exterior Algebras) Let R be a commutative ring with idenity, in which $1 + 1 \neq 0$ holds. A $\mathbb{Z}$-*graded* R-algebra A is a R-algebra, which can be written as a direct sum $\bigoplus_{n \in \mathbb{Z}} A_n$ of R-submodules $A_n \subseteq A$, which additionally satisfy

$$\forall n, m \in \mathbb{Z} : \quad A_n \cdot A_m \subseteq A_{n+m}.$$

A $\mathbb{Z}$-graded R-algebra is called *alternating*, if $A_n = 0$ for $n < 0$ and

$$\forall n, m \in \mathbb{N}_0, x \in A_n, y \in A_m : \quad xy = (-1)^{nm} yx.$$

(i) Define a subcategory $\mathbf{Alg}_R^{\mathrm{alt}}$ of alternating R-algebras from $\mathbf{Alg}_R$ and show that the functor $\Lambda : \mathbf{Mod}_R \to \mathbf{Alg}_R$ from Example 4.23 can be considered as $\Lambda : \mathbf{Mod}_R \to \mathbf{Alg}_R^{\mathrm{alt}}$.
(ii) Consider the forgetful functor $G : \mathbf{Alg}_R^{\mathrm{alt}} \to \mathbf{Mod}_R$ and show that $\Lambda \dashv G$.

4.4 Adjoint Functors

Exercise 4.20 (Universal Enveloping Algebra) Let R be a commutative ring with identity and **LieAlg**$_R$ the category of R-Lie algebras (see Example 4.5 and Exercise 4.3).

(i) For an associative R-algebra A, define the *commutator product*
$$\forall a, b \in A : [a, b] := ab - ba.$$
Show that $(A, [\cdot, \cdot])$ is an R-Lie algebra and by replacing multiplication, the identity becomes a functor $G : \mathbf{Alg}_R \to \mathbf{LieAlg}_R$.

(ii) Let L be an R-Lie algebra and $\mathrm{T}(L)$ the tensor algebra over L. Let $I(L) \subseteq \mathrm{T}(L)$ be the ideal of $\mathrm{T}(L)$ generated by the elements of the form $x \otimes y - y \otimes y - [x, y] \in \mathrm{T}(L)$ with $x, y \in L$. Show that by
$$\mathrm{ob}(\mathbf{LieAlg}_R) \to \mathrm{ob}(\mathbf{Alg}_R) \,,\, L \mapsto \mathrm{U}(L) := \mathrm{T}(L)/I(L),$$
$$\mathbf{LieAlg}_R(L, L') \to \mathbf{Alg}_R\big(\mathrm{U}(L), \mathrm{U}(L')\big) \,,\, \varphi \mapsto \big(t + I(L) \mapsto \mathrm{T}(\varphi)(t) + I(L')\big)$$
a functor $\mathrm{U} : \mathbf{LieAlg}_R \to \mathbf{Alg}_R$ is defined, which is left adjoint to G.

$\mathrm{U}(L)$ is called the *universal enveloping* algebra of L.

Uniqueness of Adjoint Functors

We now begin preparations for proving the uniqueness of adjoint functors. To create the appropriate notational framework for this, we first introduce the category of *all* functors between two categories, even though this category presents us with new set-theoretical problems.

Definition 4.50 (Natural Transformations and Functor Category) Let **A** and **B** be categories. The *functor category* $[\mathbf{A}, \mathbf{B}]$ has as objects all functors $F : \mathbf{A} \to \mathbf{B}$. A morphism $\Phi \in [\mathbf{A}, \mathbf{B}](F, F')$ is given by a family $(\Phi_A)_{A \in \mathrm{ob}(\mathbf{A})}$ of morphisms $\Phi_A : F(A) \to F'(A)$ that satisfy

$$\forall \alpha \in \mathbf{A}(A, A') : \quad F'(\alpha) \circ \Phi_A = \Phi_{A'} \circ F(\alpha) \tag{4.2}$$

Such a Φ is also called a *natural transformation* from F to F' and we write
$$\mathbf{A} \; \underset{F'}{\overset{F}{\rightrightarrows}} \Downarrow\Phi \; \mathbf{B}$$
for Φ, if we want to convey all the information in the notation.

The set-theoretic problem with Definition 4.50 is that the natural transformations Φ from F to F' do not always form a set. However, we required this in Definition 4.11 for the morphisms between two objects. This problem can be solved in different ways. Some authors also allow categories for which the morphisms between two objects do not form sets, and call categories, as we have introduced them in Definition 4.11, *locally small* (see [Le14, Def. 1.1.1 and §3.2]). Other authors limit the validity of the axioms to a chosen *universe* of sets, which itself is a set (see [ML98, §I.6]). As in Section 4.2, we ignore the problem here and focus on the potential of the described terms to describe construction patterns.

In order for [**A**, **B**], regardless of the set-theoretic difficulties, to become a category, one must define a concatenation of natural transformations. This is done for $\Phi \in [\mathbf{A}, \mathbf{B}](F, F')$ and $\Psi \in [\mathbf{A}, \mathbf{B}](F', F'')$ by the formula $(\Psi \circ \Phi)_A := \Psi_A \circ \Phi_A$ for $A \in \mathrm{ob}(\mathbf{A})$. A category also requires an identity transformation 1_F, which is simply given by $(1_F)_A := 1_{F(A)}$.

With these settings, we now get the concept of the *natural isomorphism* of functors, as an isomorphism in the category [**A**, **B**] (see Definition 4.17), which can also be explicitly formulated without recourse to this category.

Definition 4.51 (Natural Isomorphism of Functors) Let **A** and **B** be categories and $F, F' : \mathbf{A} \to \mathbf{B}$ functors. A *natural isomorphism* from F to F' is a natural transformation $\Phi = (\Phi_A)_{A \in \mathrm{ob}(\mathbf{A})}$ from F to F', for which there is a natural transformation $\Psi = (\Psi_A)_{A \in \mathrm{ob}(\mathbf{A})}$ from F' to F that satisfies

$$\forall A \in \mathrm{ob}(\mathbf{A}): \quad \Psi_A \circ \Phi_A = 1_{F(A)} \text{ and } \Phi_A \circ \Psi_A = 1_{F'(A)}$$

We write $F \simeq F'$, if there is a natural isomorphism from F to F'.

Exercise 4.21 (Natural Isomorphism of Functors) Let **A** and **B** be categories and $F, F' : \mathbf{A} \to \mathbf{B}$ functors. Let $\Phi = (\Phi_A)_{A \in \mathrm{ob}(\mathbf{A})}$ be a natural transformation from F to F', for which each $\Phi_A \in \mathbf{B}(F(A), F'(A))$ is an isomorphism. Show that the $\Phi_A^{-1} \in \mathbf{B}(F(A), F'(A))$ form a natural transformation from F' to F and Φ is a natural isomorphism.

With the help of functor categories, one can summarize the represented functors H^X and H_X from Example 4.25.

Example 4.52 (Represented Functors) Let **C** be a category.

(i) One defines a functor $H^{\bullet} : \mathbf{C}^{\mathrm{op}} \to [\mathbf{C}, \mathbf{Set}]$ by

$$\mathrm{ob}(\mathbf{C}) \to \mathrm{ob}([\mathbf{C}^{\mathrm{op}}, \mathbf{Set}]), \ X \mapsto H_X,$$

$$\mathbf{C}(X, X') \to [\mathbf{C}^{\mathrm{op}}, \mathbf{Set}](H_X, H_{X'}), \ \psi \mapsto H_\psi := \mathbf{C}^{\mathrm{op}} \underset{H_{X'}}{\overset{H_X}{\rightrightarrows}} \mathbf{Set},$$

where $\Phi_C : \mathbf{C}(X, C) = H^X(C) \to H^{X'}(C) = \mathbf{C}(X', C), \ f \mapsto f \circ \varphi$.

(ii) One defines a functor $H_\bullet : \mathbf{C} \to [\mathbf{C}^{op}, \mathbf{Set}]$ by

$$\mathrm{ob}(\mathbf{C}) \to \mathrm{ob}([\mathbf{C}^{op}, \mathbf{Set}]) \,, \; X \mapsto H_X,$$

$$\mathbf{C}(X, X') \to [\mathbf{C}^{op}, \mathbf{Set}](H_X, H_{X'}) \,, \; \psi \mapsto H_\psi := \quad \mathbf{C}^{op} \underset{H_{X'}}{\overset{H_X}{\Downarrow \psi}} \mathbf{Set} \,,$$

where $\Psi_C : \mathbf{C}(C, X) = H_X(C) \to H_{X'}(C) = \mathbf{C}(C, X'), \; h \mapsto \psi \circ h$. □

In Example 4.52, part (ii) is actually redundant. If one considers the functor $H^\bullet : \mathbf{C}^{op} \to [\mathbf{C}, \mathbf{Set}]$ for the category $\mathbf{D} = \mathbf{C}^{op}$, one gets (exercise!) nothing other than the functor $H_\bullet : \mathbf{D} \to [\mathbf{D}^{op}, \mathbf{Set}]$.

Example 4.53 (Evaluation Functor) Let $\mathbf{C}$ be a category. One defines a functor $\mathrm{ev} : \mathbf{C}^{op} \times [\mathbf{C}^{op}, \mathbf{Set}] \to \mathbf{Set}$ by

$$\mathrm{ob}(\mathbf{C}^{op} \times [\mathbf{C}^{op}, \mathbf{Set}]) \to \mathrm{ob}(\mathbf{Set}) \,, \; (C, F) \mapsto F(C)$$

and

$$(\mathbf{C}^{op} \times [\mathbf{C}^{op}, \mathbf{Set}])\big((C, F), (C', F')\big) \to \mathbf{Set}\big(F(C), F'(C')\big)$$

$$\left(\varphi, \quad \mathbf{C}^{op} \underset{F'}{\overset{F}{\Downarrow \Phi}} \mathbf{Set} \right) \mapsto F'(\varphi) \circ \Phi_C.$$

□

A comparison of Definitions 4.42, 4.50, and 4.51 suggests that the adjointness of functors can be expressed by natural isomorphisms of functors. For this, we consider a pair $\mathbf{A} \underset{G}{\overset{F}{\rightleftarrows}} \mathbf{B}$ of functors and form the functors (see Example 4.16 and Exercise 4.8)

$$F^{op} \times 1_\mathbf{B} : \mathbf{A}^{op} \times \mathbf{B} \to \mathbf{B}^{op} \times \mathbf{B} \quad \text{and} \quad 1_{\mathbf{A}^{op}} \times G : \mathbf{A}^{op} \times \mathbf{B} \to \mathbf{A}^{op} \times \mathbf{A}.$$

If we combine these functors with the Hom-functor $\mathrm{Hom}_\mathbf{A} : \mathbf{A}^{op} \times \mathbf{A} \to \mathbf{Set}$ respectively its analogue $\mathrm{Hom}_\mathbf{B}$ for $\mathbf{B}$ (see Exercise 4.7 and Exercise 4.9), we find the two functors relevant in the adjunction of F and G

$$\mathbf{A}\big(\cdot, G(\cdot)\big), \mathbf{B}\big(F(\cdot), \cdot\big) : \mathbf{A}^{op} \times \mathbf{B} \to \mathbf{Set}.$$

From the definitions it follows immediately (exercise!), that $\mathbf{A} \underset{G}{\overset{F}{\rightleftarrows}} \mathbf{B}$ holds if and only if

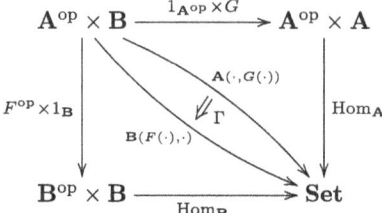

The central tool for proving the uniqueness of adjoint functors is the Yoneda lemma.

Lemma 4.54 (Yoneda) *Let $\mathbf{A}$ be a category. Then the evaluation functor*

$$\mathrm{ev} : \mathbf{A}^{\mathrm{op}} \times [\mathbf{A}^{\mathrm{op}}, \mathbf{Set}] \to \mathbf{Set}$$

is naturally isomorphic to the composition of the functors

$$H_{\bullet}^{\mathrm{op}} \times 1_{[\mathbf{A}^{\mathrm{op}}, \mathbf{Set}]} : \mathbf{A}^{\mathrm{op}} \times [\mathbf{A}^{\mathrm{op}}, \mathbf{Set}] \to [\mathbf{A}^{\mathrm{op}}, \mathbf{Set}]^{\mathrm{op}} \times [\mathbf{A}^{\mathrm{op}}, \mathbf{Set}]$$

and

$$\mathrm{Hom}_{[\mathbf{A}^{\mathrm{op}}, \mathbf{Set}]} : [\mathbf{A}^{\mathrm{op}}, \mathbf{Set}]^{\mathrm{op}} \times [\mathbf{A}^{\mathrm{op}}, \mathbf{Set}] \to \mathbf{Set}.$$

Proof We are looking for a natural isomorphism Φ from $F := \mathrm{ev}$ to $F' := \mathrm{Hom}_{[\mathbf{A}^{\mathrm{op}}, \mathbf{Set}]} \circ (H_{\bullet}^{\mathrm{op}} \times 1_{[\mathbf{A}^{\mathrm{op}}, \mathbf{Set}]})$. With $\mathbf{C} := \mathbf{A}^{\mathrm{op}} \times [\mathbf{A}^{\mathrm{op}}, \mathbf{Set}]$ we are thus looking for a natural isomorphism $\mathbf{C} \underset{F'}{\overset{F}{\rightrightarrows}}{\Downarrow \Phi} \mathbf{Set}$. For $(A, X) \in \mathrm{ob}(\mathbf{C})$ we have

$$F(A, X) = X(A) \quad \text{and} \quad F'(A, X) = [\mathbf{A}^{\mathrm{op}}, \mathbf{Set}](H_A, X).$$

So we are looking for a natural bijection for $(A, X) \in \mathrm{ob}(\mathbf{C})$

$$\Phi_{(A,X)} : X(A) \to [\mathbf{A}^{\mathrm{op}}, \mathbf{Set}](H_A, X), \quad x \mapsto \widetilde{x},$$

where $\widetilde{x} := (\widetilde{x}_B)_{B \in \mathrm{ob}(\mathbf{A}^{\mathrm{op}})} = \mathbf{A}^{\mathrm{op}} \underset{X}{\overset{H_A}{\rightrightarrows}}{\Downarrow \widetilde{x}} \mathbf{Set}$. That is, each $\widetilde{x}_B$ is a mapping $\widetilde{x}_B : H_A(B) = \mathbf{A}^{\mathrm{op}}(A, B) \to X(B)$. Note that for $\varphi \in \mathbf{A}^{\mathrm{op}}(A, B)$ the morphism $X(\varphi) \in \mathbf{Set}(X(A), X(B))$ is simply a mapping $X(A) \to X(B)$.

4.4 Adjoint Functors

Claim 1: The mapping $\tilde{x}_B(\varphi) := X(\varphi)(x)$ defines a natural transformation
$$\mathbf{A}^{\mathrm{op}} \underset{X}{\overset{H_A}{\rightrightarrows}} \Downarrow \tilde{x} \quad \mathbf{Set}.$$

With Claim 1, the mappings $\Phi_{(A,X)}$ have been determined.

Claim 2: The mappings $\Phi_{(A,X)}$ define a natural transformation
$$\mathbf{C} \underset{F'}{\overset{F}{\rightrightarrows}} \Downarrow \Phi \quad \mathbf{Set}.$$

We construct the inverse mapping to $\Phi_{(A,X)}$: For
$$\alpha = (\alpha_B)_{B \in \mathrm{ob}(\mathbf{A}^{\mathrm{op}})} = \mathbf{A}^{\mathrm{op}} \underset{X}{\overset{H_A}{\rightrightarrows}} \Downarrow \tilde{x} \quad \mathbf{Set}$$

we set $\widehat{\alpha} := \alpha_A(1_A)$. This makes sense because α_A is a mapping $H_A(A) = \mathbf{A}^{\mathrm{op}}(A, A) = \mathbf{A}(A, A) \to X(A)$ and $1_A \in \mathbf{A}(A, A)$ holds. Thus, $\widehat{\alpha} \in X(A)$, and we have constructed a mapping
$$\Psi_{(A,X)} : [\mathbf{A}^{\mathrm{op}}, \mathbf{Set}](H_A, X) \to X(A), \quad \alpha \mapsto \widehat{\alpha}.$$

Claim 3: We have $\Psi_{(A,X)} = \Phi_{(A,X)}^{-1}$.

Claim 4: The mappings $\Psi_{(A,X)}$ define a natural transformation
$$\mathbf{C} \underset{F}{\overset{F'}{\rightrightarrows}} \Downarrow \Psi \quad \mathbf{Set}.$$

The combination of the two natural transformations Ψ and Φ results in the identity natural transformation according to Claim 3. This reduces the Yoneda lemma to the four listed claims. The verification of these claims consists in each case of a short explicit calculation, using only the relevant definitions. These calculations are left to the reader as an exercise (they can be found, e.g., in [Le14, §4.2]). □

The most important application of the Yoneda lemma for us is the realization that a category can always be "embedded" into a functor category (analogous to the embedding of an integral domain into its quotient field). To be able to define the concept of embedding more precisely, we make the following definitions.

Definition 4.55 (Full and Faithful Functors) A functor $F : \mathbf{A} \to \mathbf{B}$ is called *full/faithful*, if for every pair of objects $A, A' \in \mathrm{ob}(\mathbf{A})$ the mapping $F : \mathbf{A}(A, A') \to \mathbf{B}(F(A), F(A'))$ is surjective/injective.

The announced *Yoneda embedding* of a category is a functor i.e., full and faithful.

Corollary 4.56 (Yoneda Embedding) *Let* **A** *be a category. Then the functor* $H_\bullet$: **A** → [**A**^{op}, **Set**] *from Example 4.52 is full and faithful.*

Proof We have to show that for $A, A' \in \text{ob}(\mathbf{A})$ the mapping

$$\mathbf{A}(A, A') \to [\mathbf{A}^{op}, \mathbf{Set}](H_A, H_{A'}), \quad \psi \mapsto H_\psi$$

is bijective. The mapping

$$\mathbf{A}(A, A') = H_{A'}(A) \to [\mathbf{A}^{op}, \mathbf{Set}](H_A, H_{A'}), \quad x \mapsto \tilde{x}$$

is bijective according to Lemma 4.54. It is therefore sufficient to show that the two mappings coincide. In the notation of the proof of Lemma 4.54 we therefore have to show that for $\psi \in \mathbf{A}(A, A')$ we have $\tilde{\psi} = H_\psi$. Equivalent to this is $\widehat{H_\psi} = \psi$. The latter follows from the calculation $\widehat{H_\psi} = (H_\psi)_A(1_A) = \psi \circ 1_A = \psi$. □

The duality between $H_\bullet$ and $H^\bullet$ explained in Example 4.52 via opposite categories now immediately yields that the functor $H^\bullet : \mathbf{A}^{op} \to [\mathbf{A}, \mathbf{Set}]$ is also full and faithful (exercise!).

The decisive property *fully faithful* (i.e., full and faithful) functors for us is that one can test the isomorphism of objects optionally in one of the two involved categories.

Exercise 4.22 (Fully Faithful Functors) Let $J : \mathbf{A} \to \mathbf{B}$ be a fully faithful functor and $A, A' \in \text{ob}(\mathbf{A})$. Then:

(i) $\varphi \in \mathbf{A}(A, A')$ is an isomorphism if and only if $J(\varphi) \in \mathbf{B}(J(A), J(A'))$ is an isomorphism.
(ii) For every isomorphism $\psi \in \mathbf{B}(J(A), J(A'))$ there is exactly one isomorphism $\varphi \in \mathbf{A}(A, A')$ with $J(\varphi) = \psi$.
(iii) A and A' are isomorphic in **A** if and only if $J(A)$ and $J(A')$ are isomorphic in **B**.

We can now conclude our program and show that adjoint functors are uniquely determined up to natural isomorphism. For existence statements about adjoint functors, we refer to [ML98], [HS73] and [Le14].

Theorem 4.57 (Uniqueness of the Adjoint Functor) *Let*

$$\mathbf{A} \underset{G}{\overset{F}{\rightleftarrows}} \bot \ \mathbf{B}$$

4.4 Adjoint Functors

be a pair of adjoint functors. Then the following holds:

(i) If $\mathbf{A} \underset{G}{\overset{F'}{\rightleftarrows}} \mathbf{B}$, then F' is naturally isomorphic to F.

(ii) If $\mathbf{A} \underset{G'}{\overset{F}{\rightleftarrows}} \mathbf{B}$, then G' is naturally isomorphic to G.

Proof

(i) Let $A \in \mathrm{ob}(\mathbf{A})$ be fixed. If we insert the object A or the morphism 1_A into the functors $\mathbf{B}(F(\cdot), \cdot)$ and $\mathbf{B}(F'(\cdot), \cdot)$, we obtain the functors

$$H^{F(A)} = \mathbf{B}(F(A), \cdot), \; H^{F'(A)} = \mathbf{B}(F'(A), \cdot) : \mathbf{B} \to \mathbf{Set}.$$

By assumption, these functors are both naturally isomorphic to the functor $\mathbf{A}(A, G(\cdot))$, which is obtained when A and 1_A are inserted into the functor $\mathbf{A}(\cdot, G(\cdot))$. In particular, $H^{F(A)}$ and $H^{F'(A)}$ are naturally isomorphic. But since $H^\bullet$ is faithful, the objects $F(A)$ and $F'(A)$ are also isomorphic according to Exercise 4.9. Since the isomorphisms of the functors

$$\mathbf{B}(F(\cdot), \cdot) \cong \mathbf{A}(\cdot, G(\cdot)) \cong \mathbf{B}(F'(\cdot), \cdot)$$

from $\mathbf{A}^{\mathrm{op}} \times \mathbf{B}$ to $\mathbf{Set}$ are natural in both variables, the isomorphisms $F(A) \to F'(A)$ combine to form a natural transformation, which then according to Exercise 4.21 is a natural isomorphism $F \cong F'$. This proves (i).

(ii) Let $B \in \mathrm{ob}(\mathbf{B})$ be fixed. If we insert the object B or the morphism 1_B into the functors $\mathbf{A}(\cdot, G(\cdot))$ and $\mathbf{A}(\cdot, G'(\cdot))$, we obtain the functors

$$H_{G(B)} = \mathbf{A}(\cdot, G(B)), \; H_{G'(B)} = \mathbf{A}(\cdot, G(B)) : \mathbf{A} \to \mathbf{Set}.$$

By assumption, these functors are both naturally isomorphic to the functor $\mathbf{B}(F(\cdot), B)$, which is obtained when B and 1_B are inserted into the functor $\mathbf{B}(F(\cdot), \cdot)$. In particular, $H_{G(B)}$ and $H_{G'(B)}$ are naturally isomorphic. But since $H_\bullet$ is faithful, the objects $G(B)$ and $G'(B)$ are also isomorphic according to Exercise 4.9. Since the isomorphisms of the functors

$$\mathbf{A}(\cdot, G(\cdot)) \cong \mathbf{B}(F(\cdot), \cdot) \cong \mathbf{A}(\cdot, G'(\cdot))$$

from $\mathbf{A}^{\mathrm{op}} \times \mathbf{B}$ to $\mathbf{Set}$ are natural in both variables, the isomorphisms $G(B) \to G'(B)$ combine to form a natural transformation, which then according to Exercise 4.21 is a natural isomorphism $G \cong G'$. This completes the proof of (ii).

□

Exercise 4.23 (Right Adjoint Functors Commute with Limits) Let **A** and **B** be categories and $F: \mathbf{A} \to \mathbf{B}$ and $G: \mathbf{B} \to \mathbf{A}$ be adjoint functors: $F \dashv G$. Show:

(i) G commutes with limits: $G\bigl(\varprojlim_{i \in I} D(i)\bigr) = \varprojlim_{i \in I} G(D(i))$.
(ii) F commutes with colimits: $F\bigl(\varinjlim_{i \in I} D(i)\bigr) = \varinjlim_{i \in I} F(D(i))$.

Equivalence of Categories

The definitions compiled for the proof of Theorem 4.57 allow us to introduce the concept of *equivalence* of categories, which is much more significant in practice than the obvious concept of *isomorphism* of categories.

Definition 4.58 (Isomorphism and Equivalence of Categories) Let **A** and **B** be categories.

(i) **A** and **B** are called *isomorphic*, if there are two functors $F: \mathbf{A} \to \mathbf{B}$ and $G: \mathbf{B} \to \mathbf{A}$ with $G \circ F = \mathrm{id}_\mathbf{A}$ and $F \circ G = \mathrm{id}_\mathbf{B}$.
(ii) **A** and **B** are called *equivalent*, if there are two functors $F: \mathbf{A} \to \mathbf{B}$ and $G: \mathbf{B} \to \mathbf{A}$ with $G \circ F \simeq \mathrm{id}_\mathbf{A}$ and $F \circ G \simeq \mathrm{id}_\mathbf{B}$ (see Definition 4.51).

We will see various examples of equivalent categories in the later chapters. To be able to easily handle an elementary example from linear algebra here, we give the following characterization of the equivalence of categories.

Proposition 4.59 (Equivalence of Categories) *Two categories* **A** *and* **B** *are equivalent if and only if there is a fully faithful functor* $F: \mathbf{A} \to \mathbf{B}$ *that is* essentially surjective, *i.e.*,

$$\forall B \in \mathrm{ob}(\mathbf{B}) \; \exists A \in \mathrm{ob}(\mathbf{A}): \quad F(A) \cong B.$$

Proof If **A** and **B** are equivalent, then there are functors $F: \mathbf{A} \to \mathbf{B}$ and $G: \mathbf{B} \to \mathbf{A}$ with $G \circ F \simeq \mathrm{id}_\mathbf{A}$ and $F \circ G \simeq \mathrm{id}_\mathbf{B}$. We show that F is fully faithful and essentially surjective: For $B \in \mathrm{ob}(\mathbf{B})$ choose $A := G(B)$. Then we have

$$B = \mathrm{id}_\mathbf{B}(B) \cong (F \circ G)(B) = F(A),$$

so F is essentially surjective. Let $\Phi = (\Phi_A)_{A \in \mathrm{ob}(\mathbf{A})}$ be the natural isomorphism from $G \circ F$ to $\mathrm{id}_\mathbf{A}$. Then we have

$$\forall \alpha \in \mathbf{A}(A, A'): \quad \alpha \circ \Phi_A = \Phi_{A'} \circ (G \circ F)(\alpha).$$

So for two objects $A, A' \in \mathrm{ob}(\mathbf{A})$ the mapping

$$\mathbf{A}(A, A') \to \mathbf{B}\bigl(F(A), F(A')\bigr), \quad \alpha \mapsto F(\alpha)$$

4.4 Adjoint Functors

is invertible with inverse
$$\mathbf{B}(F(A), F(A')) \to \mathbf{A}(A, A'), \quad \beta \mapsto \Phi_{A'} \circ G(\beta) \circ \Phi_A^{-1}.$$

This shows that F is also fully faithful.

Conversely, we now assume that $F: \mathbf{A} \to \mathbf{B}$ is an essentially surjective fully faithful functor. Due to the essential surjectivity of F, for each $B \in \mathrm{ob}(\mathbf{B})$ we can choose a $G(B) := A \in \mathrm{ob}(\mathbf{A})$ and an isomorphism $\Psi_B : B \to F(A)$. For each morphism $\beta \in \mathbf{B}(B, B')$ we consider the commutative diagram

$$\begin{array}{ccc} B & \xrightarrow{\Psi_B} & F(G(B)) \\ \beta \downarrow & & \downarrow \Psi_{B'} \circ \beta \circ \Psi_B^{-1} \\ B' & \xrightarrow{\Psi_{B'}} & F(G(B')) \end{array}$$

Since F is fully faithful, the mapping
$$\mathbf{A}(A, A') \to \mathbf{B}(F(A), F(A')), \quad \alpha \mapsto F(\alpha)$$
is bijective. So also
$$\mathbf{A}(A, A') \to \mathbf{B}(B, B'), \quad \alpha \mapsto \Psi_{B'} \circ F(\alpha) \circ \Psi_B^{-1}$$
is bijective. Applied to $A = G(B)$ and $A' = G(B')$ this yields a (uniquely determined) morphism $G(\beta) := \alpha \in \mathbf{A}(A, A')$ with $\Psi_{B'} \circ F(\alpha) \circ \Psi_B^{-1} = \beta$. One now verifies (details as an exercise!) that $G: \mathbf{B} \to \mathbf{A}$ is a functor and $\Psi = (\Psi_B)_{B \in \mathrm{ob}(\mathbf{B})}$ is a natural isomorphism of $\mathrm{id}_\mathbf{B}$ and $F \circ G$.

To also construct a natural isomorphism of $\mathrm{id}_\mathbf{A}$ and $G \circ F$, we consider for $A \in \mathrm{ob}(\mathbf{A})$ the isomorphism $\Psi_{F(A)}^{-1} : F(G(F(A))) \to F(A)$. Since
$$\mathbf{A}(G(F(A)), A) \to \mathbf{B}(F(G(F(A))), F(A)), \quad \alpha \mapsto F(\alpha)$$
is bijective, there is a uniquely determined morphism $\Phi_A \in \mathbf{A}(A, G(F(A)))$ with $F(\Phi_A) = \Psi_{F(A)}^{-1}$. The construction immediately shows that Φ_A is an isomorphism with $F(\Phi_A^{-1}) = \Psi_{F(A)}$. Thus, the above diagram for $\beta = F(\alpha)$ with $\alpha \in \mathbf{A}(A, A')$ yields the commutative diagram

$$\begin{array}{ccc} F(A) & \xleftarrow{F(\Phi_A)} & F(G(F(A))) \\ F(\alpha) \downarrow & & \downarrow F(\Phi_{A'}^{-1} \circ \alpha \circ \Phi_A) \\ F(A') & \xleftarrow{F(\Phi_{A'})} & F(G(F(A'))) \end{array}$$

Again from the full faithfulness of F it follows that $\Phi_{A'}^{-1} \circ \alpha \circ \Phi_A = \alpha$. Thus, we have the commutative diagram

$$\begin{array}{ccc} A & \xleftarrow{\Phi_A} & G(F(A)) \\ {\scriptstyle \alpha}\downarrow & & \downarrow{\scriptstyle \Phi_{A'}^{-1} \circ \alpha \circ \Phi_A} \\ A' & \xleftarrow{\Phi_{A'}} & G(F(A')) \end{array}$$

and $\Phi = (\Phi_A)_{A \in \mathrm{ob}(\mathbf{A})}$ is a natural isomorphism of $\mathrm{id}_{\mathbf{A}}$ and $G \circ F$. □

Example 4.60 (Equivalence of Categories) Let $\mathbb{K}$ be a field and $\mathbf{Mat}_{\mathbb{K}}$ the category, whose objects are the vector spaces $\mathbb{K}^n$ with $n \in \mathbb{N}_0$ ($\mathbb{K}^0 := \{0\}$) and whose morphisms are defined by $\mathbf{Mat}_{\mathbb{K}}(\mathbb{K}^n, \mathbb{K}^m) := \mathrm{Mat}_{n \times m}(\mathbb{K})$ with matrix multiplication as composition. Then $\mathbf{Mat}_{\mathbb{K}}$ is equivalent to the category $\mathbf{Vect}_{\mathbb{K}}^{\mathrm{fin}}$ of finite-dimensional $\mathbb{K}$-vector spaces with $\mathbb{K}$-linear mappings as morphisms.

To see this, we consider the functor $F : \mathbf{Mat}_{\mathbb{K}} \to \mathbf{Vect}_{\mathbb{K}}^{\mathrm{fin}}$, which is defined by $F(\mathbb{K}^n) := \mathbb{K}^n$ and

$$\forall A \in \mathrm{Mat}_{n \times m}(\mathbb{K}), v \in \mathbb{K}^n : \quad F(A)v := Av$$

It is obviously fully faithful. Since every finite-dimensional $\mathbb{K}$-vector space is isomorphic to a $\mathbb{K}^n$ (choose a basis), Proposition 4.59 provides the claim. □

Literature: The book [Gr08] is a standard reference (first edition 1968) for universal algebra, which goes far beyond the few basic concepts that we have discussed here. Classic texts on category theory are [HS73] and [ML98]. The fact that category theory plays a much larger role in current mathematics than universal algebra is also reflected in the fact that there are a number of newer books that are entirely or partially dedicated to the topic, e.g., [Le14] and [KS06]. Apart from that, most books on modern algebraic geometry, representation theory or topology have at least one chapter on categories and functors. The addition of natural transformations to objects and morphisms is a very first step towards a *higher category theory*, which plays an important role in modern homotopy theory. We do not go into this in this book, but an elementary introduction can be found in [La21].

Part II
Local Structures

The continuity of a real-valued function on a topological space is a local property: If one knows the function only in a small neighborhood of each point of the domain, one can decide whether the function is continuous or not. On the other hand, the integrability of a real-valued function on a topological space equipped with a measure is not a local property. For example, for the constant function 1 on $\mathbb{R}$, the restriction to any bounded open set is integrable, but the entire function is not.

For open subsets of $\mathbb{R}^n$, the differentiability of real-valued functions is also a local property. Given the usefulness of differentiable functions, the question arises whether the concept of differentiability can be extended to other topological spaces. What requirement must one impose on a topological space for this? One answer to this question is: "The topological space must locally look like $\mathbb{R}^n$." Thus, it is required that the topological space carries an additional structure that is of a local nature.

In this part of the book, the idea of a local structure is clarified and a number of significant local structures are introduced. We begin in Chap. 5 with the abstract concept of a sheaf, which on the one hand clarifies the idea of a local property and on the other hand is a very versatile and powerful conceptual tool that is indispensable in modern mathematics.

As a first concrete local structure, we consider the concept of a manifold, whose local model is a vector space with a suitable sheaf of functions. We restrict ourselves to finite-dimensional vector spaces over $\mathbb{R}$ or $\mathbb{C}$, which then leads to finite-dimensional real or complex manifolds. For the associated sheaves of functions, there are still various choices. For continuous functions, one obtains the concept of a topological manifold. If at least one continuous derivative is required, one obtains differentiable manifolds of various degrees of regularity. At the end of the scale, there are the analytic manifolds, for which the functions of the local model are the power series with positive radius of convergence. For real manifolds, a good part of the local theory is covered in advanced calculus courses, in which differential and integral calculus in one and several variables is treated. Therefore, the focus in Chap. 6 is on the construction of derived local structures such as the tangent bundle and the sheaf of differential forms. With the help of differential forms, one can develop an integration theory for manifolds. The approach here is exemplary for local structures: One knows integrals in the local models and knows (from the transformation formula) how they behave under diffeomorphisms. This knowledge is used to define a global integral of differential forms of maximum degree by gluing. Also exemplary for this context is Stokes' theorem, which is obtained by gluing from the fundamental theorem of calculus.

The local theory of complex manifolds, i.e., the differential and integral calculus of complex-valued functions in one or more complex variables, is usually not covered in calculus courses. Since a central result for this, the Cauchy integral theorem, can be easily derived from Stokes' theorem, we prove some basic local results and thus show as an exemplary global result that every complex-differentiable function on a compact complex manifold is constant.

In Chap. 7 we turn to the second family of local structures to be discussed in more detail in this part: algebraic varieties and schemes. Here, the local theory requires further algebraic preparations. In particular, we show Hilbert's Nullstellensatz, which generalizes the fundamental theorem of algebra and allows to describe the zero sets of polynomials with coefficients in an algebraically closed field by ideals in polynomial rings. These zero sets are the local models for algebraic varieties, whose points can then be identified with maximal ideals of a ring. The local models for schemes are even more diverse and serve the purpose of also being able to make statements about zero sets of polynomials with coefficients in arbitrary fields (or even rings).

Sheaves 5

Sheaves are structures on topological spaces that can be described by specifying local data. One should imagine local data as objects that are assigned to open subsets, e.g., the set of all continuous functions defined on the open subset. In order for such an assignment to be considered local data, the objects assigned to two open sets should be compatible with each other when the open sets intersect. For continuous functions, this is the case since restrictions of continuous functions to the intersection are also continuous. A refinement of this idea is the concept of a presheaf, which we will examine more closely in Section 5.1. However, one only has a truly local structure when it is sufficient to specify compatible local data, and it is then guaranteed that they come from a uniquely determined global object, as is the case with continuous or differentiable functions. Making this additional condition on presheaves precise, one arrives at the concept of a sheaf.

We will repeatedly consider sheaves of functions with specified differentiability properties on open subsets of $\mathbb{K}^n$ with $\mathbb{K}$ equal to $\mathbb{R}$ or $\mathbb{C}$. These sheaves will play a central role in Ch. 6. We assume that the readers are familiar with calculus in one or more real variables. This theory has a complex counterpart in many parts, which is often not explicitly mentioned in calculus courses, but can be literally transferred from the real case by simply replacing $\mathbb{R}$ with $\mathbb{C}$ and small intervals with small circular disks. This applies in particular to the definitions of differentiability and power series expansions of functions (see e.g. [Di69] for uniform presentations).

5.1 Presheaves and Sheaves

The most economical way to introduce the concept of a presheaf is to define it as a functor. However, one should always keep the idea of local data in mind.

Definition 5.1 (Presheaves) Let $(X, \mathfrak{T})$ be a topological space, i.e., $\mathfrak{T}$ is the set of open subsets of X. Consider the category $\mathfrak{T}^{op}$, whose objects are the elements of $\mathfrak{T}$, and in which there is exactly one (single) morphism $U \to V$ between two open subsets U and V of X if $V \subseteq U$ holds (see Exercise 4.10). Furthermore, let **C** be a category. A **C**-*presheaf* over X is a functor $\mathcal{F}\colon \mathfrak{T}^{op} \to \mathbf{C}$. That is, a presheaf assigns an object $\mathcal{F}(U)$ from **C** to each open subset U of X, and to each pair (U, V) of open subsets of X with $V \subseteq U$ a **C**-morphism $\rho_{V,U} := \rho^{\mathcal{F}}_{V,U} \colon \mathcal{F}(U) \to \mathcal{F}(V)$ is assigned. $\rho_{V,U}$ is called the *restriction* from U to V. If $W \subseteq V \subseteq U$ are open in X, then $\rho_{W,V} \circ \rho_{V,U} = \rho_{W,U}$.

In this generality, the objects $\mathcal{F}(U) \in \mathrm{ob}(\mathbf{C})$ need not be sets. Then it does not make sense to speak of elements of $\mathcal{F}(U)$. Mostly, one considers **C**-presheaves for categories $\mathbf{C} = \mathbf{C}_{\Phi,\Gamma}$ of algebraic structures of a given type (Φ, Γ) (see Exercise 4.3), e.g., presheaves of sets (when $\Phi = \emptyset$), abelian groups, rings, or modules. For this type of presheaves, the elements of $\mathcal{F}(U)$ are called *sections* of $\mathcal{F}$ over U.

The simplest presheaves are given by function spaces, where the restriction is given by limiting the functions to subsets.

Example 5.2 (Presheaves) If $(X, \mathfrak{T})$ is a topological space, one can take $\mathcal{F}(U)$ as the space of continuous functions on U with values in a fixed topological space R. This gives the presheaf $\mathcal{C}^0_{X,R} := \mathcal{C}_{X,R} := \mathcal{F}$ of continuous R-valued functions on X. If R is a topological vector space, e.g., $\mathbb{R}$ or $\mathbb{C}$, then $\mathcal{C}_{X,R}$ is a presheaf of vector spaces. If R is a normed vector space, one can take $\mathcal{F}(U)$ as the set of bounded functions on U and thus obtain the presheaf $\mathcal{B}_{X,R}$ of bounded R-valued functions on X.

Let $\mathbb{K}$ be equal to $\mathbb{R}$ or $\mathbb{C}$. If X is an open subset of $\mathbb{K}^n$ and V is a complete $\mathbb{K}$-vector space, one can also take $\mathcal{F}(U)$ as the set of k-times continuously differentiable V-valued functions ($k \in \mathbb{N} \cup \{\infty\}$) and thus obtain the presheaf $\mathcal{C}^{\mathbb{K},k}_{X,V} := \mathcal{F}$. One can also take $\mathcal{F}(U)$ as the $\mathbb{K}$-analytic V-valued functions, i.e., functions that can be represented around each point by a power series with coefficients in V and a positive radius of convergence. We denote this presheaf with $\mathcal{C}^{\mathbb{K},\omega}_{X,V}$. If $V = \mathbb{K}$, we usually omit the V from the notation. The $\mathbb{K}$ is also often omitted from the notation if it is clear from the context whether real or complex differentiability is meant.

For $\mathbb{K} = \mathbb{C}$, the presheaves $\mathcal{C}^{\mathbb{C},k}_{X,V}$ for $k \in \mathbb{N} \cup \{\infty\} \cup \{\omega\}$ all coincide (see Theorem 6.82), but we will not use this result from local complex analysis anywhere in this book. One often also writes $\mathcal{H}ol_{X,V}$ for this presheaf and calls its sections *holomorphic functions*.

Another alternative for $\mathcal{F}(U)$ is the set of integrable functions on U. This gives the presheaf $\mathcal{L}^1_X$. □

Note the restrictive conditions made in Example 5.2 for X, where presheaves of differentiable or holomorphic functions are discussed. To achieve our goal of

5.1 Presheaves and Sheaves

defining suitable local structures on topological spaces that allow us to consider differentiable or holomorphic functions on more general spaces, we need to be able to compare presheaves. Then we can consider, e.g., spaces that "look locally like" $\mathbb{R}^n$ or $\mathbb{C}^n$. In a first step, we compare presheaves that are defined over the same space.

Definition 5.3 (Morphisms of Presheaves) Let $(X, \mathfrak{T})$ be a topological space, **C** a category and $\mathcal{F}, \mathcal{G}$ two **C**-presheaves over X. A *morphism* is a natural transformation $\varphi \colon \mathcal{F} \to \mathcal{G}$. That is, for each $U \in \mathfrak{T}$ a **C**-morphism $\varphi_U \colon \mathcal{F}(U) \to \mathcal{G}(U)$ is assigned, and for $V \subseteq U$ in $\mathfrak{T}$ the diagram

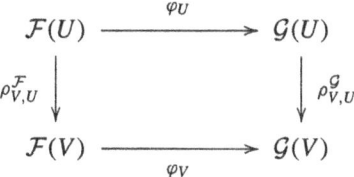

commutative. An *isomorphism* from $\mathcal{F}$ to $\mathcal{G}$ is a morphism $\varphi \colon \mathcal{F} \to \mathcal{G}$, for which there is a morphism $\psi \colon \mathcal{G} \to \mathcal{F}$ that satisfies the identities $\psi_U \circ \varphi_U = 1_{\mathcal{F}(U)}$ and $\varphi_U \circ \psi_U = 1_{\mathcal{G}(U)}$ for each $U \in \mathfrak{T}$.

Exercise 5.1 (Category of C-Presheaves) Let $(X, \mathfrak{T})$ be a topological space and **C** a category. Show that the **C**-presheaves together with the **C**-presheaf morphisms form a category $\mathbf{PS}(X, \mathbf{C})$. □

For presheaves of abelian groups, morphisms also provide new examples of presheaves.

Example 5.4 (Kernel and Image of a Morphism) Let $(X, \mathfrak{T})$ be a topological space and $\mathcal{F}, \mathcal{G}$ presheaves of abelian groups over X. Further, let $\varphi \colon \mathcal{F} \to \mathcal{G}$ be a morphism. Then the kernels and images of the φ_U for $U \in \mathfrak{T}$ each form a presheaf of abelian groups over X.

To have a concrete example in mind, consider $X = \mathbb{C}$ with the presheaf $U \mapsto \mathcal{H}ol_{\mathbb{C}}(U)$ of holomorphic functions. Then the family $\varphi_U \colon \mathcal{H}ol_{\mathbb{C}}(U) \to \mathcal{H}ol_{\mathbb{C}}(U), s \mapsto \exp \circ s$ for open $U \subseteq \mathbb{C}$ defines a morphism of presheaves of abelian groups. $\ker(\varphi_U) = \{s \in \mathcal{H}ol_{\mathbb{C}}(U) \mid s(U) \subseteq 2\pi i \mathbb{Z}\}$, so $\ker(\varphi_U)$ consists of the locally constant (i.e., constant on connected components) functions on U with values in $2\pi i \mathbb{Z}$. The image $\operatorname{im}(\varphi_U)$, on the other hand, consists of those holomorphic functions $f \colon U \to \mathbb{C}^\times := \mathbb{C} \setminus \{0\}$ that have a holomorphic logarithm, i.e., can be written as $f = e^h$ with $h \in \mathcal{H}ol_{\mathbb{C}}(U)$. For a given f, one can find a small neighborhood U' around each point and a logarithm of $f|_{U'}$, because the exponential function $\exp \colon \mathbb{C} \to \mathbb{C}^\times$ is locally invertible around each point according to the theorem of the local inverse. □

Example 5.5 (Differential Operators as Presheaf Morphisms) Let X be an open subset of $\mathbb{R}$ and $\mathcal{C}_X^\infty$ the presheaf of infinitely differentiable functions on X (see Example 5.2). Then the derivative $C^\infty(U, \mathbb{R}) \ni f \mapsto \varphi_U(f) = \frac{df}{dx}$ for open $U \subseteq X$ defines a morphism of presheaves of real vector spaces. Note that each $C^\infty(U, \mathbb{R})$ also has a multiplication that makes it an $\mathbb{R}$-algebra. Thus, $\mathcal{C}_X^\infty$ can also be interpreted as a sheaf of $\mathbb{R}$-algebras. However, since the derivative of a product is not the product of the derivatives, φ is not a morphism of presheaves of $\mathbb{R}$-algebras.

Now let X be an open subset of $\mathbb{R}^n$ and D a differential operator on X (see Example 1.7). Then also $C^\infty(U, \mathbb{R}) \ni f \mapsto \varphi_U(f) := D(f)$ for open $U \subseteq X$ defines a morphism of presheaves of real vector spaces. □

In order to compare presheaves over different topological spaces, we introduce the concept of the direct image of a presheaf.

Definition 5.6 (Direct Image) Let $(X, \mathfrak{T})$ and $(X', \mathfrak{T}')$ be topological spaces and $f \colon X \to X'$ a continuous mapping. If $\mathcal{F}$ is a **C**-presheaf over X, then

$$\forall U' \in \mathfrak{T}' : \quad (f_*\mathcal{F})(U') := \mathcal{F}\big(f^{-1}(U')\big)$$

defines a **C**-presheaf $f_*\mathcal{F}$ over X', which is called the *direct image* of $\mathcal{F}$ under f.

We want to consider two presheaves $\mathcal{F}$ and $\mathcal{F}'$ over X and X' as equivalent if there is a homeomorphism $f \colon X \to X'$ for which $f_*\mathcal{F}$ and $\mathcal{F}'$ are isomorphic. The approach for a local structure that allows differential calculus is then a presheaf that is locally equivalent to a presheaf of differentiable functions in this sense. In this context, it should be noted that a presheaf $\mathcal{F}$ over a topological space X can be restricted to a presheaf $\mathcal{F}|_U$ over U on any open subset $U \subseteq X$. Local equivalence of $\mathcal{F}$ with a presheaf of differentiable functions then means nothing more than that for every point in X there is an open neighborhood U for which the restriction $\mathcal{F}|_U$ is equivalent to a presheaf of differentiable functions.

Example 5.7 (Restriction of a Presheaf) Let $(X, \mathfrak{T})$ be a topological space, **C** a category and $\mathcal{F}$ a **C**-presheaf over X. For each open subset $U \subseteq X$ setting $V \mapsto \mathcal{F}(V)$ for open $V \subseteq U$ defines a **C**-presheaf $\mathcal{F}|_U$ over U. □

Remark 5.8 (Inverse Image) Analogous to Definition 5.6 of the direct image of a **C**-presheaf, one could also try to construct a presheaf $\mathcal{F}$ over X for a continuous mapping $f \colon X \to X'$ and a presheaf $\mathcal{F}'$ over X'. However, since images of open sets under continuous functions are not automatically open, it is not sufficient to simply set $\mathcal{F}(U) := \mathcal{F}'\big(f(U)\big)$. Instead, one will have to work with open subsets $V \subseteq X'$ that contain $f(U)$ and then form a limit in **C**. However, this only works if **C** also contains the corresponding limits. We will return to this construction and its connection with the construction of the inverse image in the next section (see Example 5.19). □

5.1 Presheaves and Sheaves

In order to be able to develop the approach described above for the development of a local structure for the generalization of differential calculus, we must take the step from presheaves to sheaves, which was already announced in the introduction to this chapter.

Sheaves

We define **C**-sheaves as **C**-presheaves whose sections can be uniquely obtained from compatible sections over arbitrarily small open subsets. For categories of the form $\mathbf{C} = \mathbf{C}_{\Phi,\Gamma}$ of algebraic structures of a given type (Φ, Γ) (see Exercise 4.3), this can be done without problems. In particular, one obtains sheaves of sets. The definitions used in the following can be formulated if the category **C** allows limits and colimits. For very general categories, one can use the "Yoneda philosophy" and call a **C**-presheaf $\mathcal{F}$ a sheaf if for each object T of **C** the assignment $U \mapsto \mathrm{Hom}_{\mathbf{C}}(T, \mathcal{F}(U))$ is a sheaf of sets. If arbitrary products exist in **C**, the sheaf condition can then be expressed again as in the definition of sheaves of sets. For simplicity, in this book we only consider $\mathbf{C}_{\Phi,\Gamma}$-sheaves.

Definition 5.9 (Sheaves) Let $(X, \mathfrak{T})$ be a topological space and $\mathbf{C} = \mathbf{C}_{\Phi,\Gamma}$ a category of algebraic structures. A **C**-presheaf $\mathcal{F}$ over X is called a **C**-*sheaf*, if for every open cover $\{V_i\}_{i \in I}$ of $U \in \mathfrak{T}$ the following two conditions are met:

(a) If $s, s' \in \mathcal{F}(U)$ satisfy the identity $\rho_{V_i, U}(s) = \rho_{V_i, U}(s') \in \mathcal{F}(V_i)$ for each $i \in I$, then $s = s' \in \mathcal{F}(U)$.
(b) If the $s_i \in \mathcal{F}(V_i)$ satisfy the identity $\rho_{V_i \cap V_j, V_i}(s_i) = \rho_{V_i \cap V_j, V_j}(s_j) \in \mathcal{F}(V_i \cap V_j)$ for each pair $i, j \in I$, then there exists an $s \in \mathcal{F}(U)$ that satisfies the identity $\rho_{V_i, U}(s) = s_i \in \mathcal{F}(V_i)$ for each $i \in I$.

Example 5.10 (Sheaves) The presheaves $\mathcal{C}_{X,R}$ and $\mathcal{C}_{X,V}^{\mathbb{K},k}$ from Example 5.2 are sheaves. On the other hand, the presheaves $\mathcal{B}_{X,V}$ and $\mathcal{L}_X^1$ are not sheaves, because functions are not necessarily bounded or integrable if X can be covered by small open sets on which the functions have these properties. □

Example 5.11 (Constant Sheaves) Let $(X, \mathfrak{T})$ be a topological space and A an algebraic structure of type (Φ, Γ), as in Exercise 4.3, e.g., an abelian group. For $U \in \mathfrak{T}$, let $\mathcal{F}(U) := A$. Then $\mathcal{F}$ defines a **C**-presheaf over X, where $\mathbf{C} = \mathbf{C}_{\Phi,\Gamma}$ is the category of algebraic structures of type (Φ, Γ). If one interprets the elements of $\mathcal{F}(U)$ as the set of constant A-valued functions on U, one realizes that the presheaf $\mathcal{F}$ is not a sheaf, e.g., when X is not connected. It can be shown that $\mathcal{F}$ is a sheaf exactly when every open subset of X is connected. If, on the other hand, one defines $A_X(U)$ as the set of locally constant A-valued functions on U (i.e., functions for which there is a neighborhood at each point on which they are constant), one obtains a sheaf A_X, which is called the *constant sheaf* with values in A. □

Example 5.12 (Skyscraper Sheaves) Let $(X, \mathfrak{T})$ be a topological space, $x \in X$ and A an (additively written) abelian group. By

$$\mathfrak{T} \ni U \mapsto \mathcal{F}(U) := \begin{cases} A & \text{if } x \in U \\ \{0\} & \text{if } x \notin U \end{cases}$$

a sheaf $\mathcal{F}$ over X is defined, which is called a *skyscraper sheaf*. □

Let $(X, \mathfrak{T})$ be a topological space, $\mathbf{C} = \mathbf{C}_{\Phi,\Gamma}$ a category of algebraic structures and $\mathcal{F}, \mathcal{G}$ two **C**-sheaves over X. A *morphism* $\varphi \colon \mathcal{F} \to \mathcal{G}$ is nothing other than a morphism between the **C**-presheaves $\mathcal{F}$ and $\mathcal{G}$, i.e., the **C**-sheaves form a full subcategory $\mathbf{S}(X, \mathbf{C})$ of the category $\mathbf{PS}(X, \mathbf{C})$ of **C**-presheaves (see Exercise 5.1).

Example 5.13 (Gluing of Sheaves) Let $(X, \mathfrak{T})$ be a topological space, $\{U_i \in \mathfrak{T} \mid i \in I\}$ an open cover of X, and $\mathbf{C} = \mathbf{C}_{\Phi,\Gamma}$ a category of algebraic structures. Further, let $\mathcal{F}_i \in \mathrm{ob}(\mathbf{S}(U_i, \mathbf{C}))$ for $i \in I$ be sheaves over U_i, where U_i is equipped with the subspace topology $\mathfrak{T}_i := \{U \in \mathfrak{T} \mid U \subseteq U_i\}$. A family of sheaf isomorphisms

$$\{\varphi_{ij} \colon \mathcal{F}_j|_{U_i \cap U_j} \to \mathcal{F}_i|_{U_i \cap U_j} \mid i, j \in I\}$$

is called a set of *gluing data*, if

$$\forall i, j, k \in I : \quad \varphi_{ij} \circ \varphi_{jk} = \varphi_{ik} \colon \mathcal{F}_k|_{U_i \cap U_j \cap U_k} \to \mathcal{F}_i|_{U_i \cap U_j \cap U_k},$$

where we also denote the isomorphism induced by φ_{ij} by restriction to U_k as $\mathcal{F}_j|_{U_i \cap U_j \cap U_k} \to \mathcal{F}_i|_{U_i \cap U_j \cap U_k}$ with φ_{ij}. In particular, gluing data satisfy $\varphi_{ii} = 1_{\mathcal{F}_i}$ for each $i \in I$.

For a set $\Phi := \{\varphi_{ij} \mid i, j \in I\}$ of gluing data and $U \in \mathfrak{T}$, we denote the set

$$\left\{ (s_i) \in \prod_{i \in I} \mathcal{F}_i(U \cap U_i) \,\middle|\, \forall i, j \in I : \rho^{\mathcal{F}_i}_{U \cap U_i \cap U_j, U \cap U_i}(s_i) = \varphi_{ij}\left(\rho^{\mathcal{F}_j}_{U \cap U_i \cap U_j, U \cap U_j}(s_j)\right) \right\}$$

by $\mathcal{F}(U)$. Then the assignments $\mathfrak{T} \ni U \mapsto \mathcal{F}(U)$ define a **C**-sheaf (exercise), which we call the *gluing* of the $\mathcal{F}_i$ by means of Φ. From the construction, it follows that the restrictions $\mathcal{F}|_{U_i}$ are isomorphic to the original sheaves $\mathcal{F}_i$ (see Example 5.7). □

Exercise 5.2 (Sheaves) Let $\varphi \colon \mathcal{F} \to \mathcal{G}$ be a morphism of sheaves of abelian groups over X. Show that the kernel presheaf (see Example 5.4) of φ is itself a sheaf. Further, find an example that shows that the image presheaf of φ is in general not a sheaf.

Exercise 5.3 (Sheaves) Let $(X, \mathfrak{T})$ and $(X', \mathfrak{T}')$ be topological spaces and $f \colon X \to X'$ a continuous mapping. Show: If $\mathcal{F}$ is a **C**-sheaf over X, then the direct image $f_*\mathcal{F}$ (see Definition 5.6) is a **C**-sheaf over X'.

5.1 Presheaves and Sheaves

Sheafification of Presheaves

The conditions from Definition 5.9 suggest that sheaves are very special presheaves. However, it turns out that all presheaves of algebraic structures can be "completed" to sheaves. The procedure is similar to the transition from the presheaf of constant functions to the sheaf of locally constant functions with values in an algebraic structure (see Example 5.11).

Definition 5.14 (Stalks of Presheaves) Let $(X, \mathfrak{T})$ be a topological space, $\mathbf{C} = \mathbf{C}_{\Phi,\Gamma}$ a category of algebraic structures, and $\mathcal{F}$ a $\mathbf{C}$-presheaf over X. For $x \in X$, consider the relation

$$(U, s) \sim_x (U', s') \quad :\Leftrightarrow \quad \left(\exists x \in V \in \mathfrak{T}, V \subseteq U \cap U' : \quad \rho_{V,U}(s) = \rho_{V,U'}(s') \right)$$

defined on the set $\{(U, s) \mid x \in U \in \mathfrak{T}, s \in \mathcal{F}(U)\}$. It is easy to show (exercise!) that $\sim_x$ is an equivalence relation. The set $\mathcal{F}_x$ of equivalence classes of $\sim_x$ is called the *stalk* of $\mathcal{F}$ in x. The elements of $\mathcal{F}_x$ are called the *germs* of $\mathcal{F}$ in x.

Remark 5.15 (Algebraic Structure of Stalks) Since the $\rho_{V,U}$ are morphisms of the category $\mathbf{C}_{\Phi,\Gamma}$, the defining condition for $\sim_x$ is compatible with the operations from Φ. It follows (exercise!) that the stalks $\mathcal{F}_x$ of a $\mathbf{C}_{\Phi,\Gamma}$-presheaf $\mathcal{F}$ over X themselves are objects of $\mathbf{C}_{\Phi,\Gamma}$ and the mapping $\mathcal{F}(U) \to \mathcal{F}_x, s \mapsto s_x := [s]_{\sim_x}$ for each pair (x, U) with $x \in U \in \mathfrak{T}$ is a $\mathbf{C}_{\Phi,\Gamma}$-morphism. Here, $[s]_{\sim_x}$ denotes the equivalence class of s with respect to $\sim_x$. □

There is a connection between the stalks of a presheaf and the inductive limits from Ch. 4.

Exercise 5.4 (Algebraic Structure of Stalks) Let $(X, \mathfrak{T})$ be a topological space, $\mathbf{C} = \mathbf{C}_{\Phi,\Gamma}$ a category of algebraic structures and $\mathcal{F}$ a $\mathbf{C}$-presheaf over X. For $x \in X$, let $I_x := \{U \in \mathfrak{T} \mid x \in U\}$ be ordered by inclusion. Show:

(i) The $\rho_{V,U} : \mathcal{F}(U) \to \mathcal{F}(V)$ for $x \in V \subseteq U$ form an inductive system in $\mathbf{C}_{\Phi,\Gamma}$ for the small category $\mathcal{I}_x$ associated with $(I_x, \subseteq)$ (see Definition 4.38).
(ii) The inductive system from (i) has an inductive limit, and this is given by the mappings $\mathcal{F}(U) \to \mathcal{F}_x, s \mapsto s_x$.
(iii) If $\varphi : \mathcal{F} \to \mathcal{G}$ is a morphism of $\mathbf{C}$-presheaves over X, then for each $x \in X$ the mapping

$$\forall x \in U \in \mathfrak{T}, s \in \mathcal{F}(U) : \quad s_x \mapsto \left(\varphi_U(s) \right)_x$$

defines a $\mathbf{C}$-morphism $\varphi_x : \mathcal{F}_x \to \mathcal{G}_x$.
(iv) The assignments $\mathcal{F} \to \mathcal{F}_x$ and $\varphi \mapsto \varphi_x$ define for each $x \in X$ a *stalk functor* $H_x : \mathbf{PS}(X, \mathbf{C}) \to \mathbf{C}$.

Theorem 5.16 (Sheafification of a Presheaf) *Let $(X, \mathfrak{T})$ be a topological space, $\mathbf{C} = \mathbf{C}_{\Phi,\Gamma}$ a category of algebraic structures and $\mathcal{F}$ a $\mathbf{C}$-presheaf over X. Then there exists a $\mathbf{C}$-sheaf $\mathcal{F}^+$ over X and a morphism $\theta := \theta^\mathcal{F} : \mathcal{F} \to \mathcal{F}^+$ of $\mathbf{C}$-presheaves* with the following universal property: *If $\mathcal{G}$ is a **C**-sheaf, for every **C**-presheaf morphism $\varphi: \mathcal{F} \to \mathcal{G}$ there exists exactly one **C**-sheaf morphism $\psi: \mathcal{F}^+ \to \mathcal{G}$ with $\psi \circ \theta = \varphi$*, i.e., the diagram

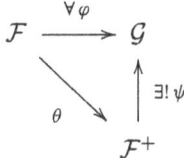

*is commutative. Furthermore, it holds that for every $x \in X$ the stalk mapping $\theta_x: \mathcal{F}_x \to \mathcal{F}_x^+$ is a **C**-isomorphism.*

Proof For every $U \in \mathfrak{T}$ we set

$$\mathcal{F}^+(U) := \left\{ s: U \to \coprod_{x \in U} \mathcal{F}_x \,\middle|\, \text{(A) and (B)} \right\},$$

where

(A) $\forall x \in U: \quad s(x) \in \mathcal{F}_x$,
(B) $\forall x \in U: \quad \exists x \in V \subseteq U$ open, $t \in \mathcal{F}(V)$ with $\big(\forall y \in V: t_y = s(y)\big)$.

Together with the restriction mappings $\rho_{V,U}^{\mathcal{F}^+}: \mathcal{F}^+(U) \to \mathcal{F}^+(V), s \mapsto s|_V$ for $V \subseteq U$ in $\mathfrak{T}$, $\mathcal{F}^+$ is a **C**-presheaf, where the algebraic operations are taken pointwise in the respective $\mathcal{F}_x$.

Now let $U \in \mathfrak{T}$ and $\{V_i\}_{i \in I}$ be an open cover of U. If $s, s' \in \mathcal{F}^+(U)$ agree on all V_i, then s and s' as elements of $\mathcal{F}^+(U)$ are equal. Thus, condition (a) from Definition 5.9 is fulfilled. Now let $s_i \in \mathcal{F}^+(V_i)$ be compatible, i.e., $s_i|_{V_i \cap V_j} = s_j|_{V_i \cap V_j}$ holds for every pair (i, j) of elements in I. For $x \in V_i$ we set $s(x) := s_i(x) \in \mathcal{F}_x$. The compatibility of the s_i shows that in this way a mapping $s: U \to \coprod_{x \in X} \mathcal{F}_x$ is defined, which fulfills (A). Because $s_i \in \mathcal{F}^+(V_i)$, for every $x \in V_i$ there is an open neighborhood V_x of x, which is contained in V_i, and a $t \in \mathcal{F}(V_x)$ with

$$\forall y \in V_x: \quad t_y = s_i(y) = s(y).$$

Thus, s also fulfills (B), so it is an element of $\mathcal{F}^+(U)$. Since obviously $s|_{V_i} = s_i$ holds, we have also verified condition (b) from Definition 5.9 for $\mathcal{F}^+$. Therefore, $\mathcal{F}^+$ is a **C**-sheaf.

5.1 Presheaves and Sheaves

Now we set for every $U \in \mathfrak{T}$

$$\theta_U : \mathcal{F}(U) \to \mathcal{F}^+(U), \quad t \mapsto (x \mapsto t_x). \tag{$\dagger$}$$

Since each of the mappings $t \mapsto t_x$ is a **C**-morphism, θ_U is also a **C**-morphism.

It remains only to show the universal property, because the uniqueness statement of the theorem is an immediate consequence of this property. Given a **C**-sheaf $\mathcal{G}$ over X and a **C**-presheaf morphism $\varphi: \mathcal{F} \to \mathcal{G}$. For $U \in \mathfrak{T}$ we want to construct a **C**-morphism $\psi_U : \mathcal{F}^+(U) \to \mathcal{G}(U)$. For $s \in \mathcal{F}^+(U)$ there is a cover $\{V_i\}_{i \in I}$ of U by open subsets $V_i \subseteq U$ and elements $t_i \in \mathcal{F}(V_i)$ with

$$\forall i \in I, x \in V_i : \quad (t_i)_x = s(x).$$

We set $\mathcal{G}(V_i) \ni \psi_i(s) := \varphi_{V_i}(t_i) : V_i \to \coprod_{x \in V_i} \mathcal{G}_x$ and choose an $x \in V_i \cap V_j$. Because $(t_i)_x = s(x) = (t_j)_x$, there is an open neighborhood V_x of x in $V_i \cap V_j$ with $\rho^{\mathcal{F}}_{V_x, V_i}(t_i) = \rho^{\mathcal{F}}_{V_x, V_j}(t_j)$. The calculation

$$\rho^{\mathcal{G}}_{V_x, V_i \cap V_j}\left(\rho^{\mathcal{G}}_{V_i \cap V_j, V_i}(\psi_i(s))\right) = \rho^{\mathcal{G}}_{V_x, V_i}(\psi_i(s)) = \rho^{\mathcal{G}}_{V_x, V_i}(\varphi_{V_i}(t_i))$$
$$= \varphi_{V_x}\left(\rho^{\mathcal{F}}_{V_x, V_i}(t_i)\right) = \varphi_{V_x}\left(\rho^{\mathcal{F}}_{V_x, V_j}(t_j)\right)$$
$$= \rho^{\mathcal{G}}_{V_x, V_i \cap V_j}\left(\rho^{\mathcal{G}}_{V_i \cap V_j, V_j}(\psi_j(s))\right)$$

provides, because $\mathcal{G}$ is a sheaf and the V_x completely cover $V_i \cap V_j$, that

$$\rho^{\mathcal{G}}_{V_i \cap V_j, V_i}(\psi_i(s)) = \rho^{\mathcal{G}}_{V_i \cap V_j, V_j}(\psi_j(s)).$$

Again with the sheaf property of $\mathcal{G}$, one thus finds an element $\psi_U(s) \in \mathcal{G}(U)$ with

$$\forall i \in I : \quad \rho^{\mathcal{G}}_{V_i, U}(\psi_U(s)) = \psi_U(s)|_{V_i} = \psi_i(s) = \varphi_{V_i}(t_i).$$

Because φ is a morphism, it follows from the construction that for $V \subseteq U$ in $\mathfrak{T}$: $\psi_V \circ \rho^{\mathcal{F}^+}_{V,U} = \rho^{\mathcal{G}}_{V,U} \circ \psi_U$. This shows that ψ is a **Set**-sheaf morphism. Since all morphisms used to construct ψ are **C**-presheaf morphisms, ψ is also a **C**-presheaf morphism, i.e., a **C**-sheaf morphism (exercise!).

If $s = \theta_U(t)$ for $t \in \mathcal{F}(U)$, then $(t_i)_x = s(x) = t_x \in \mathcal{F}_x$ for all $x \in V_i$. So for each $x \in V_i$ there is an open neighborhood $V_x \subseteq V_i$ with $\rho^{\mathcal{F}}_{V_x, V_i}(t_i) = \rho^{\mathcal{F}}_{V_x, U}(t)$. Because φ is a morphism, it follows

$$\rho^{\mathcal{G}}_{V_x, U}(\psi_U(s)) = \rho^{\mathcal{G}}_{V_x, V_i}(\psi_i(s)) = \rho^{\mathcal{G}}_{V_x, V_i}(\varphi_{V_i}(t_i)) = \rho^{\mathcal{G}}_{V_x, U}(\varphi_U(t)).$$

So, again because $\mathcal{G}$ is a sheaf, $\psi_U(s) = \varphi_U(t)$. That is, we have shown that $\psi \circ \theta = \varphi$.

To show the uniqueness of ψ, assume that ψ' is a **C**-sheaf morphism that satisfies $\psi' \circ \theta = \varphi$. For $s \in \mathcal{F}^+(U)$ and t_i, $i \in I$ as above, it then follows from $s|_{V_i} = \theta_{V_i}(t_i)$ that

$$\rho^{\mathcal{G}}_{V_i,U}(\psi'_U(s)) = \psi'_{V_i}(\rho^{\mathcal{F}^+}_{V_i,U}(s)) = \psi'_{V_i}(s|_{V_i}) = \psi'_{V_i}(\theta_{V_i}(t_i))$$
$$= (\psi' \circ \theta)_{V_i}(t_i) = \varphi_{V_i}(t_i) = (\psi \circ \theta)_{V_i}(t_i)$$
$$= \rho^{\mathcal{G}}_{V_i,U}(\psi_U(s)).$$

This shows $\psi'_U(s) = \psi_U(s)$, again because $\mathcal{G}$ is a sheaf. Since U was arbitrary, it follows $\psi' = \psi$. This proves the universal property of $\mathcal{F}^+$, and it remains only to show the last statement of the theorem.

So let $x \in X$ and $\theta_x \colon \mathcal{F}_x \to \mathcal{F}^+_x$ be the stalk mapping to θ (see Exercise 5.4). The characterization (†) of θ shows that for $t \in \mathcal{F}(U)$: $\theta_x(t_x) = (y \mapsto t_y)_x$, where the germ of $(y \mapsto t_y) \in \mathcal{F}^+(U)$ in x is to be taken for the presheaf $\mathcal{F}^+(U)$. If for $t' \in \mathcal{F}(U)$ it holds that $t_x = t'_x$, then there exists a neighborhood $x \in U' \in \mathfrak{T}$ with $\rho^{\mathcal{F}}_{U',U}(t) = \rho^{\mathcal{F}}_{U',U}(t')$, so also $t_y = t'_y$ for all $y \in U'$. Thus, θ_x is injective. The surjectivity of θ_x is, however, an immediate consequence of condition (B). □

If one applies the sheafification construction from the proof of Theorem 5.16 to a presheaf of constant functions with values in an algebraic structure, one obtains the locally constant functions with property (B). Similarly, one finds the locally bounded functions when starting with the presheaf of bounded functions. For the presheaf $\mathcal{L}^1_X$ of integrable functions from Example 5.2, one finds the locally integrable functions. All these examples illustrate that the concept of sheaves is suitable for characterizing local properties.

Remark 5.17 (Sheafification as an Adjoint Functor) Theorem 5.16 also has an interesting categorical interpretation: If $\mathcal{F}$ and $\mathcal{G}$ are two **C**-presheaves over X and $\alpha \colon \mathcal{F} \to \mathcal{G}$ is a morphism, Theorem 5.16 first provides the morphisms $\theta^{\mathcal{G}} \colon \mathcal{G} \to \mathcal{G}^+$ and $\theta^{\mathcal{F}} \colon \mathcal{F} \to \mathcal{F}^+$ and then for $\varphi := \theta^{\mathcal{G}} \circ \alpha \colon \mathcal{F} \to \mathcal{G}^+$ a morphism $\alpha^+ \colon \mathcal{F}^+ \to \mathcal{G}^+$ with $\alpha^+ \circ \theta^{\mathcal{F}} = \theta^{\mathcal{G}} \circ \alpha$. Thus, the diagram

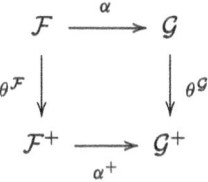

is commutative. The uniqueness statements in Theorem 5.16 show that $\mathcal{F} \mapsto \mathcal{F}^+$ and $\alpha \mapsto \alpha^+$ provide a sheafification functor Sheaf: $\mathbf{PS}(X, \mathbf{C}) \to \mathbf{S}(X, \mathbf{C})$. It is immediately clear from the definitions that there is also a forgetful functor $V \colon \mathbf{S}(X, \mathbf{C}) \to \mathbf{PS}(X, \mathbf{C})$.

5.1 Presheaves and Sheaves

If one applies Theorem 5.16 to a **C**-sheaf $\mathcal{F}$, the uniqueness statement also shows that $\mathcal{F} = \mathcal{F}^+$ and $\theta^{\mathcal{F}} = 1_{\mathcal{F}}$.

Theorem 5.16 further provides that for $\mathcal{F} \in \mathrm{ob}(\mathbf{PS}(X, \mathbf{C}))$ and $\mathcal{G} \in \mathrm{ob}(\mathbf{S}(X, \mathbf{C}))$ there is a well-defined mapping

$$\Gamma_{\mathcal{F},\mathcal{G}} \colon \mathrm{Hom}_{\mathbf{PS}(X,\mathbf{C})}(\mathcal{F}, V(\mathcal{G})) \to \mathrm{Hom}_{\mathbf{S}(X,\mathbf{C})}(\mathcal{F}^+, \mathcal{G}), \quad \varphi \mapsto \psi$$

where $\Gamma_{\mathcal{F},\mathcal{G}}(\varphi) = \psi$ is uniquely determined by the identity $\varphi = \psi \circ \theta^{\mathcal{F}}$. In particular, $\Gamma_{\mathcal{F},\mathcal{G}}$ is injective. For each $\varphi \in \mathrm{Hom}_{\mathbf{PS}(X,\mathbf{C})}(\mathcal{F}, V(\mathcal{G}))$, the diagram

$$\begin{array}{ccc} \mathcal{F} & \xrightarrow{\varphi} & \mathcal{G} \\ {\scriptstyle \theta^{\mathcal{F}}}\downarrow & & \downarrow{\scriptstyle 1_{\mathcal{G}}} \\ \mathcal{F}^+ & \xrightarrow{\varphi^+} & \mathcal{G} \end{array}$$

is commutative, so $\varphi^+ = \psi = \Gamma_{\mathcal{F},\mathcal{G}}(\varphi)$. For $\widetilde{\psi} \in \mathrm{Hom}_{\mathbf{S}(X,\mathbf{C})}(\mathcal{F}^+, \mathcal{G})$ we define $\widetilde{\varphi} := \widetilde{\psi} \circ \theta^{\mathcal{F}} \in \mathrm{Hom}_{\mathbf{PS}(X,\mathbf{C})}(\mathcal{F}, V(\mathcal{G}))$ and obtain

$$\Gamma_{\mathcal{F},\mathcal{G}}(\widetilde{\varphi}) \circ \theta^{\mathcal{F}} = \widetilde{\varphi}^+ \circ \theta^{\mathcal{F}} = \widetilde{\varphi} = \widetilde{\psi} \circ \theta^{\mathcal{F}}.$$

It follows that $\Gamma_{\mathcal{F},\mathcal{G}}(\widetilde{\varphi}) = \widetilde{\psi}$, i.e., $\Gamma_{\mathcal{F},\mathcal{G}}$ is also surjective. Directly from the definitions, it is now derived that $\Gamma_{\bullet,\bullet}$ fulfills the naturality condition from Definition 4.42 (Exercise!). This then shows that the sheafification functor Sheaf is left adjoint to the forgetful functor $V \colon \mathbf{S}(X, \mathbf{C}) \to \mathbf{PS}(X, \mathbf{C})$. □

The following example shows how to construct sheaves using sheafification as limits and colimits from sheaf-valued diagrams.

Example 5.18 (Limits and Colimits of Sheaves) Let $(X, \mathfrak{T})$ be a topological space and $\mathbf{C} = \mathbf{C}_{\Phi, \Gamma}$ a category of algebraic structures. Furthermore, let $\mathbf{I}$ be a small category and $D \colon \mathbf{I} \to \mathbf{S}(X, \mathbf{C})$ a diagram. For each $U \in \mathfrak{T}$ we define the *section functor* $\rho_U \colon \mathbf{S}(X, \mathbf{C}) \to \mathbf{C}$ by

$$\mathrm{ob}(\mathbf{S}(X, \mathbf{C})) \ni \mathcal{F} \mapsto \mathcal{F}(U) \in \mathrm{ob}(\mathbf{C}),$$
$$\mathbf{S}(X, \mathbf{C})(\mathcal{F}, \mathcal{F}') \ni \varphi \mapsto \varphi_U \in \mathbf{C}(\mathcal{F}(U), \mathcal{F}'(U)).$$

By combining with D we obtain a diagram $D_U := \rho_U \circ D \colon \mathbf{I} \to \mathbf{C}$. By

$$U \mapsto D_U(i), \quad (U \to V) \mapsto \left(\rho_{V,U}^{D(i)} \colon D_U(i) \to D_V(i)\right)$$

for $V \subseteq U$ in $\mathfrak{T}$, for each $i \in \mathrm{ob}(\mathbf{I})$ a functor $\widehat{D}(i) \colon \mathfrak{T}^{\mathrm{op}} \to \mathbf{C}$ is defined (see Exercise 4.10). Because $\rho_V \circ D(i) = \rho_{V,U} \circ \rho_U \circ D(i)$, the $\widehat{D}(i)$ are presheaves.

(i) We assume that for each $U \in \mathfrak{T}$ the limit $L_U := \varprojlim_{\mathbf{I}} D_U$ exists. Then the universal property provides

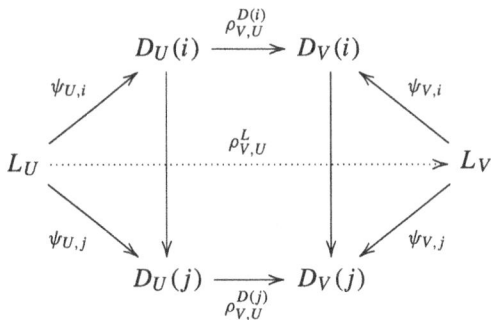

and $U \mapsto L_U$ becomes a presheaf with the $\rho^L_{V,U}$, which is sheafified to a sheaf $\mathcal{L}_D \in \mathbf{S}(X, \mathbf{C})$. Then one verifies (exercise) that $\mathcal{L}_D$ is the categorical limit $\varprojlim_{\mathbf{I}} D$ of D in $\mathbf{S}(X, \mathbf{C})$.

(ii) We assume that for each $U \in \mathfrak{T}$ the limit $K_U := \varinjlim_{\mathbf{I}} D_U$ exists. Then the universal property provides

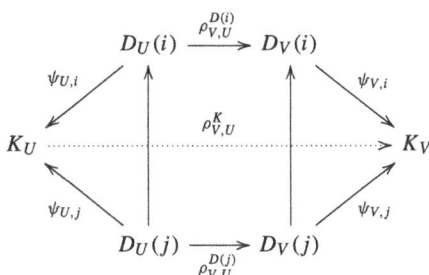

and $U \mapsto K_U$ becomes a presheaf with the $\rho^K_{V,U}$, which is sheafified to a sheaf $\mathcal{K}_D \in \mathbf{S}(X, \mathbf{C})$. Then one verifies (exercise) that $\mathcal{K}_D$ is the categorical limit $\varinjlim_{\mathbf{I}} D$ of D in $\mathbf{S}(X, \mathbf{C})$. □

If one applies the constructions of Example 5.18 to direct products (see Example 4.28), one obtains the concept of the *direct image* of sheaves.

Exercise 5.5 (Direct Image) With the notations from Definition 5.6, show:

(i) The assignment $f_* \colon \mathrm{ob}(\mathbf{PS}(X, \mathbf{C})) \to \mathrm{ob}(\mathbf{PS}(X', \mathbf{C}))$ can be extended to a functor $f_* \colon \mathbf{PS}(X, \mathbf{C}) \to \mathbf{PS}(X', \mathbf{C})$.
(ii) The restriction of the functor f_* to $\mathbf{S}(X, \mathbf{C})$ yields a functor $f_* \colon \mathbf{S}(X, \mathbf{C}) \to \mathbf{S}(X', \mathbf{C})$.

5.1 Presheaves and Sheaves

(iii) If $X \xrightarrow{f} Y \xrightarrow{g} Z$ are continuous mappings, then

$$(g \circ f)_* = g_* \circ f_* \colon \mathbf{S}(X, \mathbf{C}) \to \mathbf{S}(Z, \mathbf{C}).$$

Example 5.19 (Inverse Image) Let $(X, \mathfrak{T})$ and $(X', \mathfrak{T}')$ be topological spaces, $f \colon X \to X'$ a continuous mapping, $\mathbf{C} = \mathbf{C}_{\Phi, \Gamma}$ a category of algebraic structures and $\mathcal{F}'$ a $\mathbf{C}$-presheaf over X'. For an open set $U \in \mathfrak{T}$, we consider the set $I_U := \{V \in \mathfrak{T}' \mid f(U) \subseteq V\}$ ordered by inclusion and consider the category $\mathbf{I}_U$ corresponding to the reversed order. Then

$$D_U \colon \mathbf{I}_U \to \mathbf{C}, \ V \mapsto \mathcal{F}'(V)$$

is a diagram in the sense of Definition 4.33. This diagram has a colimit $L_U \in \mathrm{ob}(\mathbf{C})$ in the sense of Definition 4.34 (see also Example 4.40): Analogous to Definition 5.14 of stalks, we define by

$$(V, s) \sim_U (\widetilde{V}, \widetilde{s}) \quad :\Leftrightarrow \quad \left(\exists\, I_U \ni W \subseteq V \cap \widetilde{V} \colon \rho^{\mathcal{F}'}_{W, V}(s) = \rho^{\mathcal{F}'}_{W, \widetilde{V}}(\widetilde{s}) \right)$$

an equivalence relation on the set $\{(V, s) \mid V \in I_U, s \in \mathcal{F}'(V)\}$. The set L_U of equivalence classes of $\sim_U$ is the sought colimit (see Remark 5.15 and Exercise 5.4—details as exercise!). The assignment

$$\mathfrak{T} \ni U \mapsto L_U \in \mathrm{ob}(\mathbf{C})$$

is a presheaf $f^{<-1>}(\mathcal{F}')$, whose sheafification is called the *inverse image* of the presheaf $\mathcal{F}'$ under f and is denoted by $f^{-1}(\mathcal{F}')$.

To calculate the stalk $\left(f^{-1}(\mathcal{F}')\right)_x$ of $f^{-1}(\mathcal{F}')$ at $x \in X$, one can refer to the presheaf $f^{<-1>}(\mathcal{F}')$ according to Theorem 5.16 and use the isomorphism $\theta_x^{f^{<-1>}(\mathcal{F}')}$ to identify the stalks at x. Then one considers small open neighborhoods U of x. Due to the continuity of f, one can also choose arbitrarily small open neighborhoods of $f(x)$ as V. This gives the inductive limit for determining the stalk of $f^{<-1>}(\mathcal{F}')$ at x just the inductive limit for determining the stalk of $\mathcal{F}'$ at $f(x)$. That is, one obtains

$$\left(f^{-1}(\mathcal{F}')\right)_x = \theta_x^{f^{<-1>}(\mathcal{F}')}\left((f^{<-1>}(\mathcal{F}'))_x\right) \equiv (f^{<-1>}(\mathcal{F}'))_x = \mathcal{F}'_{f(x)}.$$

If $f \colon X \to X'$ is the inclusion of an open subset $X \subseteq X'$, it immediately follows from the construction that $f^{-1}(\mathcal{F}) = \mathcal{F}|_X$. $\square$

In general, the inverse sheaves in the form presented here are not very inviting. However, they have a number of useful properties that one would not want to do without (see e.g., Theorem 5.20). Fortunately, there is a much simpler description of the inverse image sheaf, which only becomes apparent when one looks at sheaves in general from a different perspective (see Definition 5.26 and Theorem 5.27).

Exercise 5.6 (Inverse Image) With the notations from Example 5.19, show:

(i) The assignment $f^{<-1>}: \operatorname{ob}(\mathbf{PS}(X', \mathbf{C})) \to \operatorname{ob}(\mathbf{PS}(X, \mathbf{C}))$ can be extended to a functor $f^{<-1>}: \mathbf{PS}(X', \mathbf{C}) \to \mathbf{PS}(X, \mathbf{C})$.
(ii) The assignment $f^{-1}: \operatorname{ob}(\mathbf{S}(X', \mathbf{C})) \to \operatorname{ob}(\mathbf{S}(X, \mathbf{C}))$ can be extended to a functor $f^{-1}: \mathbf{S}(X', \mathbf{C}) \to \mathbf{S}(X, \mathbf{C})$.
(iii) If $X \xrightarrow{f} Y \xrightarrow{g} Z$ are continuous mappings, then

$$(g \circ f)^{-1} = f^{-1} \circ g^{-1}: \mathbf{S}(Z, \mathbf{C}) \to \mathbf{S}(X, \mathbf{C}).$$

If $f: X \to X'$ is a homeomorphism, then the functors "inverse image" and "direct image" are inverse to each other, because then by construction $f^{<-1>}\mathcal{F}'$ is nothing other than $(f^{-1})_*\mathcal{F}'$. In general, however, the two functors are still adjoint.

Theorem 5.20 (Inverse and Direct Images) *Let $(X, \mathfrak{T})$ and $(X', \mathfrak{T}')$ be topological spaces, $f: X \to X'$ a continuous mapping and $\mathbf{C} = \mathbf{C}_{\Phi, \Gamma}$ a category of algebraic structures. Then $f_*: \mathbf{S}(X, \mathbf{C}) \to \mathbf{S}(X', \mathbf{C})$ and $f^{-1}: \mathbf{S}(X', \mathbf{C}) \to \mathbf{S}(X, \mathbf{C})$ are adjoint functors. More precisely, it holds that $f^{-1} \dashv f_*$.*

Proof Let $\mathcal{F}$ and $\mathcal{F}'$ be two $\mathbf{C}$-sheaves over X and X' respectively, and $\psi: \mathcal{F}' \to f_*\mathcal{F}$ a morphism in $\mathbf{S}(X', \mathbf{C})$. We want to define a morphism $\varphi := \Gamma_{\mathcal{F}, \mathcal{F}'}(\psi): f^{-1}\mathcal{F}' \to \mathcal{F}$ in $\mathbf{S}(X, \mathbf{C})$ corresponding to ψ. According to Remark 5.17, it is sufficient to define a morphism $\varphi: f^{<-1>}\mathcal{F}' \to \mathcal{F}$ in $\mathbf{PS}(X, \mathbf{C})$. Let $U \in \mathfrak{T}$ and $s \in f^{<-1>}\mathcal{F}'(U)$. Then there is an open neighborhood V of $f(U)$ in X' and an $s_V \in \mathcal{F}'(V)$ with (V, s_V) in the equivalence class s (see Example 5.19). Then $\psi_V(s_V) \in f_*\mathcal{F}(V) = \mathcal{F}(f^{-1}(V))$, and we set

$$\varphi_U(s) := \rho^{\mathcal{F}}_{U, f^{-1}(V)}(\psi_V(s_V)) \in \mathcal{F}(U).$$

This construction is compatible with restrictions, i.e.,

$$\mathfrak{T} \ni U \mapsto \varphi_U: f^{<-1>}\mathcal{F}'(U) \to \mathcal{F}(U)$$

is indeed a morphism $\varphi: f^{<-1>}\mathcal{F}' \to \mathcal{F}$ in $\mathbf{PS}(X, \mathbf{C})$. We denote the associated morphism $f^{-1}\mathcal{F}' \to \mathcal{F}$ in $\mathbf{S}(X, \mathbf{C})$ also with φ.

Conversely, if $\varphi: f^{-1}\mathcal{F}' \to \mathcal{F}$ is a morphism in $\mathbf{S}(X, \mathbf{C})$, we consider $V \in \mathfrak{T}'$. Because $f(f^{-1}(V)) \subseteq V$ we have a $\mathbf{C}$-morphism $\mathcal{F}'(V) \to f^{<-1>}\mathcal{F}'(f^{-1}(V))$, which can be combined with the natural morphism $f^{<-1>}\mathcal{F}'(f^{-1}(V)) \to$

$f^{-1}\mathcal{F}'(f^{-1}(V))$. Note that

$$\varphi_{f^{-1}(V)}\colon f^{-1}\mathcal{F}'(f^{-1}(V)) \to \mathcal{F}(f^{-1}(V)) = f_*\mathcal{F}(V).$$

Concatenating these mappings we obtain a **C**-morphism $\psi_V\colon \mathcal{F}'(V) \to f_*\mathcal{F}(V)$. Again, the construction is compatible with restrictions, i.e., $\mathfrak{T}' \ni V \mapsto \psi_V\colon \mathcal{F}'(V) \to f_*\mathcal{F}(V)$ is indeed a morphism in $\mathbf{S}(X', \mathbf{C})$. One verifies (exercise!) that the two constructions mentioned are inverse to each other and the mappings $\Gamma_{\bullet,\bullet}$ with

$$\Gamma_{\mathcal{F},\mathcal{F}'}\colon \mathbf{S}(X', \mathbf{C})(\mathcal{F}', f_*(\mathcal{F})) \to \mathbf{S}(X, \mathbf{C})(f^{-1}(\mathcal{F}'), \mathcal{F}))$$

also fulfill the naturality condition from Definition 4.42. This proves the claim. □

Remark 5.21 (Mappings of Stalks) In the situation of Theorem 5.20 we write $\Gamma_{\mathcal{F},\mathcal{F}'}(\psi) =: \psi^\sharp$ and $\Gamma^{-1}_{\mathcal{F},\mathcal{F}'}(\varphi) =: \varphi^\flat$. For a given $\psi \in \mathbf{S}(X', \mathbf{C})(\mathcal{F}', f_*(\mathcal{F}))$ and $x \in X$ we want to describe the stalk mapping

$$\psi^\sharp_x\colon (f^{-1}(\mathcal{F}'))_x \to \mathcal{F}_x$$

The construction in the proof of Theorem 5.20 provides for $U \in \mathfrak{T}$ and $\tilde{s} \in (f^{<-1>}(\mathcal{F}'))(U)$ an open neighborhood V of $f(U)$ and a $s_V \in \mathcal{F}'(V)$ with (V, s_V) in the equivalence class of $\tilde{s}$ and

$$\psi^\sharp_U\big(\theta^{f^{<-1>}(\mathcal{F}')}(\tilde{s})\big) = \rho^{\mathcal{F}}_{U, f^{-1}(V)}(\psi_V(s_V)) \in \mathcal{F}(U).$$

Now let U be a neighborhood of x and $s \in (f^{-1}(\mathcal{F}'))(U)$. If U is small enough, we can assume that $s = \theta^{f^{<-1>}(\mathcal{F}')}(\tilde{s})$ for an $\tilde{s} \in (f^{<-1>}(\mathcal{F}'))(U)$. Using Theorem 5.16 we find

$$\psi^\sharp_x(s_x) = \psi_{f(x)}(t_{f(x)}) \in (f^{-1}(\mathcal{F}'))_x = \mathcal{F}'_{f(x)},$$

if $s \in (f^{-1}(\mathcal{F}'))(U)$ is represented by $t \in \mathcal{F}'(V)$ in a small neighborhood of x. □

5.2 Étalé Spaces

We introduced sheaves over X as special presheaves over X because we wanted to encode local properties. In some books, one finds a completely different (but equivalent) definition, namely as special topological spaces with a continuous mapping to X. In reference to the usual name for the spaces associated with a sheaf, the corresponding type of space here will be called *étalé space* over X.

Definition 5.22 (Étalé Spaces) Let $(E, \mathfrak{S})$ and $(X, \mathfrak{T})$ be topological spaces, $\mathbf{C} = \mathbf{C}_{\Phi, \Gamma}$ a category of algebraic structures, and $\pi: E \to X$ a continuous mapping that satisfies the following conditions:

(a) π is a local homeomorphism, i.e., for each point in E there is an open neighborhood W for which the mapping $\pi|_W: W \to \pi(W) \in \mathfrak{T}$ is a homeomorphism.
(b) For each $x \in X$, the fiber $E_x := \pi^{-1}(x)$ over x is an object of $\mathbf{C}$.
(c) The fiber-wise operations

$$\{(e_1, \ldots, e_k) \in E \times \ldots \times E \mid \pi(e_1) = \ldots = \pi(e_k)\} \to E$$

are continuous mappings.

Then the triple (E, X, π) is called a $\mathbf{C}$-*étalé space* over X.

Let $U \in \mathfrak{T}$. A continuous function $s: U \to E$ is called a *section* of π over U if $\pi \circ s = \mathrm{id}_U$.

Property (a) shows that there are plenty of sections: For each point $e \in E$, one finds an open neighborhood W for which $\pi|_W: W \to U := \pi(W) \in \mathfrak{T}$ is a homeomorphism, i.e., $s := (\pi|_W)^{-1}: U \to W \subseteq E$ is a section. Thus, the images of sections give E the shape of a puff pastry. Continuous paths exist only within the layers (see Figure 5.1).

Together, properties (a)–(c) provide that

$$U \mapsto \mathcal{F}_\pi(U) := \{s: U \to E \mid s \text{ is a section of } \pi \text{ over } U\}$$

is a $\mathbf{C}$-sheaf (exercise). We call $\mathcal{F}_\pi$ the *section sheaf* of (E, X, π).

It will turn out that every $\mathbf{C}$-sheaf is isomorphic to the section sheaf of an étalé space that is uniquely determined up to isomorphism. This space will then be called *the* étalé space of the sheaf.

Fig. 5.1 Puff pastry structure of étalé spaces

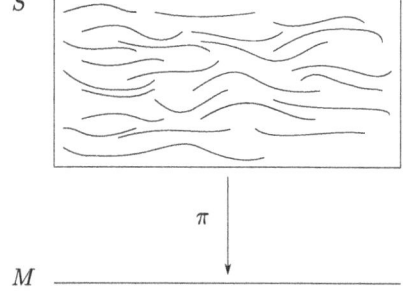

5.2 Étalé Spaces

Fig. 5.2 Fibers and germs in étalé spaces

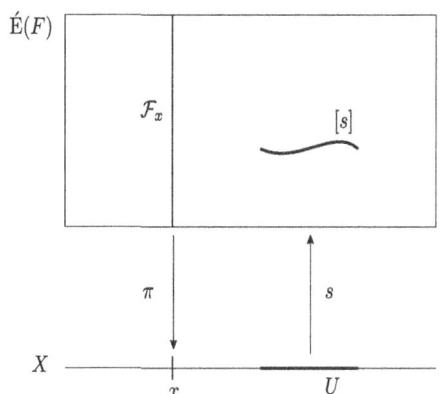

Construction 5.23 (Étalé Space of a Sheaf) Let $(X, \mathfrak{T})$ be a topological space, $\mathbf{C} = \mathbf{C}_{\Phi,\Gamma}$ a category of algebraic structures, and $\mathcal{F}$ a $\mathbf{C}$-sheaf over X. We set $\acute{\mathrm{E}}(\mathcal{F}) := \coprod_{x \in X} \mathcal{F}_x$ and define $\pi \colon \acute{\mathrm{E}}(\mathcal{F}) \to X$ by $\pi(\mathcal{F}_x) = \{x\}$.

For $U \in \mathfrak{T}$ and $s \in \mathcal{F}(U)$ we also set $[s] := \{s_x \in \mathcal{F}_x \mid x \in U\}$ (see Figure 5.2). Then the set $\mathfrak{S}$ of all subsets of $\acute{\mathrm{E}}(\mathcal{F})$ that can be written as a union of sets of the form $[s]$ with $s \in \mathcal{F}(U)$ and $U \in \mathfrak{T}$ is a topology on $\acute{\mathrm{E}}(\mathcal{F})$:

By definition, $\mathfrak{S}$ is closed under arbitrary unions. Also, $\emptyset \in \mathfrak{S}$, because $\emptyset \in \mathfrak{T}$ holds. Since for every $p \in \acute{\mathrm{E}}(\mathcal{F})$ there is an $x \in X$ with $p \in \mathcal{F}_x$, there is also a $U \in \mathfrak{T}$ and an $s \in \mathcal{F}(U)$ with $s_x = p$. Thus, $p \in [s]$, and $\acute{\mathrm{E}}(\mathcal{F}) \in \mathfrak{S}$ holds. It remains to prove that $\mathfrak{S}$ is closed under finite intersections. To do this, we first note that for $p \in [s] \cap [t]$ with $s \in \mathcal{F}(U)$ and $t \in \mathcal{F}(V)$ there is an open $W \subseteq V \cap U$ and an $r \in \mathcal{F}(W)$ with $[r] \subseteq [s] \cap [t]$, because one finds an $x \in V \cap U$ with $p = s_x = t_x$. Thus, $[s] \cap [t]$ is the union of sets of the form $[r]$, i.e., in $\mathfrak{S}$. The assertion follows from De Morgan's law.

We want to show that $(\acute{\mathrm{E}}(\mathcal{F}), X, \pi)$ is a $\mathbf{C}$-étalé space. Condition (b) from Definition 5.22 is fulfilled according to Remark 5.15, because $\acute{\mathrm{E}}(\mathcal{F})_x = \mathcal{F}_x$ holds.

To show (a), we note that for $U \in \mathfrak{T}$ and $s \in \mathcal{F}(U)$ the mapping $\pi|_{[s]} \colon [s] \to U$ is continuous, because for open $V \subseteq U$ the set

$$(\pi|_{[s]})^{-1}(V) = \{s_x \in [s] \mid x \in V\} = [\rho_{V,U}(s)]$$

is open. The inverse mapping $f \colon U \to [s]$, $x \mapsto s_x$ of $\pi|_{[s]}$ is also continuous, because for each subset of the form $[t]$ of $[s]$ with $t \in \mathcal{F}(V)$ and open $V \subseteq U$ the set $f^{-1}([t]) = \pi([t]) = V$ is open. Since $\acute{\mathrm{E}}(\mathcal{F})$ is covered by open subsets of the form $[s]$ with $s \in \mathcal{F}(U)$ and $U \in \mathfrak{T}$, property (a) is thus proven.

Finally, let $\varphi \in \Phi$ be a k-ary operation, $(p_1, \ldots, p_k) \in \mathcal{F}_x^k$ and $p_0 = \varphi(p_1, \ldots, p_k)$. We choose $U_0, \ldots, U_k \in \mathfrak{T}$ and $s_j \in \mathcal{F}(U_j)$ for $j = 0, \ldots, k$ with $p_j \in [s_j]$. Then there is a $U \in \mathfrak{T}$, which is contained in all U_j and

$$\rho_{U,U_0}(s_0) = \varphi\bigl(\rho_{U,U_1}(s_1), \ldots, \rho_{U,U_k}(s_k)\bigr)$$

holds. But then

$$\varphi\big(\{(e_1,\ldots,e_k) \in [\rho_{U,U_1}(s_1)] \times \ldots \times [\rho_{U,U_k}(s_k)] \mid \pi(e_1) = \ldots = \pi(e_k)\}\big) \subseteq [s_0],$$

and because U_0 could be chosen arbitrarily small, this proves the continuity statement (c). □

We have now assigned a $\mathbf{C}_{\Phi,\Gamma}$-sheaf $\mathcal{F}_\pi$ over X to each $\mathbf{C}_{\Phi,\Gamma}$-étalé space (E, X, π) over X as well as a $\mathbf{C}_{\Phi,\Gamma}$-étalé space $(\acute{E}(\mathcal{F}), X, \pi)$ to each $\mathbf{C}_{\Phi,\Gamma}$-sheaf $\mathcal{F}$ over X. Our next goal is to show that the combination of the two assignments (in both possible orders) yields an object that is isomorphic to the original object. To do this, we first need to introduce a concept of isomorphism for $\mathbf{C}_{\Phi,\Gamma}$-étalé spaces.

Definition 5.24 (Morphisms of Étalé Spaces) Let $\mathbf{C} = \mathbf{C}_{\Phi,\Gamma}$ be a category of algebraic structures and let (E, X, π) and (E', X, π') be two $\mathbf{C}$-étalé spaces over X. A *morphism* $(E, X, \pi) \to (E', X, \pi')$ is a continuous mapping $\varphi : E \to E'$ with $\pi' \circ \varphi = \pi$, and

$$\forall x \in X: \quad \varphi|_{E_x} E_x \to E'_x \text{ is a } \mathbf{C}\text{-morphism.}$$

One verifies (exercise) that the $\mathbf{C}$-étalé spaces over X together with the morphisms from Definition 5.24 form a category $\acute{E}(X, \mathbf{C})$. This also gives us a concept of isomorphisms of $\mathbf{C}$-étalé spaces over X.

Theorem 5.25 (Étalé Spaces versus Sheaves) *Let* $\mathbf{C} = \mathbf{C}_{\Phi,\Gamma}$ *be a category of algebraic structures and* $(X, \mathfrak{T})$ *a topological space.*

(i) *The assignments* $(E, X, \pi) \mapsto S(E, X, \pi) := \mathcal{F}_\pi$ *and*

$$\mathrm{Hom}_{\acute{E}(X,\mathbf{C})}\big((E, X, \pi), (E', X, \pi')\big) \ni \varphi \mapsto \widetilde{\varphi} \in \mathrm{Hom}_{S(X,\mathbf{C})}(\mathcal{F}_\pi, \mathcal{F}_{\pi'})$$

with

$$\forall U \in \mathfrak{T}: \quad (S\varphi)_U := \widetilde{\varphi}_U : \mathcal{F}_\pi(U) \to \mathcal{F}_{\pi'}(U), \ s \mapsto \varphi \circ s$$

define a functor $S \colon \acute{E}(X, \mathbf{C}) \to \mathbf{S}(X, \mathbf{C})$.

(ii) *The assignments* $\mathcal{F} \mapsto \acute{E}(\mathcal{F})$ *and*

$$\mathrm{Hom}_{\mathbf{S}(X,\mathbf{C})}(\mathcal{F}, \mathcal{F}') \ni \widetilde{\varphi} \mapsto \varphi \in \mathrm{Hom}_{\acute{E}(X,\mathbf{C})}\big(\acute{E}(\mathcal{F}), \acute{E}(\mathcal{F}')\big)$$

with

$$\forall p = s_x \in [s]: \quad (\acute{E}\widetilde{\varphi})(p) := \varphi(p) := \big(\widetilde{\varphi}_U(s)\big)_x,$$

where $x \in U \in \mathfrak{T}$ *and* $s \in \mathcal{F}(U)$ *is, define a functor* $\acute{E} \colon \mathbf{S}(X, \mathbf{C}) \to \acute{E}(X, \mathbf{C})$.

5.2 Étalé Spaces

(iii) For all $\mathcal{F} \in \text{ob}(\mathbf{S}(X, \mathbf{C}))$, $\mathcal{F}$ and $S(\text{É}(\mathcal{F}))$ are isomorphic.
(iv) For all $(E, X, \pi) \in \text{ob}(\text{É}(X, \mathbf{C}))$, (E, X, π) and $\text{É}(S(E, X, \pi)) = \text{É}(\mathcal{F}_\pi)$ are isomorphic.

Proof

(i) Let $\varphi \colon (E, X, \pi) \to (E', X, \pi')$ and $\varphi' \colon (E', X, \pi') \to (E'', X, \pi'')$ be morphisms in $\text{É}(X, \mathbf{C})$. Then the calculation

$$\begin{aligned}
\bigl(S(\varphi' \circ \varphi)\bigr)_U(s) &= (\varphi' \circ \varphi) \circ s = \varphi' \circ (\varphi \circ s) \\
&= (S\varphi')_U\bigl((S\varphi)_U(s)\bigr) = \bigl((S\varphi')_U \circ (S\varphi)_U\bigr)(s) \\
&= \bigl((S\varphi' \circ S\varphi)_U\bigr)(s)
\end{aligned}$$

for $U \in \mathfrak{T}$ and $s \in \mathcal{F}_\pi(U)$, shows that $S(\varphi' \circ \varphi) = (S\varphi') \circ (S\varphi)$. Thus, condition (b) from Definition 4.18 for S is fulfilled. Since the identities $1_{\{E,X,\pi\}}$ are obviously mapped to the identities $1_{\mathcal{F}_\pi}$, i.e., condition (a) from Definition 4.18 is also fulfilled, S is indeed a functor.

(ii) Let $\widetilde{\varphi} \colon \mathcal{F} \to \mathcal{F}'$ and $\widetilde{\varphi}' \colon \mathcal{F}' \to \mathcal{F}''$ be morphisms in $\mathbf{S}(X, \mathbf{C})$. Then the calculation

$$\begin{aligned}
\text{É}(\widetilde{\varphi}' \circ \widetilde{\varphi})(s_x) &= \bigl((\widetilde{\varphi}' \circ \widetilde{\varphi})_U(s)\bigr)_x = \bigl((\widetilde{\varphi}'_U \circ \widetilde{\varphi}_U)(s)\bigr)_x \\
&= \bigl((\widetilde{\varphi}'_U)((\widetilde{\varphi}_U)(s))\bigr)_x = \text{É}(\widetilde{\varphi}')\bigl((\widetilde{\varphi}_U)(s))_x\bigr) \\
&= \text{É}(\widetilde{\varphi}')\bigl(\text{É}\widetilde{\varphi}(s_x)\bigr) = \bigl(\text{É}(\widetilde{\varphi}') \circ \text{É}(\widetilde{\varphi})\bigr)(s_x)
\end{aligned}$$

for $x \in U \in \mathfrak{T}$ and $s \in \mathcal{F}(U)$, shows that $\text{É}(\widetilde{\varphi}' \circ \widetilde{\varphi}) = (\text{É}\widetilde{\varphi}') \circ (\text{É}\widetilde{\varphi})$. Again, it is clear that the identities are mapped onto each other, so É is also a functor.

(iii) Let $U \in \mathfrak{T}$. Then

$$S(\text{É}\mathcal{F})(U) = \{t \colon U \xrightarrow{\text{continuous}} \text{É}\mathcal{F} \mid \forall x \in U : t(x) \in \mathcal{F}_x\}.$$

The continuity of t at $x \in U$ means that there is an open neighborhood V of x in U and a $s \in \mathcal{F}(V)$ with $t(x) = s_x$. This makes it clear that the mapping

$$\widetilde{\varphi}_U \colon \mathcal{F}(U) \to S(\text{É}\mathcal{F})(U), \quad s \mapsto (x \mapsto s_x)$$

is well-defined. Since the formation of germs is a $\mathbf{C}$-morphism and the compositions on $S(\text{É}\mathcal{F})(U)$ are defined pointwise, the $\widetilde{\varphi}_U$ also define a $\mathbf{C}$-morphism.

It is now sufficient to show that $\widetilde{\varphi}_U$ is invertible, because then $\widetilde{\varphi}_U^{-1}$ is automatically a $\mathbf{C}$-morphism, and because $U \in \mathfrak{T}$ was arbitrary, $U \mapsto \widetilde{\varphi}_U$ and $U \mapsto \widetilde{\varphi}_U^{-1}$ define mutually inverse $\mathbf{S}(X, \mathbf{C})$-morphisms.

So let $t \in S(\acute{E}\mathcal{F})(U)$ and $\{V_i\}_{i \in I}$ be an open cover of U, for which $s_i \in \mathcal{F}(V_i)$ with

$$\forall x \in V_i : \quad t(x) = (s_i)_x$$

exist. Then the s_i are compatible on the intersections of the V_i and define, because $\mathcal{F}$ is a sheaf, an $s \in \mathcal{F}(U)$ with

$$\forall x \in U : \quad t(x) = s_x.$$

Then $\widetilde{\varphi}_U(s) = t$, i.e., $\widetilde{\varphi}_U$ is surjective. The injectivity of $\widetilde{\varphi}_U$ also follows from the sheaf property, because two sections that agree everywhere locally also agree. This proves the claim.

(iv) As a set, the étalé space $\acute{E}(\mathcal{F}_\pi)$ is just $\coprod_{x \in X} (\mathcal{F}_\pi)_x = \coprod_{x \in X} E_x = E$. The projection $\acute{E}(\mathcal{F}_\pi) \to X$ is given by $(\mathcal{F}_\pi)_x \ni p \mapsto x$, so it coincides with π. It remains only to show that the topologies of $\acute{E}(\mathcal{F}_\pi)$ and E coincide, because then the identity is an $\acute{E}(X, \mathbf{C})$-isomorphism between $\acute{E}(\mathcal{F}_\pi)$ and E.

Let $W \subseteq E$ be open. Then W is the union of a family $\{W_i\}_{i \in I}$ of open subsets of W, for which $\pi|_{W_i} : W_i \to U_i := \pi(W_i) \in \mathfrak{T}$ are homeomorphisms. The inverse mappings $s_i = (\pi|_{W_i})^{-1} : U_i \to W_i \subseteq E$ are sections of $S(E, X, \pi)(U_i)$, and in the notation of Construction 5.23 we have $W_i = [s_i]$. Thus, W is open in $\acute{E}(\mathcal{F}_\pi)$ as a union of the W_i. Conversely, if $W \subseteq \acute{E}(\mathcal{F}_\pi)$ is open, one finds a family $\{U_i\}_{i \in I}$ of open subsets of X and sections $s_i \in \mathcal{F}_\pi(U_i)$, for which W is the union of the $[s_i] = s_i(U_i)$. It is therefore sufficient to show that sections of $\mathcal{F}_\pi$ are automatically open mappings. This in turn is immediately clear, because $\pi : E \to X$ is a local homeomorphism.

□

Exercise 5.7 (Category Equivalence of Sheaves and Étalé Spaces) Let $(X, \mathfrak{T})$ be a topological space and $\mathbf{C} = \mathbf{C}_{\Phi,\Gamma}$ a category of algebraic structures. Show that the categories $\mathbf{S}(X, \mathbf{C})$ of C-sheaves over X and the category $\acute{E}(X, \mathbf{C})$ of C-étalé spaces over X are equivalent.

Exercise 5.8 (Products of Sheaves and Étalé Spaces) Let $(X, \mathfrak{T})$ be a topological space and $\mathbf{C} = \mathbf{C}_{\Phi,\Gamma}$ a category of algebraic structures. Transfer the concept of product sheaves (see Example 5.18) over X to the category $\acute{E}(X, \mathbf{C})$ of C-étalé spaces over X.

Hint: For two étalé spaces (E, X, π) and (E', X, π') over X, consider the set-theoretic fiber product $E \times_X E' := \{(e, e') \in E \times E' \mid \pi(e) = \pi'(e')\}$ from Example 4.36.

Definition 5.26 (Inverse Image) Let $(X, \mathfrak{T})$ and $(X', \mathfrak{T}')$ be topological spaces, $f : X \to X'$ a continuous mapping, $\mathbf{C} = \mathbf{C}_{\Phi,\Gamma}$ a category of algebraic structures

5.2 Étalé Spaces

and (E', X', π') a **C**-étalé space over X'. Then

$$\pi : E := \{(e', x) \in E' \times X \mid f(x) = \pi'(e')\} \to X, \quad (e', x) \mapsto x$$

defines an étalé space (E, X, π), which is called the *inverse image* of (E', X', π') under f and denoted by $f^{-1}(E', X', \pi')$. The topology of E is given as the subspace topology of E as a (closed) subset of $E' \times X$, which in turn is endowed with the product topology (details as exercise).

Theorem 5.27 (Inverse Image) *The constructions of inverse images of sheaves and étalé spaces correspond to each other when applying the functors S and $\acute{E}$.*

Proof In light of Theorem 5.25, it suffices to show that the **C**-presheaves

$$\mathfrak{T} \ni U \mapsto S(f^{-1}(E', X', \pi'))(U) \quad \text{and} \quad \mathfrak{T} \ni U \mapsto f^{-1}(S(E', X', \pi'))(U)$$

over X are isomorphic. The space $S(f^{-1}(E', X', \pi'))(U)$ can be identified with

$$\{s : U \xrightarrow{\text{continuous}} E' \mid \pi' \circ s = f\}$$

because the étalé space of $f^{-1}(E', X', \pi')$ is given by $E = \{(e', x) \in E' \times X \mid \pi'(e') = f(x)\}$ and thus for a section $s : U \to E$ the second component is redundant. The continuity of s means that for each $x \in U$ there is an open neighborhood U_1 in U, a $f(x) \in V \in \mathfrak{T}'$ and a $t \in S(E', X', \pi')(V)$ with $s|_{U_1} = t \circ f|_{U_1}$.

Furthermore,

$$f^{-1}(S(E', X', \pi'))(U) = \left\{ \widetilde{s} : U \xrightarrow{\text{continuous}} \coprod_{x \in U} f^{-1}(S(E', X', \pi'))_x \right\}$$

holds, where the continuity of $\widetilde{s}$ means that for each $x \in U$ there is an open neighborhood U_1 in U and an $r \in f^{-1}(S(E', X', \pi'))(U_1)$ with

$$\forall y \in U_1 : \quad \widetilde{s}(y) = r_{f(y)}.$$

According to the definition of $f^{-1}(S(E', X', \pi'))(U_1) = \varinjlim_{V \supseteq f(U_1)} S(E', X', \pi')(V)$, this is equivalent to the fact that for $x \in U$ there is a neighborhood U_1 of x in U and an open neighborhood V_1 of $f(U_1)$ in X' as well as a $t \in S(E', X', \pi')(V_1)$ with $s|_{U_1} = t \circ f|_{U_1}$.

To complete the proof, it is now sufficient to prove that

$$f^{-1}(S(E', X', \pi'))_x = S(E', X', \pi')_{f(x)}.$$

However, this can be verified directly from the definitions (exercise). $\square$

5.3 Ringed Spaces

In this section, we consider a special class of sheaves that will always play the role of the sheaf of functions relevant to the respective structure in later descriptions of local structures. For a topological space, this is, e.g., the sheaf of continuous functions, for a (real or complex) differentiable manifold the sheaf of (real or complex) differentiable functions, and for an algebraic variety the sheaf of regular functions. All these examples have in common that the respective functions can be multiplied with each other, thus creating a ring structure.

Definition 5.28 (Ringed Spaces) Let $(X, \mathfrak{T})$ be a topological space and $\mathcal{O}_X$ a sheaf of commutative rings over X, then the pair $(X, \mathcal{O}_X)$ is called a *ringed space*. If R is a fixed commutative ring and $\mathcal{O}_X$ a sheaf of commutative R-algebras (i.e., $\mathbf{C} = \mathbf{CAlg}_R$; see Section 4.2), then the pair $(X, \mathcal{O}_X)$ is called an *R-ringed space* and $\mathcal{O}_X$ the *structure sheaf* of X.

A *morphism* $(X, \mathcal{O}_X) \to (Y, \mathcal{O}_Y)$ of ringed spaces $(X, \mathcal{O}_X)$ and $(Y, \mathcal{O}_Y)$ is a pair $(f, f^\flat)$, where $f: X \to Y$ is a continuous mapping and $f^\flat: \mathcal{O}_Y \to f_*\mathcal{O}_X$ is a $\mathbf{C}$-sheaf morphism. □

Every ringed space is a $\mathbb{Z}$-ringed space, so from now on we only talk about R-ringed spaces.

Exercise 5.9 (Category of Ringed Spaces) Let $R \in \mathrm{ob}(\mathbf{CRing}_1)$. Show that the R-ringed spaces $(X, \mathcal{O}_X)$ together with the morphisms $(f, f^\flat): (X, \mathcal{O}_X) \to (Y, \mathcal{O}_Y)$ form a category $\mathbf{RSp}_R$ (*R-ringed spaces*), for which

$$(g, g^\flat) \circ (f, f^\flat) = (g \circ f, f^\flat \circ g^\flat): (X, \mathcal{O}_X) \to (Z, \mathcal{O}_Z)$$

holds, if $(f, f^\flat): (X, \mathcal{O}_X) \to (Y, \mathcal{O}_Y)$ and $(g, g^\flat): (Y, \mathcal{O}_Y) \to (Z, \mathcal{O}_Z)$.

The R-ringed spaces, together with the morphisms described in Definition 5.28, form a category (see Exercise 5.9). The question immediately arises as to why, in the definition of a morphism of ringed spaces, one works with the direct image and not with the inverse image of the corresponding structure sheaves. Indeed, one could just as well replace the morphism $f^\flat: \mathcal{O}_Y \to f_*\mathcal{O}_X$ with a morphism $f^\sharp: f^{-1}\mathcal{O}_Y \to \mathcal{O}_X$, because according to Theorem 5.20 there is a natural isomorphism between $\mathrm{Hom}_{\mathbf{S}(X, \mathbf{CAlg}_R)}(f^{-1}\mathcal{O}_Y, \mathcal{O}_X)$ and $\mathrm{Hom}_{\mathbf{S}(Y, \mathbf{CAlg}_R)}(\mathcal{O}_Y, f_*\mathcal{O}_X)$.

Example 5.29 (Ringed Function Spaces) Let $X \subseteq \mathbb{K}^n$ be an open subset. Then the pairs $(X, \mathcal{C}_X^{\mathbb{K},k})$ with the sheaves $\mathcal{C}_X^{\mathbb{K},k}$ from Example 5.10 are $\mathbb{K}$-ringed spaces.

If Y is an open subset of $\mathbb{K}^m$ and $f: X \to Y$ is a k-times continuously $\mathbb{K}$-differentiable mapping, then $(f, f^\flat)$ with

$$f_V^\flat: \mathcal{C}_Y^{\mathbb{K},k}(V) = C^{\mathbb{K},k}(V, \mathbb{K}) \to C^{\mathbb{K},k}(f^{-1}(V), \mathbb{K}) = (f_*\mathcal{C}_X^{\mathbb{K},k})(V), \quad h \mapsto h \circ f$$

for an open subset $V \subseteq Y$ is a morphism $(X, \mathcal{C}_X^{\mathbb{K},k}) \to (Y, \mathcal{C}_Y^{\mathbb{K},k})$. □

5.3 Ringed Spaces

The next example is of central importance in algebraic geometry.

Example 5.30 (Spectrum of a Ring) Let R be a commutative ring with identity and $\text{Spec}(R) := \{P \triangleleft R \mid P \text{ is prime }\}$ the set of prime ideals of R (see Definition 1.19). $\text{Spec}(R)$ is called the *spectrum* of R. The spectrum of R becomes a topological space when the subsets

$$V(I) := \{P \in \text{Spec}(R) \mid I \subseteq P\}$$

with $I \triangleleft R$ are defined to be the closed subsets of $\text{Spec}(R)$. To see this, we prove the following three claims:

(i) $\forall I, J \triangleleft R: \quad V(IJ) = V(I) \cup V(J)$.
(ii) For a family $\{I_\alpha\}_{\alpha \in A}$ of ideals in R, it holds: $V(\sum_{\alpha \in A} I_\alpha) = \bigcap_{\alpha \in A} V(I_\alpha)$.
(iii) $V(R) = \emptyset$ and $V(\{0\}) = \text{Spec}(R)$.

Claim (i) shows that the union of two (and thus finitely many) closed sets is closed. Claim (ii) shows that intersections of arbitrary families of closed sets are closed. Claim (iii) provides the closedness of $\emptyset$ and $\text{Spec}(R)$. It follows that the set $\mathfrak{Z}$ of $\text{Spec}(R) \setminus V(I)$ with $I \triangleleft R$ indeed defines a topology on $\text{Spec}(R)$. This topology is called the *Zariski topology*.

To prove (i), we first establish that $I \subseteq P$ for $I, J \triangleleft R$ and $P \in \text{Spec}(R)$ immediately yields $IJ \subseteq P$, thus $V(IJ) \supseteq V(I) \cap V(J)$. Conversely, if $IJ \subseteq P$ and $J \not\subseteq P$, then there exists an $r \in J \setminus P$, and because $Ir \subseteq P$ it follows that $I \subseteq P$, since P is prime. This shows $V(IJ) \subseteq V(I) \cap V(J)$. Statement (ii) is clear, because $\sum_{\alpha \in A} I_\alpha \subseteq P$ is equivalent to $I_\alpha \subseteq P$ for all $\alpha \in A$. The last statement follows because no prime ideal contains the identity, but every prime ideal contains zero.

For $P \in \text{Spec}(R)$, let R_P be the ring that arises from R when we localize in $S := R \setminus P$. Note that S meets the conditions from Exercise 1.4 precisely because P is a prime ideal.

Now we can define the structure sheaf $\mathcal{O} := \mathcal{O}_{\text{Spec}(R)}$ for $\text{Spec}(R)$: For $U \in \mathfrak{Z}$ open, we set

$$\mathcal{O}(U) := \left\{ t : U \to \coprod_{P \in U} R_P \;\middle|\; t(P) \in R_P \text{ and (A)} \right\},$$

where

(A) $\forall P \in U \; \exists P \in U_P \in \mathfrak{Z}, r, s \in R, s \notin \bigcup_{Q \in U_P} Q :$
$\left(\forall Q \in U_P : t(Q) = \frac{r}{s} \in R_Q \right).$

Note that for $s \in R \setminus P$, we have: $P \notin V(sR)$. Thus,
$$U(s) := \text{Spec}(R) \setminus V(sR) \in \mathfrak{Z}$$
is an open neighborhood of P. For $Q \in U(s)$, we have $Q \notin V(sR)$, which means there exists an $r \in R$ with $sr \notin Q$. Because Q is an ideal, it follows that $s \notin Q$, which means we have $s \notin \bigcup_{Q \in U(s)} Q$. Therefore, $\mathcal{O}(U) \neq \emptyset$.

The $\mathcal{O}(U)$ are rings with respect to pointwise operations, since if $s, s' \in R \setminus Q$ for $Q \in \text{Spec}(R)$, then $s, s' \in R \setminus Q$ also follows. The construction is compatible with restriction mappings, therefore $\mathfrak{Z} \ni U \mapsto \mathcal{O}(U)$ is a presheaf of rings. Since condition (A) is of local nature, $\mathcal{O}$ is even a sheaf (exercise). Thus, $(\text{Spec}(R), \mathcal{O}_{\text{Spec}(R)})$ is a ringed space.

If R' is another commutative ring with identity and $\varphi \colon R \to R'$ is a ring homomorphism that preserves the one, then we find a morphism $(f, f^\flat) \colon (\text{Spec}(R'), \mathcal{O}_{\text{Spec}(R')}) \to (\text{Spec}(R), \mathcal{O}_{\text{Spec}(R)})$ as follows. The mapping $f \colon \text{Spec}(R') \to \text{Spec}(R)$ is given by
$$\forall P' \in \text{Spec}(R') : \quad f(P') = \varphi^{-1}(P') \in \text{Spec}(R)$$
(well-defined due to Exercise 1.7). Now let $U \in \mathfrak{Z}_R$ and $t \in \mathcal{O}_{\text{Spec}(R)}(U)$ and $P' \in f^{-1}(U) \subseteq \text{Spec}(R')$. By
$$\forall r \in R, s \in R \setminus \varphi^{-1}(P') : \quad \varphi_{P'}\left(\frac{r}{s}\right) := \frac{\varphi(r)}{\varphi(s)} \in R'_{P'}$$
a ring homomorphism $\varphi_{P'} \colon R_{f(P')} \to R'_{P'}$ is defined, so we obtain a mapping
$$f_U^\flat(t) \colon f^{-1}(U) \to \coprod_{P' \in f^{-1}(U)} R'_{P'}, \quad P' \mapsto \varphi_{P'} \circ t \circ f(P').$$
One verifies (exercise) that $f_U^\flat(t) \in \mathcal{O}_{\text{Spec}(R')}(f^{-1}(U)) = f_* \mathcal{O}_{\text{Spec}(R)}(U)$ and that the mapping
$$f_U^\flat \colon \mathcal{O}_{\text{Spec}(R)}(U) \to f_* \mathcal{O}_{\text{Spec}(R)}(U), \quad t \mapsto f_U^\flat(t)$$
is a ring homomorphism. Since the construction is compatible with restrictions, the $f_U^\flat$ together define a sheaf morphism $f^\flat$.

One further verifies (exercise) that the assignments $R \mapsto (\text{Spec}(R), \mathcal{O}_{\text{Spec}(R)})$ and $(\varphi \colon R \to R') \mapsto (f, f^\flat) \colon (\text{Spec}(R'), \mathcal{O}_{\text{Spec}(R')}) \to (\text{Spec}(R), \mathcal{O}_{\text{Spec}(R)})$ define a functor

$$\text{Spec} \colon \mathbf{CRing}_1 \to \mathbf{RSp}$$

5.3 Ringed Spaces

Finally, let R_0 be a commutative ring with identity. Then the above constructions can be transferred to commutative R_0-algebras, and one obtains a functor

$$\text{Spec}\colon \mathbf{CAlg}_{R_0} \to \mathbf{RSp}_{R_0}.$$

□

Let $X \subseteq \mathbb{R}^n$ be as in Example 5.29. If a continuous function $s\colon X \to \mathbb{R}$ does not vanish at a point $x \in X$, then there is a small neighborhood U of x where the function does not vanish anywhere. Then one can form the inverse function $U \to \mathbb{R}$, $x \mapsto \frac{1}{f(x)}$, whose germ at x is then the inverse of the germ s_x of s at x in the ring $(\mathcal{C}_X^{\mathbb{R},k})_x$. Thus, every germ $s_x \in (\mathcal{C}_X^{\mathbb{R},k})_x$ that lies in a proper ideal of $(\mathcal{C}_X^{\mathbb{R},k})_x$ is contained in the kernel of the evaluation mapping

$$\text{ev}_x \colon (\mathcal{C}_X^{\mathbb{R},k})_x \to \mathbb{R}, \quad s_x \mapsto s(x).$$

In other words, $\ker(\text{ev}_x)$ is the only maximal ideal in $(\mathcal{C}_X^{\mathbb{R},k})_x$. Rings with only one maximal ideal are called *local rings*. If one factors out the maximal ideal from such a ring, one obtains a field, which is called the *residue field* of the ring. The evaluation mapping shows that in the present case the residue field is simply $\mathbb{R}$.

Now if $Y \subseteq \mathbb{R}^m$ is open and $f\colon X \to Y$ is a continuous mapping, a sheaf morphism $f^\flat \colon \mathcal{C}_Y^{\mathbb{R},k} \to f_*(\mathcal{C}_X^{\mathbb{R},k})$ yields a sheaf morphism $f^\sharp \colon f^{-1}(\mathcal{C}_Y^{\mathbb{R},k}) \to \mathcal{C}_X^{\mathbb{R},k}$ and this, according to Exercise 5.4 and Theorem 5.27 (see also Remark 5.21), provides the ring homomorphism

$$f_x^\sharp \colon (\mathcal{C}_Y^{\mathbb{R},k})_{f(x)} = (f^{-1}\mathcal{C}_Y^{\mathbb{R},k})_x \to (\mathcal{C}_X^{\mathbb{R},k})_x, \quad s_{f(x)} \mapsto (f^\flat(s))_x. \qquad (*)$$

Since all ring homomorphisms considered here preserve the identity, we find $\text{ev}_x \circ f_x^\sharp = \text{ev}_{f(x)}$, so that $f_x^\sharp(\ker(\text{ev}_{f(x)})) \subseteq \ker(\text{ev}_x)$, and $f^\flat(s) = s \circ f$. If one applies $(*)$ to the projections $s(x_1, \dots, x_n) = x_j$, one sees that the component functions of $f\colon X \to Y \subseteq \mathbb{R}^m$ are all in $\mathcal{C}_X^{\mathbb{R},k}(X)$, i.e., $f\colon X \to \mathbb{R}^m$ is k-times continuously differentiable. Conversely, $s \mapsto s \circ f$ provides a sheaf morphism $f^\flat \colon \mathcal{C}_Y^{\mathbb{R},k} \to f_*(\mathcal{C}_X^{\mathbb{R},k})$ for each k-times continuously differentiable $f\colon X \to \mathbb{R}^m$.

The argument just demonstrated can be transferred to other sheaves of continuous functions, as long as the considered family of functions remains closed under the transition from $f|_U$ to $\frac{1}{f|_U}$, as for example is the case for differentiable functions. For spectra of rings, one has analogous results, although the proofs look quite different (see Theorem 5.34). This leads to the following definitions.

Definition 5.31 (Locally Ringed Spaces) An R-ringed space $(X, \mathcal{O}_X)$ is called a *locally R-ringed space*, if the stalks $(\mathcal{O}_X)_x$ for $x \in X$ are all local rings. Let $(X, \mathcal{O}_X)$ and $(Y, \mathcal{O}_Y)$ be locally ringed spaces. A morphism

$$(f, f^\flat) \colon (X, \mathcal{O}_X) \to (Y, \mathcal{O}_Y)$$

of ringed spaces is a *morphism* of *locally* ringed spaces, if for each $x \in X$ the ring homomorphism

$$f_x^\sharp : (f^{-1}\mathcal{O}_Y)_x = (\mathcal{O}_Y)_{f(x)} \to (\mathcal{O}_X)_x$$

(see Exercise 5.4 and Theorem 5.20) is *local*, i.e., it maps the maximal ideals into each other. □

The locally R-ringed spaces form a subcategory $\mathbf{RSp}_{\mathrm{loc},R}$ of the category $\mathbf{RSp}_R$ of R-ringed spaces (exercise).

Proposition 5.32 (Locally Ringed Function Spaces) *Let $X \subseteq \mathbb{K}^n$ and $Y \subseteq \mathbb{K}^m$ be open, and $\mathcal{O}_X := \mathcal{C}_X^{\mathbb{K},k}$ and $\mathcal{O}_Y := \mathcal{C}_Y^{\mathbb{K},k}$ for $k \in \mathbb{N}_0 \cup \{\infty\} \cup \{\omega\}$. Then:*

(i) *$(X, \mathcal{O}_X)$ is a locally $\mathbb{K}$-ringed space.*
(ii) *For each $x \in X$, the residue field of $\mathcal{O}_{X,x}$ is equal to $\mathbb{K}$.*
(iii) *Every morphism $(f, f^\flat): (X, \mathcal{O}_X) \to (Y, \mathcal{O}_Y)$ is automatically local, i.e., a morphism of locally $\mathbb{K}$-ringed spaces.*
(iv) *For every morphism $(f, f^\flat): (X, \mathcal{O}_X) \to (Y, \mathcal{O}_Y)$, the components of the vector-valued mapping $f: X \to Y$ are in $\mathcal{O}_X(X)$, and the sheaf morphism $f^\sharp: f^{-1}\mathcal{O}_Y \to \mathcal{O}_X$ associated with $f^\flat: \mathcal{O}_Y \to f_*(\mathcal{O}_X)$ is given by*

$$\forall x \in X: \quad f_x^\sharp : (\mathcal{O}_Y)_{f(x)} = (f^{-1}\mathcal{O}_Y)_x \to (\mathcal{O}_X)_x, \quad s_{f(x)} \mapsto (s \circ f)_x.$$

Proof The argument before Definition 5.31 can be applied to each of the considered ringed spaces. □

We want to show that ring spectra are locally ringed spaces. For this, we need to show in particular that the localized rings R_P from the construction of $\mathcal{O}_{\mathrm{Spec}(R)}$ in Example 5.30 are local rings.

Lemma 5.33 (Localizations are Local) *Let R be a commutative ring with identity and $P \in \mathrm{Spec}(R)$. Then the localized ring R_P is a local ring. The uniquely determined maximal ideal of R_P is*

$$\left\{ \frac{r}{s} \in R_P \,\middle|\, r \in P, s \in R \setminus P \right\}.$$

Proof Consider the ideal $I_P := \left\{ \frac{r}{s} \in R_P \,\middle|\, r \in P, s \in R \setminus P \right\}$ in R_P. If $\frac{r}{s} \in R_P \setminus I_P$, then $r \notin P$, i.e., $\frac{r}{s}$ is a unit in R_P. Therefore, I_P is a maximal ideal and the complement of I_P is exactly the set of units in R. Since every proper ideal and in particular every non-unit is contained in a maximal ideal, the union of all maximal ideals is exactly the complement of the units. Therefore, there can be no other maximal ideals than I_p. □

5.3 Ringed Spaces

Theorem 5.34 (Ring Spectra are Locally Ringed Spaces) *The functor* Spec: $\mathbf{CAlg}_{R_0} \to \mathbf{RSp}_{R_0}$ *factors through the category* $\mathbf{RSp}_{loc, R_0}$ *for every* $R_0 \in \mathrm{ob}(\mathbf{CRing})$.

Proof The following two statements need to be shown:

(i) $(\mathrm{Spec}(R), \mathcal{O}_{\mathrm{Spec}(R)})$ is a locally R_0-ringed space for every $R \in \mathrm{ob}(\mathbf{CAlg}_{R_0})$.
(ii) For every identity-preserving ring homomorphism $\varphi : R \to R'$ and corresponding morphism

$$\mathrm{Spec}(\varphi) = (f, f^\flat) : (\mathrm{Spec}(R'), \mathcal{O}_{\mathrm{Spec}(R')}) \to (\mathrm{Spec}(R), \mathcal{O}_{\mathrm{Spec}(R)}),$$

the ring homomorphism

$$f^\sharp_{P'} : (f^{-1} \mathcal{O}_{\mathrm{Spec}(R)})_{P'} = (\mathcal{O}_{\mathrm{Spec}(R)})_{f(P')} \to (\mathcal{O}_{\mathrm{Spec}(R')})_{P'}$$

is local for every $P' \in \mathrm{Spec}(R')$.

To show (i), it suffices, according to Lemma 5.33, to prove that for every $P \in \mathrm{Spec}(R)$ the stalk $(\mathcal{O}_{\mathrm{Spec}(R)})_P$ is isomorphic to R_P.

The evaluation mappings $\mathrm{ev}_{U,P} : \mathcal{O}_{\mathrm{Spec}(R)}(U) \to R_P$, $t \mapsto t(P)$ for $P \in U \in \mathfrak{Z}_R$ are compatible with the restrictions and therefore provide ring homomorphisms $\varphi_P : (\mathcal{O}_{\mathrm{Spec}(R)})_P \to R_P$. We want to show that φ_P is bijective.

Every germ $x \in (\mathcal{O}_{\mathrm{Spec}(R)})_P$ is represented on an open neighborhood U_P of P by a pair (r, s) with $r \in R$ and $s \in R \setminus P$ according to Example 5.30. Then $\varphi_P(x) = \frac{r}{s} \in R_P$. If $x, x' \in (\mathcal{O}_{\mathrm{Spec}(R)})_P$ with $\varphi_P(x) = \varphi_P(x')$ are represented on a neighborhood U_P of P by (r, s) and (r', s') respectively, then $\frac{r}{s} = \frac{r'}{s'} \in R_P$. There is then an $s'' \in R \setminus P$ with $rs's'' = r'ss'' \in R$. Let the open neighborhoods $U(s), U(s')$ and $U(s'')$ of P be defined as in Example 5.30. Then for every $Q \in U_P \cap U(s) \cap U(s') \cap U(s'')$, $\frac{r}{s} = \frac{r'}{s'} \in R_Q$, where the fractions are to be taken in R_Q. But this shows $x = x'$, and φ_P is injective. The argument also shows that for $\frac{r}{s} \in R_P$ there is a neighborhood U of P and a $t \in \mathcal{O}_{\mathrm{Spec}(R)}(U)$ with $\mathrm{ev}_{U,P}(t) = \frac{r}{s}$. It follows that $\varphi(t_P) = \frac{r}{s}$ for $t_P \in (\mathcal{O}_{\mathrm{Spec}(R)})_P$, so φ_P is also surjective. This proves (i).

To show (ii), we first note that Remark 5.21 allows us to determine $f^\sharp_{P'}$ explicitly. Under the identifications $(\mathcal{O}_{\mathrm{Spec}(R)})_{f(P')} \cong R_{f(P')} = R_{\varphi^{-1}(P')}$ and $(\mathcal{O}_{\mathrm{Spec}(R')})_{P'} \cong R'_{P'}$ and with the formula for $f^\flat_U(t)$ in Example 5.30, we get

$$f^\sharp_{P'} : R_{\varphi^{-1}(P')} \to R'_{P'}, \quad \frac{r}{s} \mapsto \frac{\varphi(r)}{\varphi(s)}.$$

Because $\varphi(\varphi^{-1}(P')) \subseteq P'$, $f^\sharp_{P'}$ maps the maximal ideals into each other according to the formula in Lemma 5.33, so it is local. □

5.4 Module Sheaves

Ringed spaces provide the conceptual framework for generalized spaces of scalar functions that fit a topological space with a given local additional structure. Module sheaves are the vector-valued extension of this concept. Therefore, the basic principles of the theory of module sheaves also look like elementary linear algebra—in the variant of modules over commutative rings from Chaps. 2 and 3.

Definition 5.35 ($\mathcal{O}_X$-Modules) Let $(X, \mathcal{O}_X)$ be an R-ringed space. A sheaf $\mathcal{F}$ of abelian groups over X is called a $\mathcal{O}_X$-*module* or a $\mathcal{O}_X$-*module sheaf*, if the following conditions are fulfilled

(a) $\forall U \subseteq X$ open: $\mathcal{F}(U)$ is a $\mathcal{O}_X(U)$-module.
(b) $\forall V \subseteq U \subseteq X$ open and $r \in \mathcal{O}_X(U)$ and $v \in \mathcal{F}(U)$:

$$\rho^{\mathcal{F}}_{V,U}(rv) = \rho^{\mathcal{O}_X}_{V,U}(r)\rho^{\mathcal{F}}_{V,U}(v),$$

i.e., the restrictions are compatible with the module structures.

A *morphism* $\varphi: \mathcal{F} \to \mathcal{G}$ between two $\mathcal{O}_X$-modules is a morphism of sheaves of abelian groups, for which all $\varphi_U: \mathcal{F}(U) \to \mathcal{G}(U)$ with open $U \subseteq X$ are $\mathcal{O}_X(U)$-module homomorphisms. We denote the resulting category of $\mathcal{O}_X$-modules with **Mod**$_{\mathcal{O}_X}$.

If $(X, \mathcal{O}_X)$ is an R-ringed space, then the stalks $\mathcal{O}_{X,x}$ are R-algebras according to Remark 5.15. The stalks $\mathcal{F}_x$ of a $\mathcal{O}_X$-module $\mathcal{F}$ are R-modules according to the same remark. But there is more: The compatibility of the scalar multiplications with the restrictions shows that $\mathcal{F}_x$ is even an $\mathcal{O}_{X,x}$-module.

Remark 5.36 (Étalé Version of Module Sheaves) Let $(X, \mathcal{O}_X)$ be an R-ringed space and $\mathcal{F}$ a $\mathcal{O}_X$-module. The $\mathcal{O}_{X,x}$-module structure of the stalks $\mathcal{F}_x$ shows that on the categorical product $\acute{E}(\mathcal{O}_X) \times \acute{E}(\mathcal{F})$ of the associated étalé spaces $(\acute{E}(\mathcal{O}_X), X, \pi)$ and $(\acute{E}(\mathcal{F}), X, \pi_\mathcal{F})$ (see Exercise 5.8) an $\acute{E}(X, \textbf{Set})$-morphism $\acute{E}(\mathcal{O}_X) \times \acute{E}(\mathcal{F}) \to \acute{E}(\mathcal{F})$ is defined, which is given on the fibers by the scalar multiplication. Similarly, we have an $\acute{E}(X, \textbf{Set})$-morphism $\acute{E}(\mathcal{F}) \times \acute{E}(\mathcal{F}) \to \acute{E}(\mathcal{F})$, which is given fiberwise by the addition.

Conversely, if $(X, \mathfrak{T})$ is a topological space and (A, X, π) an object of $\acute{E}(X, \textbf{Alg}_R)$, then X together with the section sheaf $S(A, X, \pi) \in \textbf{S}(X, \textbf{Alg}_R)$ is an R-ringed space $(X, \mathcal{O}_X)$, i.e., an object of **RSp**$_R$. If in addition $(E, X, \pi_E) \in$ ob$(\acute{E}(X, \textbf{Mod}_R))$, then X together with the section sheaf $\mathcal{F} := S(E, X, \pi) \in$ $\textbf{S}(X, \textbf{Mod}_R)$ is a sheaf of R-modules. If additionally a $\acute{E}(X, \textbf{Set})$-morphism $(A, X, \pi) \times (E, X, \pi_E) \to (E, X, \pi_E)$ is given, which defines for each $x \in X$ an A_x-module structure on the R-module E_x, then $\mathcal{F}$ is even a $\mathcal{O}_X$-module. To see this, one realizes that $A_x = \mathcal{O}_{X,x}$ and $E_x = \mathcal{F}_x$, and multiplies local sections

5.4 Module Sheaves

pointwise (details as exercise). Thus, it is concluded that under the equivalence of the categories $\mathbf{S}(X, \mathbf{Mod}_R)$ and $\acute{\mathbf{E}}(X, \mathbf{Mod}_R)$ (see Exercise 5.7), the subcategory $\mathbf{Mod}_{\mathcal{O}_X}$ corresponds to the following subcategory $\acute{\mathbf{E}}\mathbf{Mod}_{\mathcal{O}_X}$ of étalé spaces in $\acute{\mathbf{E}}(X, \mathbf{Mod}_R)$: The *objects* (E, X, π_E) of $\acute{\mathbf{E}}\mathbf{Mod}_{\mathcal{O}_X}$ are equipped with a $\acute{\mathbf{E}}(X, \mathbf{Set})$-morphism

$$\acute{\mathbf{E}}(\mathcal{O}_X) \times (E, X, \pi_E) \to (E, X, \pi_E)$$

that makes the R-module E_x a $\mathcal{O}_{X,x}$-module for each $x \in X$. A *morphism* $(E, X, \pi_E) \to (E', X, \pi_{E'})$ between two objects of $\acute{\mathbf{E}}\mathbf{Mod}_{\mathcal{O}_X}$ is a $\acute{\mathbf{E}}(X, \mathbf{Mod}_R)$-morphism $\varphi \colon E \to E'$, for which the restrictions $\varphi|_{E_x} \colon E_x \to E'_x$ are morphisms of $\mathcal{O}_{X,x}$-modules. □

The considerations from Remark 5.36 show that the constructions of limits and colimits of sheaves from Example 5.18 yield $\mathcal{O}_X$-modules when starting with $\mathcal{O}_X$-modules. Since $\mathbf{Mod}_R$ in particular allows categorical sums and products (see Construction 2.16), there are therefore also categorical sums and products of $\mathcal{O}_X$-modules. Similarly, one also finds the existence of injective and projective limits of $\mathcal{O}_X$-modules (see Exercise 4.12).

But also functorial constructions of modules, which, like the tensor product, cannot be described as a limit or colimit, can be transferred to $\mathcal{O}_X$-modules. As for limits and colimits, there are two options. Either one applies the construction separately for each open set $U \in \mathfrak{T}$ to the section modules $\mathcal{F}(U)$ and makes a presheaf out of it, which is then sheafified, or one considers the étalé spaces associated with the $\mathcal{O}_X$-modules, applies the construction fiberwise, and shows that the result belongs to $\acute{\mathbf{E}}\mathbf{Mod}_{\mathcal{O}_X}$ by verifying that the fiberwise scalar multiplication defines a morphism of étalé spaces.

Example 5.37 (Tensor Products of $\mathcal{O}_X$-Modules) Let $(X, \mathcal{O}_X)$ be an R-ringed space and $\mathcal{F}$ and $\mathcal{G}$ two $\mathcal{O}_X$-modules.

(i) The sheafification of the presheaf $U \mapsto \mathcal{F}(U) \otimes_{\mathcal{O}_X(U)} \mathcal{G}(U)$ is a $\mathcal{O}_X$-module $\mathcal{F} \otimes_{\mathcal{O}_X} \mathcal{G}$ with respect to the $\mathcal{O}_X(U)$-module structures on $\bigl(\mathcal{F} \otimes_{\mathcal{O}_X} \mathcal{G}\bigr)(U)$, which are induced by the morphisms

$$\mathcal{O}_X(U) \times \bigl(\mathcal{F}(U) \otimes_{\mathcal{O}_X(U)} \mathcal{G}(U)\bigr) \to \mathcal{F}(U) \otimes_{\mathcal{O}_X(U)} \mathcal{G}(U), \quad (s, a \otimes b) \mapsto (sa) \otimes b$$

in sheafification as morphisms

$$\mathcal{O}_X(U) \times \bigl(\mathcal{F} \otimes_{\mathcal{O}_X} \mathcal{G}\bigr)(U) \to \bigl(\mathcal{F} \otimes_{\mathcal{O}_X} \mathcal{G}\bigr)(U).$$

(ii) The $\mathbf{S}(X, \mathbf{Set})$-morphism $\pi: \mathcal{F} \times \mathcal{G} \to \mathcal{F} \otimes_{\mathcal{O}_X} \mathcal{G}$ given by the mappings

$$\pi_U: \mathcal{F}(U) \times \mathcal{G}(U) = (\mathcal{F} \times \mathcal{G})(U) \to (\mathcal{F} \otimes_{\mathcal{O}_X} \mathcal{G})(U), \quad (a, b) \mapsto a \otimes b$$

for $U \in \mathfrak{T}$ is $\mathcal{O}_X$-bilinear in the sense that all π_U are $\mathcal{O}_X(U)$-bilinear.

(iii) The $\mathcal{O}_X$-module $\mathcal{F} \otimes_{\mathcal{O}_X} \mathcal{G}$, together with $\pi: \mathcal{F} \times \mathcal{G} \to \mathcal{F} \otimes_{\mathcal{O}_X} \mathcal{G}$, satisfies the following universal property: For every $\mathcal{O}_X$-bilinear $\mathbf{S}(X, \mathbf{Set})$-morphism $\varrho: \mathcal{F} \times \mathcal{G} \to \mathcal{H}$, there exists exactly one $\mathbf{Mod}_{\mathcal{O}_X}$-morphism $\overline{\varrho}: \mathcal{F} \otimes_{\mathcal{O}_X} \mathcal{G} \to \mathcal{H}$ with $\overline{\varrho} \circ \pi = \varrho$, i.e., we have the following commutative diagram

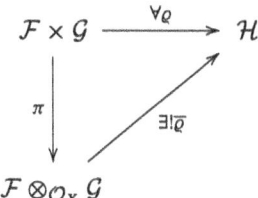

A comparison with Definition 3.3 shows why $\mathcal{F} \otimes_{\mathcal{O}_X} \mathcal{G}$ is referred to as the tensor product of the $\mathcal{O}_X$-modules $\mathcal{F}$ and $\mathcal{G}$.

(iv) The construction can be iterated, thus providing an associative tensor product (see Proposition 3.10). □

The existence of direct sums and tensor products of $\mathcal{O}_X$-modules allows us to consider tensor algebras in $\mathbf{Mod}_{\mathcal{O}_X}$, and further suggests the construction of symmetric and exterior algebras.

Exercise 5.10 ($\mathcal{O}_X$-Tensor Algebras) Define the tensor algebra of an $\mathcal{O}_X$-module and prove a characterizing universal property for it.

Exercise 5.11 (Symmetric and Exterior $\mathcal{O}_X$-Algebras) Transfer the constructions and results from Section 3.3 to $\mathcal{O}_X$-algebras.

The following example provides in particular the concept of a *dual $\mathcal{O}_X$-module*.

Example 5.38 (Hom-Spaces of $\mathcal{O}_X$-Modules) Let $(X, \mathcal{O}_X)$ be an R-ringed space and $\mathcal{F}$ and $\mathcal{G}$ two $\mathcal{O}_X$-modules.

(i) The presheaf $U \mapsto \mathrm{Hom}_{\mathcal{O}_X(U)}(\mathcal{F}(U), \mathcal{G}(U))$ is a sheaf, which we denote by $\mathcal{H}om_{\mathcal{O}_X}(\mathcal{F}, \mathcal{G})$. Each $(\mathcal{H}om_{\mathcal{O}_X}(\mathcal{F}, \mathcal{G}))(U)$ carries a $\mathcal{O}_X(U)$-module structure:

$$\mathcal{O}_X(U) \times \mathrm{Hom}_{\mathcal{O}_X(U)}(\mathcal{F}(U), \mathcal{G}(U)) \to \mathrm{Hom}_{\mathcal{O}_X(U)}(\mathcal{F}(U), \mathcal{G}(U),$$

$$(s, \varphi) \mapsto (a \mapsto s\varphi(a)).$$

This makes $\mathcal{H}om_{\mathcal{O}_X}(\mathcal{F}, \mathcal{G})$ an $\mathcal{O}_X$-module.

(ii) When $\mathcal{G}$ in (i) is equal to $\mathcal{O}_X$, then the $\mathcal{O}_X$-module $\mathcal{H}om_{\mathcal{O}_X}(\mathcal{F}, \mathcal{G}) = \mathcal{H}om_{\mathcal{O}_X}(\mathcal{F}, \mathcal{O}_X)$ is denoted by $\mathcal{F}^\vee$ and is called the *dual $\mathcal{O}_X$-module* to $\mathcal{F}$.

(iii) If I is an index set and $\mathcal{O}_X^{(I)} := \bigoplus_{i \in I} \mathcal{O}_X$ is the direct sum and $\mathcal{F}^I = \prod_{i \in I} \mathcal{F}$ is the direct product, then

$$\mathcal{H}om_{\mathcal{O}_X}(\mathcal{O}_X^{(I)}, \mathcal{F}) \cong \mathcal{H}om_{\mathcal{O}_X}(\mathcal{O}_X, \mathcal{F}^I).$$

The isomorphism is given for $U \in \mathfrak{T}$ by

$$\mathrm{Hom}_{\mathcal{O}_X(U)}\left(\mathcal{O}_X^{(I)}(U), \mathcal{F}(U)\right) \to \mathrm{Hom}_{\mathcal{O}_X(U)}\left(\mathcal{O}_X(U), \mathcal{F}^I(U)\right)$$

$$\varphi_U \mapsto \left(s \mapsto (\varphi(s_i))_{i \in I}\right)$$

where $s_i \in \mathcal{O}_X^{(I)}(U)$ is the element with s at the i-th position and all others are zeros. □

$\mathcal{O}_X$-modules, which are at least locally of the form $\mathcal{O}_X^{(I)}$, play a special role in geometry. For finite index sets, they will reappear in later chapters as *vector bundles*.

Definition 5.39 (Locally Free Sheaves) Let $(X, \mathcal{O}_X)$ be a ringed space and $\mathcal{F}$ an $\mathcal{O}_X$-module.

(i) $\mathcal{F}$ is called *free*, if there is an index set I for which the direct sum parametrized by I, $\mathcal{O}_X^{(I)} := \bigoplus_I \mathcal{O}_X$, is isomorphic to $\mathcal{F}$ as an $\mathcal{O}_X$-module.
(ii) $\mathcal{F}$ is called *locally free*, if for every $x \in X$ there is a $x \in U \in \mathfrak{T}$ for which $\mathcal{F}|_U$ is free. If for each $x \in X$ the U can be chosen such that $\mathcal{F}|_U$ is isomorphic to a finite direct sum of $\mathcal{O}_X|_U$, then $\mathcal{F}$ is called *locally free of finite type*.

The following example suggests the connection between locally free $\mathcal{O}_X$-modules and vector bundles.

Example 5.40 (Locally Free $\mathcal{O}_X$-Modules of Functions) Let $X \subseteq \mathbb{K}^n$ be an open subset and V a finite-dimensional $\mathbb{K}$-vector space. Then the sheaf $\mathcal{C}_{X,V}^{\mathbb{K},k}$ from Example 5.10 is a locally free $\mathcal{C}_X^{\mathbb{K},k}$-module of finite type. We denote it by $\mathcal{C}_X^{\mathbb{K},k} \otimes V$. The following module sheaves are obtained from the already discussed constructions for $\mathcal{O}_X$-modules. They are all locally free $\mathcal{C}_X^{\mathbb{K},k}$-modules of finite type (details as exercise).

(i) If V and W are finite-dimensional $\mathbb{K}$-vector spaces, then

$$\left(\mathcal{C}_X^{\mathbb{K},k} \otimes V\right) \otimes_{\mathcal{O}_X} \left(\mathcal{C}_X^{\mathbb{K},k} \otimes W\right) \cong \mathcal{C}_X^{\mathbb{K},k} \otimes (V \otimes W).$$

(ii) If $V = \varprojlim_{i \in I} V_i$ is a finite-dimensional $\mathbb{K}$-vector space, then

$$\varprojlim_{i \in I}(\mathcal{C}_X^{\mathbb{K},k} \otimes V_i) \cong \mathcal{C}_X^{\mathbb{K},k} \otimes \left(\varprojlim_{i \in I} V_i\right).$$

In particular, for finite-dimensional $\mathbb{K}$-vector spaces V and W

$$\left(\mathcal{C}_X^{\mathbb{K},k} \otimes V\right) \times \left(\mathcal{C}_X^{\mathbb{K},k} \otimes W\right) \cong \mathcal{C}_X^{\mathbb{K},k} \otimes (V \times W).$$

(iii) If $V = \varinjlim_{i \in I} V_i$ is a finite-dimensional $\mathbb{K}$-vector space, then

$$\varinjlim_{i \in I}(\mathcal{C}_X^{\mathbb{K},k} \otimes V_i) \cong \mathcal{C}_X^{\mathbb{K},k} \otimes \left(\varinjlim_{i \in I} V_i\right).$$

In particular, for finite-dimensional $\mathbb{K}$-vector spaces V and W

$$\left(\mathcal{C}_X^{\mathbb{K},k} \otimes V\right) \oplus \left(\mathcal{C}_X^{\mathbb{K},k} \otimes W\right) \cong \mathcal{C}_X^{\mathbb{K},k} \otimes (V \oplus W).$$

$\square$

Literature: Students usually learn about sheaves either in the theory of several complex variables, the theory of Riemann surfaces, or in algebraic geometry. This means that sheaves are rather rare in undergraduate studies. In particular, few courses use the sheaf approach to introduce manifolds, as is done e.g., in [Ra04] and [We15]. Texts on complex analysis in several variables always contain chapters on sheaves (see e.g., [Ta02] or [We79]), as do texts on modern algebraic geometry like [Ha77] and [GW10]. There are also various texts like [Br97, Go73, KS06] and [Te75], in which sheaves are the focus. In these texts, sheaf cohomology is usually also covered.

Manifolds 6

Let $\mathbb{K}$ be equal to $\mathbb{R}$ or $\mathbb{C}$. A $\mathbb{K}$-manifold is a topological space that "locally looks like" $\mathbb{K}^n$. How to understand "locally" has already been discussed in Ch. 5: The topological space should carry the structure of a ringed space, whose restriction to sufficiently small neighborhoods is isomorphic to a suitable ringed space on an open subset of $\mathbb{K}^n$. What is considered suitable here depends on which lens one wants to use to assess "looks the same". The weakest conceivable lens is when one only wants to consider continuous functions on the open pieces of $\mathbb{K}^n$. Then one considers the ringed spaces $\mathcal{C}_U$ on the open subsets U of $\mathbb{K}^n$. This is then referred to as a topological manifold. The stronger the lens, the more orders of differentiability one can recognize and accordingly considers $\mathcal{C}_U^{\mathbb{K},k}$ with $k \in \mathbb{N} \cup \{\infty, \omega\}$. In these cases, one speaks of *differentiable manifolds*.

For technical reasons, two a-priori assumptions are made about the underlying topological space $(X, \mathfrak{T})$: It should be a Hausdorff space, and the topology $\mathfrak{T}$ should have a *countable basis*, i.e., there is a countable family $\mathcal{B} := \{U_i \in \mathfrak{T} \mid i \in \mathbb{N}\}$, for which every $U \in \mathfrak{T}$ can be written as a union of elements from $\mathcal{B}$. There are two reasons for such a-priori assumptions. They enable the use of results from other areas in the theory of manifolds or in the application to questions from other areas. But the a-priori assumptions must also be fulfilled for the key examples that the theory is supposed to cover. In the case of the mentioned a-priori assumptions for manifolds, both justifications apply.

Definition 6.1 (Manifolds and Their Morphisms) Let $(X, \mathfrak{T})$ be a Hausdorff space, whose topology has a countable basis.

(i) A ringed space $(X, \mathcal{O}_X)$ is called a *manifold of type* $\mathcal{C}^{\mathbb{K},k}$ or $\mathcal{C}^{\mathbb{K},k}$-*manifold*, if every point $x \in X$ has a neighborhood $U \in \mathfrak{T}$ for which the restriction $(U, \mathcal{O}_X|_U)$ of $(X, \mathcal{O}_X)$ to U is isomorphic to $(V, \mathcal{C}_V^{\mathbb{K},k})$ for an open subset $V \subseteq \mathbb{K}^n$. We then also denote the structure sheaf $\mathcal{O}_X$ of X with $\mathcal{C}_X^{\mathbb{K},k}$.

(ii) A *morphism* between two manifolds of the same type is a morphism between the corresponding ringed spaces.
(iii) The category that the manifolds of type $\mathcal{C}^{\mathbb{K},k}$ form together with their morphisms, we denote with **Man**$_{\mathbb{K},k}$.

According to Proposition 5.32, all manifolds of type $\mathcal{C}^{\mathbb{K},k}$ are locally $\mathbb{K}$-ringed spaces, and the morphisms between these locally $\mathbb{K}$-ringed spaces are automatically local. So one could also have included the "locally" in the definition.

Example 6.2 (Open Subsets of $\mathbb{K}^n$) Let $X \subseteq \mathbb{K}^n$ be an open subset. Then $(X, \mathcal{C}_X^{\mathbb{K},k})$ for $k \in \mathbb{N}_0 \cup \{\infty, \omega\}$ is a manifold of type $\mathcal{C}^{\mathbb{K},k}$. A family of such examples are the groups $GL(m, \mathbb{K})$ which are open subsets of the space of $m \times m$ matrices with entries in $\mathbb{K}$ (which is identified with $\mathbb{K}^{m^2}$). □

Example 6.3 (Open Subsets of Manifolds) Let $(X, \mathcal{O}_X)$ be a manifold of type $\mathcal{C}^{\mathbb{K},k}$ and $U \subseteq X$ an open subset. Then $(U, \mathcal{O}_X|_U)$ is a manifold of type $\mathcal{C}^{\mathbb{K},k}$, and $(\iota, \iota^\flat): (U, \mathcal{O}_X|_U) \to (X, \mathcal{O}_X)$ with the inclusion $\iota: U \to X$ and the restriction $\iota^\flat = \rho_{U,X}$ is a morphism of $\mathcal{C}^{\mathbb{K},k}$-manifolds. □

According to Proposition 5.32, for a morphism $(f, f^\flat): (X, \mathcal{O}_X) \to (Y, \mathcal{O}_Y)$ of $\mathcal{C}^{\mathbb{K},k}$-manifolds, the sheaf morphism $f^\flat: \mathcal{O}_Y \to f_*\mathcal{O}_X$ is completely determined by f. Therefore, it is usually omitted in the notation and $f: X \to Y$ is referred to as a *mapping of class* $\mathcal{C}^{\mathbb{K},k}$.

Example 6.4 (Restriction of Morphisms) Let $(X, \mathcal{O}_X)$ and $(Y, \mathcal{O}_Y)$ be manifolds of type $\mathcal{C}^{\mathbb{K},k}$ and $f: X \to Y$ a mapping of class $\mathcal{C}^{\mathbb{K},k}$. If $U \subseteq X$ and $V \subseteq Y$ are open subsets such that $f(U) \subseteq V$, then $f|_U: U \to V$ is also a mapping of class $\mathcal{C}^{\mathbb{K},k}$. □

6.1 Charts and Parametrizations

Charts and parametrizations of manifolds are nothing more than the underlying mappings of the local isomorphisms of ringed spaces, whose existence is required in Definition 6.1 of manifolds.

Definition 6.5 (Charts and Parametrizations) Let $(X, \mathcal{O}_X)$ be a manifold of type $\mathcal{C}^{\mathbb{K},k}$. If now $U \subseteq X$ and $V \subseteq \mathbb{K}^n$ are open and there is an isomorphism $\Phi := (\varphi, \varphi^\flat): (U, \mathcal{O}_X|_U) \to (V, \mathcal{C}_V^{\mathbb{K},k})$, then $\varphi: U \to V$ is called a *chart* and $\varphi^{-1}: V \to U$ a *parametrization* of the manifold. In this case, we also call the open set U a *coordinate* or *chart neighborhood*.

If two charts $\varphi_1: U_1 \to V_1$ and $\varphi_2: U_2 \to V_2$ of a manifold $(X, \mathcal{O}_X)$ have an overlapping domain, then $\varphi_2 \circ \varphi_1^{-1}$ on its domain $\varphi_1(U_1 \cap U_2) = V_1 \cap \varphi_1 \circ \varphi_2^{-1}(V_2) \subseteq$

6.1 Charts and Parametrizations

$\mathbb{K}^n$ is in the corresponding function class according to Proposition 5.32. That is, each component function of $\varphi_2 \circ \varphi_1^{-1}$ is a local section of $\mathcal{C}_{\mathbb{K}^n}^{\mathbb{K},k}$.

Remark 6.6 (Dimension) For manifolds of type $\mathcal{C}^{\mathbb{K},k}$ with $k \neq 0$, the n in Definition 6.1 is uniquely determined for each domain U of a map, because a locally invertible differentiable mapping provides isomorphisms of the $\mathbb{K}^n$ via the derivative. We then call n the *dimension* of U. If the domains of two maps intersect, their dimensions must match. If X is *connected*, meaning it cannot be written as a disjoint union of open subsets, then the dimensions of all U must match. Otherwise, one could write X as a disjoint union of the sets X_n, which are obtained as unions of all domains of maps of dimension n. For connected X, the uniquely determined dimension of its map domains is called the dimension of X and is denoted by $\dim_{\mathbb{K}}(X)$.

For topological manifolds, i.e., manifolds of type $\mathcal{C}^{\mathbb{K},0}$, the same conclusions and definitions apply. However, they are based on a relatively difficult topological theorem (see [MT97, Cor. 7.14]), which states that a homeomorphism between two open sets $V \subseteq \mathbb{K}^n$ and $V' \subseteq \mathbb{K}^{n'}$ can only exist if $n = n'$.

If $\{U_i \in \mathfrak{T} \mid i \in I\}$ is a cover of X and there is an isomorphism $\Phi_i := (\varphi_i, \varphi_i^b) \colon (U_i, \mathcal{O}_X|_{U_i}) \to (V_i, \mathcal{C}_{\mathbb{K}^n}^k|_{V_i})$ for each $i \in I$, then the family $(U_i, \varphi_i)_{i \in I}$ of maps satisfies the following condition:

$$\forall i, j \in I \text{ is } \varphi_{ij} := \varphi_i \circ \varphi_j^{-1} \colon \varphi_j(U_i \cap U_j) \to \varphi_i(U_i \cap U_j) \text{ a } \mathcal{C}_{\mathbb{K}^n}^k\text{-morphism.}$$

Conversely, such data also provide a manifold structure on a topological space.

Example 6.7 (Gluing of Charts) Let $(X, \mathfrak{T})$ be a topological space, $\{U_i \in \mathfrak{T} \mid i \in I\}$ a cover of X and $(\varphi_i \colon U_i \to V_i)_{i \in I}$ a family of homeomorphisms onto open subsets $V_i \subseteq \mathbb{K}^n$. Consider the following condition (see Figure 6.1):

$$\forall i, j \in I \text{ is } \varphi_{ij} := \varphi_i \circ \varphi_j^{-1} \colon \varphi_j(U_i \cap U_j) \to \varphi_i(U_i \cap U_j) \text{ a } \mathcal{C}^{\mathbb{K},k}\text{-mapping.} \quad (*)$$

If condition $(*)$ holds, the $(\varphi_{ij})_{i,j \in I}$ form a set of gluing data (see Example 5.13) for the sheaf family $(\mathcal{F}_i)_{i \in I}$ with $\mathcal{F}_i := \varphi_i^{-1}(\mathcal{C}_{V_i}^{\mathbb{K},k}) = (\varphi_i^{-1})_*(\mathcal{C}_{V_i}^{\mathbb{K},k})$ over U_i. Thus, Example 5.13 provides a sheaf $\mathcal{F}$ over X, whose restriction to U_i is just $\mathcal{F}_i$. Since the $(V_i, \mathcal{C}_{V_i}^{\mathbb{K},k})$ are $\mathbb{K}$-ringed spaces, the same applies to the isomorphic $(U_i, \mathcal{F}_i)$. On the other hand, since being a $\mathbb{K}$-ringed space is a local property, $(X, \mathcal{F})$ is a $\mathbb{K}$-ringed space i.e., locally isomorphic to $(\mathbb{K}^n, \mathcal{C}_{\mathbb{K}^n}^{\mathbb{K},k})$. This shows that $(X, \mathcal{F})$ is a $\mathcal{C}^{\mathbb{K},k}$-manifold. □

Example 6.7 and our preliminary considerations show that Definition 6.1 is equivalent to the traditional definition of manifolds via charts and atlases. Here, an *atlas* is a family of charts whose domains of definition cover the entire space. In this language, a maximal atlas (which contains all charts for which the transition

Fig. 6.1 Change of charts

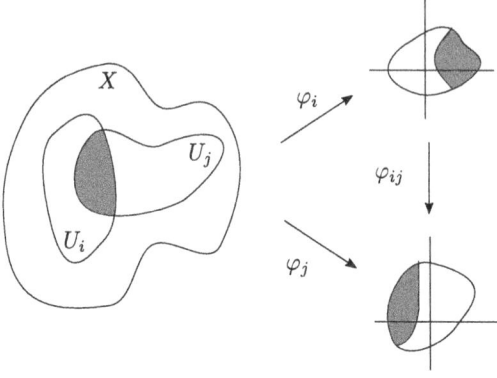

Fig. 6.2 Differentiability of mappings

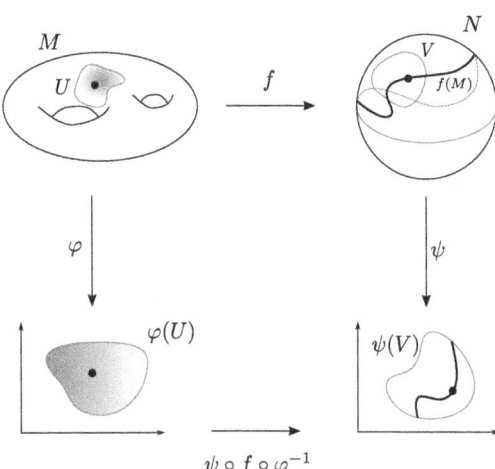

functions are $C_{\mathbb{K}^m}^{\mathbb{K},k}$ morphisms) is called a *differentiable structure* on the space. The use of the sheaf concept saves the handling of the unwieldy maximal atlases.

Remark 6.8 (Morphisms in Charts) Let M and N be manifolds of type $C^{\mathbb{K},k}$ and $f\, M \to N$ a mapping. f is exactly a **Man**$_{\mathbb{K},k}$-morphism when for every point $p \in X$ there are charts (U, φ) of M and (V, ψ) of N for which $p \in U$ and $f(p) \in V$ and also: $\psi \circ f \circ \varphi^{-1} : \varphi(f^{-1}(V) \cap U) \to \psi(V)$ is a $C^{\mathbb{K},k}$-mapping (see Figure 6.2). If $k \neq 0$, i.e., if the involved manifolds are differentiable, then we also speak of differentiable mappings. In the case $k = \infty$, we also say *smooth* instead of differentiable. For $k = \omega$, the mapping is called *analytic*. □

Fig. 6.3 Chart neighborhoods on the circle

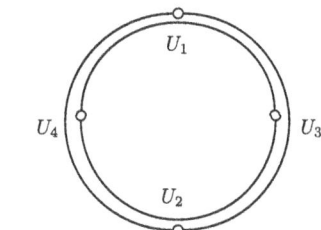

Fig. 6.4 Stereographic projections

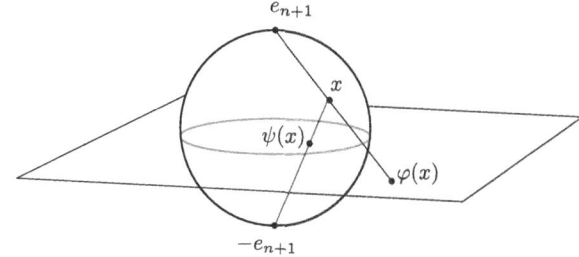

Example 6.9 (Spheres)

(i) Let $X := \mathbb{S}^1 := \{(x_1, x_2) \in \mathbb{R}^2 \mid x_1^2 + x_2^2 = 1\}$ be the 1-sphere. For the mappings
$\varphi_1 : U_1 = \{(x_1, x_2) \in M \mid x_2 > 0\} \to\,]-1, 1[\,, \varphi_1(x_1, x_2) = x_1,$
$\varphi_2 : U_2 = \{(x_1, x_2) \in M \mid x_2 < 0\} \to\,]-1, 1[\,, \varphi_2(x_1, x_2) = x_1,$
$\varphi_3 : U_3 = \{(x_1, x_2) \in M \mid x_1 > 0\} \to\,]-1, 1[\,, \varphi_3(x_1, x_2) = x_2,$
$\varphi_4 : U_4 = \{(x_1, x_2) \in M \mid x_1 < 0\} \to\,]-1, 1[\,, \varphi_4(x_1, x_2) = x_2$
we have $\varphi_3 \circ \varphi_1^{-1} : \varphi_1(U_1 \cap U_3) \to \varphi_3(U_1 \cap U_3)$, $x_1 \mapsto \sqrt{1 - x_1^2}$ and an analogous formula for $\varphi_4 \circ \varphi_2^{-1}$ (see Figure 6.3).

(ii) Let $X := \mathbb{S}^n := \{x \in \mathbb{R}^{n+1} \mid |x| = 1\}$ be the n-sphere. For $e_{n+1} := (0, \ldots, 0, 1)$ we set $V = X \setminus \{e_{n+1}\}$ and $U = X \setminus \{-e_{n+1}\}$ and

$$\psi : V \to \mathbb{R}^n, \quad \psi(x) = \frac{1}{1 + x_{n+1}}(x_1, \ldots, x_n),$$

$$\varphi : U \to \mathbb{R}^n, \quad \varphi(x) = \frac{1}{1 - x_{n+1}}(x_1, \ldots, x_n)$$

(stereographic projections, see Figure 6.4). Then

$$t\,\varphi(x) + (1 - t)e_{n+1} = \left(\frac{t x_1}{1 - x_{n+1}}, \ldots, \frac{t x_n}{1 - x_{n+1}}, 1 - t\right) \overset{t=1-x_{n+1}}{=} x.$$

For $(t_1, \ldots, t_n) := \varphi(x_1, \ldots, x_{n+1})$ we find

$$\sum_{k=1}^n x_k^2 = 1 - x_{n+1}^2 = (1 - x_{n+1})(1 + x_{n+1}),$$

$$\sum_{k=1}^n t_k^2 = \frac{\sum x_k^2}{(1 - x_{n+1})^2} = \frac{(1 + x_{n+1})}{(1 - x_{n+1})},$$

so $x_{n+1} = (\alpha - 1)(\alpha + 1)^{-1}$ with $\alpha = \sum t_k^2$ and $x_k = t_k(1 - x_{n+1})$. Thus we obtain:

$$\psi \circ \varphi^{-1}(t_1, \ldots, t_n) = \psi\left(t_1 \cdot 1 - x_{n+1}, \ldots, t_n \cdot 1 - x_{n+1}, \frac{\sum t_k^2 - 1}{\sum t_k^2 + 1}\right)$$

$$= \frac{1}{1 + x_{n+1}} (t_1 \cdot 1 - x_{n+1}, \ldots, t_n \cdot 1 - x_{n+1})$$

$$= \frac{1}{\sum t_k^2} (t_1, \ldots, t_n).$$

□

Note that in Example 6.9 two different families of maps for $\mathbb{S}^1$ have been given (through linear projections or stereographic projections). It is easy to verify that the union of these families of maps still fulfills condition (∗) in Example 6.7. Thus, both variants define the same sheaf on $\mathbb{S}^1$, i.e., the same differentiable structure.

Exercise 6.1 (Tori) Let X be the n-dimensional torus $\mathbb{R}^n/\mathbb{Z}^n$ with the quotient topology (see Figure 6.5): $\pi : \mathbb{R}^n \to X$, $\pi(x) = x + \mathbb{Z}^n$.
Provide a family of maps that make X a real C^∞-manifold.

Exercise 6.2 (Products of Manifolds) Let $(X, \mathcal{O}_X)$ and $(Y, \mathcal{O}_Y)$ be manifolds of type $C^{\mathbb{K},k}$ and $X \times Y$ be equipped with the product topology. If (U, φ) and (V, ψ) are maps for X and Y respectively, then consider

$$\varphi \times \psi : U \times V \to \mathbb{K}^n \times \mathbb{K}^m = \mathbb{K}^{n+m}$$

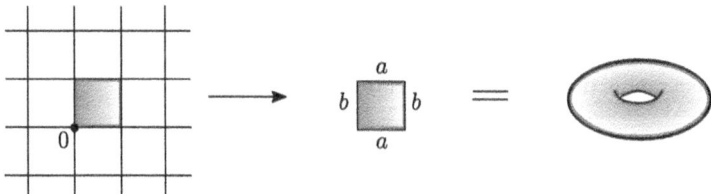

Fig. 6.5 Two-dimensional torus

6.1 Charts and Parametrizations

Fig. 6.6 Regular hyperplanes

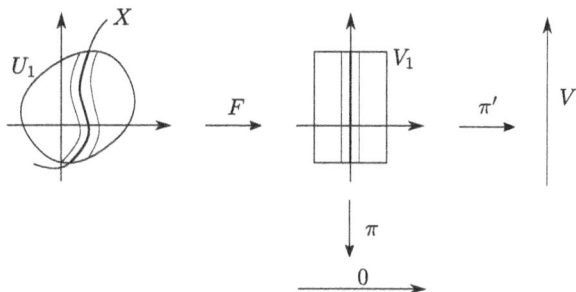

as a homeomorphism with an open image. Show that the entirety of all such maps can be glued according to Example 6.7. The manifold thus constructed, $(X \times Y, \mathcal{O}_{X \times Y})$, is called the *product* of $(X, \mathcal{O}_X)$ and $(Y, \mathcal{O}_Y)$. Further show that it is the categorical product in $\mathbf{Man}_{\mathbb{K},k}$.

Example 6.10 ($\mathbb{K}$-Hyperplanes) Let $D \subseteq \mathbb{K}^{n+1}$ be open and $f : D \to \mathbb{K}$ in $\mathcal{C}^{\mathbb{K},k}_{\mathbb{K}^{n+1}}(D)$ with $k \neq 0$. Furthermore, let $\mathrm{grad}(f(p)) \neq 0$ for all $p \in D$ with $f(p) = 0$, where $\mathrm{grad} f(p) := \left(\frac{\partial f}{\partial x_1}(p), \ldots, \frac{\partial f}{\partial x_{n+1}}(p)\right)$. Then $X := \{x \in D \mid f(x) = 0\}$ is an n-dimensional manifold of type $\mathcal{C}^{\mathbb{K},k}$: Let $p \in X$ be fixed. By renumbering, we can achieve that $\frac{\partial f}{\partial x_1}(p) \neq 0$. We consider the function $F : \mathbb{K}^{n+1} \to \mathbb{K}^{n+1}$, which is given by

$$x = (x_1, x_2, \ldots, x_{n+1}) \mapsto (f(x), x_2, \ldots, x_{n+1}).$$

The Jacobian matrix of F at the point p is:

$$\begin{pmatrix} \frac{\partial f}{\partial x_1}(p) & \frac{\partial f}{\partial x_2}(p) & \frac{\partial f}{\partial x_3}(p) & \cdots & \frac{\partial f}{\partial x_{n+1}}(p) \\ 0 & 1 & 0 & \cdots & 0 \\ \vdots & 0 & \ddots & \ddots & \vdots \\ \vdots & \vdots & \ddots & \ddots & 0 \\ 0 & 0 & \cdots & 0 & 1 \end{pmatrix}$$

So, according to the inverse function theorem, F is locally invertible at p. This means, there is a neighborhood U_1 of p in $\mathbb{K}^{n+1}$ and a neighborhood V_1 of 0 in $\mathbb{K} \times \mathbb{K}^n$, such that $F : U_1 \to V_1$ is a $\mathcal{C}^{\mathbb{K},k}$-morphism. Note that $f = \pi \circ F$ holds, where $\pi : \mathbb{K} \times \mathbb{K}^n$ is the projection onto the first component. By reducing the neighborhoods, we can assume that V_1 is of the form $]-\varepsilon, \varepsilon[\times V$. We set $U = X \cap U_1$ and achieve that $V = \pi'(V_1)$, where $\pi' : \mathbb{K} \times \mathbb{K}^n \to \mathbb{K}^n$ is the canonical projection onto the second component (see Figure 6.6). Let $\varphi = \pi' \circ F|_U$, then $\varphi(U) = V$. With $\varphi(x) = \varphi(y)$ and $\pi \circ F(y) = f(y) = 0 = f(x) = \pi \circ F(x)$ it also follows that $F(x) = F(y)$ and thus $x = y$. So φ is bijective and continuous. Since the projection π' is an open mapping, φ is even a homeomorphism. If now (U', φ') is another chart of X that was constructed in the same way as (U, φ), then the mapping $\varphi' \circ \varphi^{-1}$ as a restriction of a mapping of the type $F' \circ F^{-1}$ is a $C^{\mathbb{K},k}$-mapping. □

Example 6.11 (Grassmann Manifolds)

(a) The *projective space* $\mathbb{P}_{\mathbb{K}}^n$ (the set of lines in $\mathbb{K}^{n+1}$ through the origin). For $x \in \mathbb{K}^{n+1} \setminus \{0\}$ set $[x] := \{\mathbb{K}x\}$. Then $[x] = [y]$, if $x = \lambda y$ with $\lambda \in \mathbb{K} \setminus \{0\}$. We define $\mathbb{P}_{\mathbb{K}}^n := (\mathbb{K}^{n+1} \setminus \{0\})/\mathbb{K}^\times$ with the quotient topology. Note that $\mathbb{P}_{\mathbb{R}}^n = S^n/\sim$ with $x \sim y$ if $x = y$ or $x = -y$. The projective space is a special case of the next example; therefore, we refrain from giving explicit charts.

(b) The *Grassmann manifolds* $\mathbb{G}_{k,n}$ (the set of k-dimensional subspaces of $\mathbb{K}^n$). Let

$$V := \{(v_1, \ldots, v_k) \mid v_i \in \mathbb{K}^n \text{ linearly independent}\}.$$

For $x \in V$, let $[x]$ be the subspace of $\mathbb{K}^n$ generated by the v_i in x. One can identify V with the set $S_{k,n}$ of all $n \times k$ matrices of rank k (by fixing a basis of $\mathbb{K}^n$) and see that the group $GL(k, \mathbb{K})$ operates on V (by multiplication from the right). It holds $[x] = [y]$, if there is an $A \in GL(k, \mathbb{K})$ with $x = yA$. The set $S_{k,n}$ is open in the set of all $n \times k$ matrices and therefore a manifold (*Stiefel manifold*). Consider the mapping $\pi : S_{k,n} \to \mathbb{G}_{k,n}$ defined by $\pi(x) = [x]$. We equip $\mathbb{G}_{k,n}$ with the quotient topology with respect to π. To specify a set of coordinate neighborhoods, we consider arbitrary k-element subsets I of $\{1, \ldots, n\}$. For $x \in S_{k,n}$, let x_I be the $k \times k$ submatrix of x given by I. We set $U_I := \{[x] \in \mathbb{G}_{k,n} \mid \det(x_I) \neq 0\}$. (Note that this definition of U_I depends only on $[x]$, not on the choice of representative). Because $\text{rank}(x) = k$ for all $x \in S_{k,n}$, it immediately follows that $\cup_I U_I = \mathbb{G}_{k,n}$. We define $\varphi_I : U_I \to \mathbb{K}^{k(n-k)}$ as follows:

$$\varphi_I([x]) := x_{I'} \cdot x_I^{-1},$$

where I' is the complement of I in $\{1, \ldots, n\}$. The mapping φ_I is well-defined and injective, since $x = yA$ implies $x_I = y_I A$ and $x_{I'} = y_{I'} A$. The mapping φ_I is surjective, because for every $(n-k) \times k$ matrix z, the matrix $\binom{1}{z} =: x$ is mapped to z under φ_I. It remains to show that the φ_I are continuous and open (exercise). Also, it must be shown that $\varphi_J \circ \varphi_I^{-1} : \varphi_I(U_I \cap U_J) \to \varphi_J(U_I \cap U_J)$ is a diffeomorphism (exercise).

□

In andvanced calculus courses, the following definition of submanifolds of $\mathbb{K}^n$ is often presented.

Definition 6.12 ($\mathbb{K}$-Submanifolds of V) Let V be a finite-dimensional $\mathbb{K}$-vector space. A subset $M \subseteq V$ is called a d-dimensional $\mathbb{K}$-submanifold of type $C^{\mathbb{K},k}$, if for every $x \in M$ there is an open neighborhood U of x in V and a bijective

6.1 Charts and Parametrizations

$C^{\mathbb{K},k}$-mapping $\varphi: U \to \varphi(U) \subseteq V$ with open image and $\mathbb{K}$-differentiable inverse function, such that

$$\varphi(U \cap M) = \varphi(U) \cap E,$$

where E is a d-dimensional $\mathbb{K}$-linear subspace of V.

Exercise 6.3 ($\mathbb{K}$-**Submanifolds of** V) Let V be a finite-dimensional $\mathbb{K}$-vector space and $M \subseteq V$ a d-dimensional differentiable submanifold of type $C^{\mathbb{K},k}$ in the sense of Definition 6.12. Show that M is a $C^{\mathbb{K},k}$-manifold.

Hint: For each $x \in M$, there is an open neighborhood $V_x \subseteq V$ of x and an injective differentiable mapping $\Phi_x: V_x \to V$ with open image $\Phi_x(V_x)$, such that $\Phi_x^{-1}: \Phi_x(V_x) \to V_x$ is also differentiable. Furthermore, there is a d-dimensional linear subspace E_x of V with

$$\Phi_x(V_x \cap M) = \Phi_x(V_x) \cap E_x.$$

For each $x \in M$, fix a linear isomorphism $\psi_x: E_x \to \mathbb{K}^d$ and set $Y_x := V_x \cap M$. Then work with the homeomorphisms

$$\varphi_x := \psi_x \circ \Phi_x|_{Y_x}: Y_x \to U_x := \psi_x\big(\Phi_x(V_x) \cap E_x\big) \subseteq \mathbb{K}^d.$$

Finally, we present an interesting family of examples, for which we first need to provide some topological background.

Remark 6.13 (Connected Components of Topological Spaces) Let $(X, \mathfrak{T})$ be a topological space. It is called *connected*, if M cannot be written as the union of two disjoint non-empty open sets. Any subset $A \subseteq X$ is called *connected*, if it is connected as a topological space with the relative topology. Since the open subsets of A are exactly the intersections of A with open subsets of X, this means that A cannot be covered by two open subsets U_1, U_2 of X with

$$U_1 \cap A \neq \emptyset, \quad U_2 \cap A \neq \emptyset, \quad U_1 \cap U_2 \cap A = \emptyset$$

If $(X, \mathfrak{T})$ is connected, then the only subsets of X that are both open and closed are the empty set and X itself. In fact, if $Y \subseteq X$ is open and closed, then $X \setminus Y$ is also open and closed, and due to

$$Y \cup (X \setminus Y) = X, \quad Y \cap (X \setminus Y) = \emptyset$$

one of the two sets must be empty, i.e., the other is all of X.

(i) Let $(Y_i)_{i \in I}$ be a family of connected subsets of a topological space $(X, \mathfrak{T})$ with $\bigcap_{i \in I} Y_i \neq \emptyset$. Then $\bigcup_{i \in I} Y_i$ is connected. To see this, set $E := \bigcup_{i \in I} Y_i$ and choose $x \in \bigcap_{i \in I} Y_i$. Let $E \subseteq F_1 \cup F_2$ with disjoint open sets F_1 and F_2. We can assume $x \in F_1$. Since Y_i is connected and $x \in F_1$, $Y_i \subseteq F_1 \cup F_2$ implies that $F_2 \cap Y_i = \emptyset$, i.e., $Y_i \subseteq F_1$. Since i was arbitrary, $E \cap F_2 = \emptyset$. Thus, E is connected.

(ii) If $A \subseteq X$ is connected, then the closure $\overline{A}$ of A in X is also connected.

Let U_1, U_2 be two open sets in M with $\overline{A} \cap U_1 \cap U_2 = \emptyset$ and $\overline{A} \subseteq U_1 \cup U_2$. Then due to the connectedness of A, either $A \cap U_1 = \emptyset$ or $A \cap U_2 = \emptyset$ holds. Without loss of generality, we assume the case $A \cap U_2 = \emptyset$. Thus, A is contained in the closed set $X \setminus U_2$. By definition of the closure, $\overline{A} \subseteq X \setminus U_2$, i.e., $\overline{A} \cap U_2 = \emptyset$. This proves the claim.

(iii)

$$x \sim y \quad :\Leftrightarrow \quad (\exists Y \subseteq X \text{ connected}) \; x, y \in Y$$

defines an equivalence relation on X, whose equivalence classes are connected and closed.

The reflexivity and symmetry of $\sim$ are clear. If $x \sim y$ and $y \sim z$, then there are connected subsets X_1 and X_2 of X with $x, y \in X_1$ and $y, z \in X_2$. In particular, $y \in X_1 \cap X_2$, so (i) implies that $X_1 \cup X_2$ is connected. This shows $x \sim z$, hence the transitivity of $\sim$.

Let A now be an equivalence class of $\sim$. To show that A is connected, we consider two open sets U_1, U_2 in X with $A \subseteq U_1 \cup U_2$ and $A \cap U_1 \cap U_2 = \emptyset$. If $x_1 \in A \cap U_1$ and $x_2 \in A \cap U_2$, then because $x_1 \sim x_2$ there is a connected subset Y of X that contains x_1 and x_2. All elements of Y are equivalent to each other, so Y is contained in A. Thus, $Y \subseteq U_1 \cup U_2$ and $Y \cap U_1 \neq \emptyset \neq Y \cap U_2$. This contradiction shows the connectedness of A.

Finally, we note that the equivalence classes of $\sim$ are closed according to (ii).

The equivalence classes $[x]_\sim$ of $\sim$ from (iii) are called the *connected components* of X. In general, connected components are not open, as can be seen from the set $X = \{0\} \cup \{\frac{1}{n} \mid n \in \mathbb{N}\}$, when it carries the relative topology of $\mathbb{R}$.

Example 6.14 (Riemann Surfaces of Function Germs) Let $U \subseteq \mathbb{C}$ be open and $f : U \to \mathbb{C}$ a $\mathcal{C}^{\mathbb{C},k}$-function. Consider the sheaf $\mathcal{C}_\mathbb{C}^{\mathbb{C},k}$ and its Étalé space $E := \text{É}(\mathcal{C}_\mathbb{C}^{\mathbb{C},k})$ (see Construction 5.23). Then the germ $f_x \in \mathcal{C}_{\mathbb{C},x}^{\mathbb{C},k}$ of f at $x \in U$ is an element of E, and we can consider the connected component M of f_x in E. The subset $M \subseteq E$ is open, because for each $m \in M$ there is an open neighborhood of m in E i.e., homeomorphic to a disk in $\mathbb{C}$ via π. Therefore, we can cover M by open sets $U \subseteq M$ for which $\pi|_U : U \to \pi(U) \subseteq \mathbb{C}$ is a homeomorphism and $\pi(U) \subseteq \mathbb{C}$ is open. The corresponding change of charts is always the identity, so in particular of the class $\mathcal{C}^{\mathbb{C},k}$. If we can now also show that M fulfills the topological a-priori conditions that we have set in Definition 6.1, then M becomes a $\mathcal{C}^{\mathbb{C},k}$-manifold. M is called the *Riemann surface* of the function germ f_x.

The Hausdorff property of M follows because $\mathbb{C}$ is a Hausdorff space, and therefore two points that do not lie in the same π-fiber can be separated by open sets. Two different points in the same fiber are separated by open sets anyway due to the local homeomorphism of π.

The existence of a countable basis for the topology on M is a consequence of a non-trivial theorem that goes back to Poincaré and Volterra (see Proposition 6.15). □

Proposition 6.15 (Poincaré-Volterra) *Let $(X, \mathfrak{T})$ be a Hausdorff topological space with a countable base and $(M, \mathfrak{S})$ a connected topological space, that is, locally isomorphic to $\mathbb{R}^n$. If $f : M \to X$ is continuous with discrete fibers, then M also has a countable base.*

Proof Let $U_i, i \in \mathbb{N}$ be a countable base for $\mathfrak{T}$ and $\mathfrak{B} := \{V \in \mathfrak{S} \mid V$ is relatively compact and $(*)\}$, where:

$$\exists i \in \mathbb{N} : \quad V \text{ is a connected component of } f^{-1}(U_i). \tag{$*$}$$

Then $\mathfrak{B}$ is a countable base for $\mathfrak{S}$.

We first show that $\mathfrak{B}$ is a base for $\mathfrak{S}$. For $a \in V \in \mathfrak{S}$, according to the conditions on M, there is a relatively compact $W \in \mathfrak{S}$ with $a \in W \subseteq \overline{W} \subseteq V$ and $\overline{W} \cap f^{-1}(f(a)) = \{a\}$, where $\overline{W}$ is the closure of W in M. Then the boundary ∂W of W, and thus also $f(\partial W)$, is compact. Moreover, $f(a) \notin f(\partial W)$. Choose an $i \in \mathbb{N}$ with $f(a) \in U_i \subseteq X \setminus f(\partial W)$ and denote the connected component of a in $f^{-1}(U_i)$ by V'. It holds that $V' \subseteq W$, because otherwise $V' \cap \partial W \neq \emptyset$, which due to $U_i \cap f(\partial W) \supseteq f(V') \cap f(\partial W)$ contradicts $U_i \cap f(\partial W) = \emptyset$. It follows $a \in V' \in \mathfrak{B}$, and together it turns out that $\mathfrak{B}$ is indeed a base for $\mathfrak{S}$.

Let $V_0 \in \mathfrak{B}$ and $\mathfrak{B}_0 := \{V_0\}$. Inductively we set $\mathfrak{B}_k := \{V \in \mathfrak{B} \mid \exists V' \in \mathfrak{B}_{k-1} : V \cap V' \neq \emptyset\}$ and consider

$$\Omega := \bigcup \{V \in \bigcup_{k \in \mathbb{N}_0} \mathfrak{B}_k\} \in \mathfrak{S},$$

$$\Omega' := \bigcup \{V \in \mathfrak{B} \setminus \bigcup_{k \in \mathbb{N}_0} \mathfrak{B}_k\} \in \mathfrak{S}.$$

Then $M = \Omega \cup \Omega'$ and $\Omega \cap \Omega' = \emptyset$. Since M is connected, $\Omega' = \emptyset$ must hold, i.e., $\mathfrak{B} = \bigcup_{k \in \mathbb{N}_0} \mathfrak{B}_k$. It is therefore sufficient to show that each $\mathfrak{B}_k$ is countable. We do this by induction over k, where the induction start is trivial. So let $\mathfrak{B}_{k-1}$ be countable for $k \in \mathbb{N}$ and $\Omega_{k-1} := \bigcup \{V \in \mathfrak{B}_{k-1}\}$.

Since each $V \in \mathfrak{B}$ is relatively compact, it can be written as a finite union of open sets that are homeomorphic to open subsets of $\mathbb{R}^n$. Since the topology of $\mathbb{R}^n$ has a countable base, families of disjoint non-empty open subsets of V are at most countable. The same applies if V is a countable union of elements of $\mathfrak{B}$, i.e., in particular for Ω_{k-1}. For $i \in \mathbb{N}$, let $\mathfrak{F}_i$ be the set of connected components of $f^{-1}(U_i)$

that intersect Ω_{k-1}. Since M is locally homeomorphic to $\mathbb{R}^n$ and $f^{-1}(U_i)$ is open, each of these connected components is open. So each $\mathfrak{F}_i$ is countable. From the definitions we see $\mathfrak{B}_k \subseteq \bigcup_{i \in \mathbb{N}} \mathfrak{F}_i$, so $\mathfrak{B}_k$ is also countable. □

6.2 Tangent Spaces and Derivatives

For two differentiable manifolds M and N, we have defined when a mapping $f : M \to N$ should be called differentiable (smooth, analytic), but have said nothing about a derivative. In the case that M and N are vector spaces, the derivative at a point $p \in X$ is a linear mapping $f'(p) : M \to N$. This derivative is a linear approximation of f at p. In general, however, we do not have a linear structure on M and N.

Tangent Spaces

There are different ways to provide a remedy, all of which lead to the same result under sufficiently strong conditions. A geometric-physical approach is to be guided by the idea that a manifold, similar to Example 6.10, is situated in a linear space and tangent vectors should be something like velocities of movements on the manifold. This idea leads to a definition of tangent vectors as equivalence classes of curves. A computational-physical approach is provided by the observation that the images of maps are contained as open subsets in a linear space. Tangent vectors are then defined as families of vectors in the ranges of charts, which transform into each other when changing charts by applying the Jacobian matrix. This corresponds to the physicist's definition "A vector is what transforms like a vector". Since the equivalence of curves is also formulated using charts and derivatives, it is easy to see that these two approaches are essentially equivalent. The situation is different with the algebraic approach, which is based on the idea that a direction can be characterized by the associated directional derivative. To show that the tangent space obtained in this way is naturally isomorphic to the other two variants, we will have to assume that our manifold is at least smooth.

We begin with the geometric method of defining the tangent space $T_p M := T_p^{\text{geo}} M$ of a manifold M at $p \in M$. So, we want to try to assign a velocity vector

$$\gamma'(t) = \text{``} \lim_{h \to 0} \frac{1}{h} \left(\gamma(t+h) - \gamma(t) \right) \text{''}$$

to each differentiable curve $\gamma : I \to M$ parameterized by an open interval I that passes through the point p at $t \in I$ (see Figure 6.7). As long as M is a subset of a $\mathbb{R}^n$, this does not cause any problems. In general, however, for a point p in a coordinate neighborhood U, one will have to consider the curve $\varphi \circ \gamma : I \to \varphi(U)$. Since the derivative $(\varphi \circ \gamma)'(t)$ does not have a meaning independent of the map (U, φ), we

6.2 Tangent Spaces and Derivatives

Fig. 6.7 Geometric tangent space

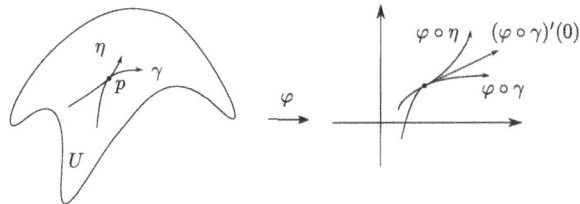

identify the differentiable curves γ through p that have the same derivative in all coordinates.

Definition 6.16 (Geometric Tangent Space) Let $(M, \mathcal{O}_M)$ be a manifold of type $C^{\mathbb{K},k}$ and $p \in M$. Let G_p be the set of $C^{\mathbb{K},k}$ mappings $\gamma \colon I \to M$, which are defined on an open neighborhood of $0 \in \mathbb{K}$ and satisfy $\gamma(0) = p$. On G_p we define an equivalence relation $\sim_p$ by

$$\gamma \sim_p \eta \quad :\Leftrightarrow \quad \forall (U, \varphi) \text{ map} : (\varphi \circ \gamma)'(0) = (\varphi \circ \eta)'(0).$$

We denote the equivalence class of γ with $[\gamma]_p$. With this, we can define the *geometric tangent space* $T_p^{\text{geo}} M$ of $p \in M$ as follows:

$$T_p^{\text{geo}} M := \{[\gamma]_p \mid \gamma \in G_p\}.$$

One can directly infer from this definition that the tangent space is a $\mathbb{K}$-vector space, but the other approaches provide this with less effort.

Exercise 6.4 (Geometric Tangent Space) Let M be an n-dimensional differentiable manifold of type $C^{\mathbb{K},k}$ and γ and η two $C^{\mathbb{K},k}$-mappings defined on the same open set $I \subseteq \mathbb{K}$. Let (U, φ) be a chart of M with $p \in U$ and $(\varphi \circ \gamma)(I) + (\varphi \circ \eta)(I) \subseteq \varphi(U)$. Consider the mapping

$$\sigma := \varphi^{-1}\big((\varphi \circ \gamma) + (\varphi \circ \eta)\big) : I \to M$$

and prove the following statements:

(i) The class $[\sigma]_p$ of σ depends only on the classes $[\gamma]_p$ and $[\eta]_p$.
(ii) One can define an addition on $T_p^{\text{geo}} M$ via $[\gamma]_p + [\eta]_p := [\sigma]_p$.
(iii) Similarly, a scalar multiplication can be defined on $T_p^{\text{geo}} M$, which, together with the addition from (ii), makes $T_p^{\text{geo}} M$ a $\mathbb{K}$-vector space.
(iv) For $j \in \{1, \ldots, n\}$ let $i_j \colon \mathbb{K} \to \mathbb{K}^n, t \mapsto \varphi(p) + (0, \ldots, 0, t, 0, \ldots, 0)$, where the t is in the j-th component, and $\gamma_j := \varphi^{-1} \circ i_j|_{i_j^{-1}(\varphi(U))} : i_j^{-1}(\varphi(U)) \to M$. Then the $[\gamma_j]_p$ form a basis for $T_p^{\text{geo}} M$.

We call the basis $[\gamma_1]_p, \ldots, [\gamma_n]_p$ the φ-*basis* for $T_p^{\text{geo}} M$.

Next, we introduce the computational physicist variant of the tangent space.

Definition 6.17 (Physicist's Tangent Space) Let $(M, \mathcal{O}_M)$ be an n-dimensional manifold of type $C^{\mathbb{K},k}$ and $p \in M$. Let Φ_p be the family of all charts $\varphi \colon U_\varphi \to V_\varphi \subseteq \mathbb{K}^n$ of M, whose domain U_φ contains the point $p \in M$. A family $(v_\varphi)_{\varphi \in \Phi_p}$ with $v_\varphi \in \mathbb{K}^n$ is called a *tangent vector* at M in p, if:

$$\forall \varphi, \psi \in \Phi_p : \quad v_\psi = \left((\psi \circ \varphi^{-1})|_{\varphi(U_\varphi \cap U_\psi)}\right)'(\varphi(p)) v_\varphi.$$

The set of all tangent vectors at M in p forms a $\mathbb{K}$-vector space $T_p^{\mathrm{phy}} M$, which we call the *physicist's tangent space*. The basis $v_\varphi^{(1)}, \ldots, v_\varphi^{(n)}$ for $T_p^{\mathrm{phy}} M$, which is created by setting $v_\varphi^{(j)} = (0, \ldots, 0, 1, 0, \ldots, 0)$ with the 1 at the j-th position, is called the φ-*basis* for $T_p^{\mathrm{phy}} M$.

Exercise 6.5 (Equivalence of Tangent Spaces) Let $(M, \mathcal{O}_M)$ be an n-dimensional manifold of type $C^{\mathbb{K},k}$ and $p \in M$. Consider the mapping

$$T_p^{\mathrm{geo}} M \to T_p^{\mathrm{phy}} M, \quad [\gamma] \mapsto \left((\varphi \circ \gamma)'(0)\right)_{\varphi \in \Phi_p}$$

and show that it is an isomorphism of $\mathbb{K}$-vector spaces that transforms the φ-bases into each other.

Our algebraic approach to the tangent space will consist in studying the effect of velocity vectors $(\varphi \circ \gamma)'(t)$ on smooth functions, rather than considering the vectors themselves: Let γ be a differentiable curve with $\gamma(t) = p \in M$ and $f \colon M \to \mathbb{R}$ a differentiable function. Furthermore, let (U, φ) again be a chart with $p \in U$. Then the value $(f \circ \gamma)'(t)$ depends only on the equivalence class $[(t, \gamma)]_p$ of (t, γ), where

$$(t, \gamma) \sim_p (s, \eta) \quad :\Leftrightarrow \quad \begin{array}{l} \gamma(t) = p = \eta(s), \\ \forall (\varphi, U_\varphi) : (\varphi \circ \gamma)'(t) = (\varphi \circ \eta)'(s). \end{array}$$

In fact, we have

$$(f \circ \gamma)'(t) = (f \circ \varphi^{-1})' \left(\varphi(\gamma(t))\right) \cdot (\varphi \circ \gamma)'(t).$$

$$\begin{array}{ccccc}
I & \xrightarrow{\gamma} & M & \xrightarrow{f} & \mathbb{R} \\
& \searrow_{\varphi \circ \gamma} & \downarrow \varphi & \nearrow_{f \circ \varphi^{-1}} & \\
& & \varphi(U) & &
\end{array}$$

6.2 Tangent Spaces and Derivatives

The product rule implies

$$\big((f \cdot g) \circ \gamma\big)'(t) = (f \circ \gamma)'(t) \cdot g(p) + (g \circ \gamma)'(t) \cdot f(p).$$

Finally, note that the value of $(f \circ \gamma)'(t)$ does not differ from $(g \circ \gamma)'(t)$ if f and g are equal in a neighborhood of p, i.e., they define the same germ at p.

Definition 6.18 (Algebraic Tangent Space) Let M be a differentiable manifold of type $\mathcal{C}^{\mathbb{K},k}$ and $p \in M$. Furthermore, let $\mathcal{C}^{\mathbb{K},k}_{M,p}$ be the $\mathbb{K}$-algebra of germs of $\mathcal{C}^{\mathbb{K},k}_M$ at p. A linear functional $\mathfrak{v} : \mathcal{C}^{\mathbb{K},k}_{M,p} \to \mathbb{R}$ is called a *derivation*, if:

$$\mathfrak{v}(f_p \cdot g_p) = \mathfrak{v}(f_p) \cdot g(p) + \mathfrak{v}(g_p) \cdot f(p).$$

The *algebraic tangent space* $T^{\mathrm{alg}}_p M$ of M at p is the set of all derivations $\mathfrak{v} : \mathcal{C}^{\mathbb{K},k}_{M,p} \to \mathbb{K}$. We equip $T^{\mathrm{alg}}_p M$ with pointwise addition and pointwise scalar multiplication, thus making the algebraic tangent space a $\mathbb{K}$-vector space.

The preliminary considerations for Definition 6.18 provide a mapping $\iota : T^{\mathrm{geo}}_p M \to T^{\mathrm{alg}}_p M$ for each $p \in M$. The elementary rules of differentiation show that this mapping ι is $\mathbb{K}$-linear. It turns out that ι is injective.

Proposition 6.19 (Tangent Spaces) *Let M be a differentiable manifold of type $\mathcal{C}^{\mathbb{K},k}$. Then for each $p \in M$ the $\mathbb{K}$-linear mapping $\iota : T^{\mathrm{geo}}_p M \to T^{\mathrm{alg}}_p M$ is injective.*

Proof To learn more about the mapping ι, we consider a chart (U, φ) for which $p \in U$ holds, and define n derivations $\frac{\partial}{\partial x_j}|_p$, $j = 1, \ldots, n$, in $T^{\mathrm{alg}}_p X$ by

$$\frac{\partial}{\partial x_j}|_p(f_p) := \frac{\partial (f \circ \varphi^{-1})}{\partial x_j}(\varphi(p)),$$

where points in $\mathbb{K}^n$ are denoted by $x = (x_1, \ldots, x_n)$. Note that in the notation of Exercise 6.4, we have: $\frac{\partial}{\partial x_j}|_p = \iota([\gamma]_p)$. If we apply the $\frac{\partial}{\partial x_j}|_p$ to the germs of the coordinate functions $x_i := \pi_i \circ \varphi$, where $\pi_i : \mathbb{K}^n \to \mathbb{K}$ is the projection onto the i-th component, we see that the $\frac{\partial}{\partial x_j}|_p$ are linearly independent. This proves the assertion. □

Proposition 6.20 (Derivations) *Let M be a differentiable manifold of type $\mathcal{C}^{\mathbb{K},k}$ and $p \in M$. Set $A_k := \{a \in \mathcal{C}^{\mathbb{K},k}_{M,p} \mid a(p) = 0\}$ and consider a linear mapping $\mathfrak{v} : \mathcal{C}^{\mathbb{K},k}_{M,p} \to \mathbb{K}$. Then the following two statements are equivalent:*

(1) $\mathfrak{v}$ *is a derivation.*

(2) $\mathfrak{v}$ *vanishes on* $A_k^2 + \mathbb{K} \cdot 1_p$, *where* A_k^2 *is the set of* $\mathbb{K}$*-linear combinations of products of the form* ab *with* $a, b \in A_k$ *and* 1_p *is the germ of the constant function* 1.

Proof Assume $\mathfrak{v}$ is a derivation. Then for $c \in \mathbb{K}$

$$\mathfrak{v}(c \cdot 1_p) = c \cdot \mathfrak{v}(1_p) = c \cdot \mathfrak{v}(1_p \cdot 1_p) = c\big(\mathfrak{v}(1_p) \cdot 1 + \mathfrak{v}(1_p) \cdot 1\big) = 2c \cdot \mathfrak{v}(1_p),$$

which proves $\mathfrak{v}(c \cdot 1_p) = 0$. If $a, b \in A_k$, then $\mathfrak{v}(ab) = a(p)\mathfrak{v}(b) + b(p)\mathfrak{v}(a) = 0$, so $\mathfrak{v}$ satisfies condition (2).

Conversely, assume that $\mathfrak{v}$ satisfies condition (2). Then we calculate for $a, b \in \mathcal{C}_{X,p}^{\mathbb{K},k}$

$$\mathfrak{v}(ab) = \mathfrak{v}\big((a - a(p))(b - b(p))\big) + \mathfrak{v}(a(p)b) + \mathfrak{v}(b(p)a) + \mathfrak{v}\big(a(p)b(p)\big)$$
$$= \mathfrak{v}(a(p)b) + \mathfrak{v}(b(p)a) = a(p)\mathfrak{v}(b) + b(p)\mathfrak{v}(a),$$

i.e., $\mathfrak{v}$ is a derivation. $\square$

For each $k \in \mathbb{N}$, one can find functions $f \in \mathcal{C}^{\mathbb{R},k}(\mathbb{R}^n)$ whose k-th derivative is not differentiable at 0. Let $f_0 \in \mathcal{C}_{\mathbb{R}^n,0}^{\mathbb{R},k}$ be the germ of such a function. We choose a linear mapping $\mathfrak{v}: \mathcal{C}_{\mathbb{R}^n,0}^{\mathbb{R},k} \to \mathbb{R}$ such that $\mathfrak{v}(f_0) = 1$ and $\mathfrak{v}(a) = 0$ if $a \in \mathcal{C}_{\mathbb{R}^n,0}^{\mathbb{R},k}$ is differentiable at 0. Using the formula

$$(gh)^{(k)} = \sum_{\ell=0}^{k} \binom{k}{\ell} g^{(\ell)} h^{(k-\ell)}$$

for the k-th derivative of the product of two functions, one sees that for $a, b \in \mathcal{C}_{\mathbb{R}^n,0}^{\mathbb{R},k}$ with $a(0) = b(0) = 0$ the k-th derivative $(ab)^{(k)}$ of the product $ab \in \mathcal{C}_{\mathbb{R}^n,0}^{\mathbb{R},k}$ is differentiable at 0. Thus, $\mathfrak{v}(ab) = 0$, and Proposition 6.20 shows that $\mathfrak{v}$ is a derivation that vanishes on $\mathcal{C}_{\mathbb{R}^n,0}^{\mathbb{R},k+1}$. It can therefore not be a linear combination of directional derivatives at 0. If one transfers this consideration using a chart into a differentiable manifold M of type $\mathcal{C}^{\mathbb{R},k}$, it follows that the mapping $\iota: T_p^{\text{geo}} M \to T_p^{\text{alg}} M$ cannot be surjective when $k \in \mathbb{N}$. Since there are even infinitely many linearly independent function germs $f \in \mathcal{C}_{\mathbb{R},0}^{\mathbb{R},k}$ whose k-th derivative at 0 is not differentiable (exercise), one can even determine that for $k \in \mathbb{N}$ the space $T_p^{\text{alg}} M$ is infinite-dimensional for every $p \in M$. The above argument does not work for $\mathbb{K} = \mathbb{C}$ because it can be shown that every complex differentiable function is analytic. This means that the types $\mathcal{C}^{\mathbb{C},k}$ all agree, and we lose nothing by always assuming $k = \omega$ in the complex case.

6.2 Tangent Spaces and Derivatives

For manifolds of type $\mathcal{C}^{\mathbb{K},k}$ with $k \in \{\infty, \omega\}$, we can show that $\iota: T_p^{\text{geo}} M \to T_p^{\text{alg}} M$ is surjective. We do this by proving that for every chart (U, φ) with $p \in U$, the $\frac{\partial}{\partial x_j}|_p$ form a basis for $T_p^{\text{alg}} M$, which we then call the φ-basis for $T_p^{\text{alg}} X$.

Proposition 6.21 (Basis for the Algebraic Tangent Space) *Let M be a differentiable manifold of type $\mathcal{C}^{\mathbb{K},k}$ with $k \in \{\infty, \omega\}$ and (U, φ) a chart on M. Let $x_j : U \to \mathbb{K}$ be the functions defined by $x_j(q) = \pi_j(\varphi(q))$ for all $q \in U$, where the $\pi_j : \mathbb{K}^n \to \mathbb{K}$ are the projections onto the j-th component. Then for every $\mathfrak{v} \in T_p^{\text{alg}} M$ the following formula holds*

$$\mathfrak{v} = \sum_{j=1}^n \left(\mathfrak{v}\big((x_j)_p\big)\right) \cdot \frac{\partial}{\partial x_j}\Big|_p.$$

In particular, the tangent vectors $\frac{\partial}{\partial x_j}|_p$, $j = 1, \ldots, n$, form a basis for $T_p^{\text{alg}} M$.

To prove this proposition, we need a lemma.

Lemma 6.22 *Let V be an open ball in $\mathbb{K}^n$, with center a, and let $F \in \mathcal{C}^{\mathbb{K},k}(V)$ with $k \in \{\infty, \omega\}$. Then there are functions $G_1, \ldots, G_n \in \mathcal{C}^{\mathbb{K},k}(V)$ that fulfill the following properties:*

(i) $F(x) = F(a) + \sum_{j=1}^n (x_j - a_j) G_j(x).$

(ii) $G_j(a) = \frac{\partial F}{\partial x_j}(a).$

Proof For fixed $x \in V$ consider the function $\xi : [-1, 1] \to \mathbb{K}$, defined by $\xi(t) = F(a + t(x - a))$:

$$F(x) = \xi(1) = \xi(0) + \int_0^1 \frac{d}{dt}\big(\xi(t)\big) dt$$

$$= \xi(0) + \int_0^1 \sum_{j=1}^n \left(\frac{\partial F}{\partial x_j}(a + t(x - a))\right)(x_j - a_j) dt$$

$$= F(a) + \sum_{j=1}^n (x_j - a_j)\left(\int_0^1 \frac{\partial F}{\partial x_j}(a + t(x - a)) dt\right)$$

$$= F(a) + \sum_{j=1}^n (x_j - a_j) G_j(x).$$

But since $F \in C^{\mathbb{K},k}(V)$ was assumed, also $G_j \in C^{\mathbb{K},k}(V)$. The second statement is clear with the above formula for the G_j. □

The example $F: \mathbb{R} \to \mathbb{R}, x \mapsto x|x|$ shows that the statement of Lemma 6.22 no longer holds if $k \in \mathbb{N}$. The formulas remain correct, but the G_j do not have to have the same degree of differentiability as F.

Proof of Proposition 6.21 We fix $\mathfrak{v} \in T_p^{\text{alg}} M$ and $f \in C^{\mathbb{K},k}_{M,p}$, and set $F = f \circ \varphi^{-1} : V \to \mathbb{K}$. By reducing the neighborhoods, we can assume that V is a ball with center $\varphi(p)$. We apply Lemma 6.22 and find functions $G_j : V \to \mathbb{K}$ that fulfill Lemma 6.22(i). With this, we calculate using Proposition 6.20

$$\mathfrak{v}(f_p) = \mathfrak{v}((F \circ \varphi)_p) = \mathfrak{v}(f(p) \cdot 1_p) + \mathfrak{v}\left(\sum_{j=1}^n (x_j - x_j(p) \cdot 1)_p \cdot (G_j \circ \varphi)_p\right)$$

$$= 0 + \sum_{j=1}^n \left(\mathfrak{v}((x_j)_p) \cdot G_j(\varphi(p)) + \mathfrak{v}((G_j \circ \varphi)_p) \cdot (x_j(p) - x_j(p) \cdot 1)\right)$$

$$= \sum_{j=1}^n \left(\mathfrak{v}((x_j)_p)\right) \cdot \frac{\partial (f \circ \varphi^{-1})}{\partial x_j}(\varphi(p)) = \sum_{j=1}^n \left(\mathfrak{v}((x_j)_p)\right) \cdot \frac{\partial}{\partial x_j}\bigg|_p (f_p).$$

Therefore, every derivation $\mathfrak{v} \in T_p M$ is a linear combination of $\frac{\partial}{\partial x_j}|_p$, $j = 1, \ldots, n$. It remains only to show that the $\frac{\partial}{\partial x_j}|_p$, $j = 1, \ldots, n$, are linearly independent. This is immediately verified by applying them to the functions x_k, $k = 1, \ldots, n$. □

We can now show that for manifolds of type $C^{\mathbb{K},k}$ with $k \in \{\infty, \omega\}$ all three tangent spaces are isomorphic to each other.

Proposition 6.23 (Equivalence of Tangent Spaces) *Let M be a manifold of type $C^{\mathbb{K},k}$ with $k \in \{\infty, \omega\}$. Then the mapping $\iota: T_p^{\text{geo}} M \to T_p^{\text{alg}} M$, $[\gamma]_p \mapsto \mathfrak{v}_\gamma$ with $\mathfrak{v}_\gamma(f_p) := (f \circ \gamma)'(t)$ is an isomorphism of $\mathbb{K}$-vector spaces, which transforms the φ-bases into each other.*

Proof Proposition 6.19 states that the mapping is well-defined and injective. We have also already seen that Proposition 6.21 provides surjectivity. To also show the last statement, we calculate for an (arbitrary) element

$$\mathfrak{v} = \sum_{j=1}^n v_j \cdot \frac{\partial}{\partial x_j}\bigg|_p$$

6.2 Tangent Spaces and Derivatives

of $T_p^{\text{alg}} M$ and the curve $\gamma : I \to M$, which for sufficiently small t is defined by

$$\gamma(t) = \varphi^{-1}\left(\varphi(p) + t \cdot \sum_{j=1}^{n} v_j \cdot (0, \ldots, 1, \ldots, 0)\right)$$

where the 1 is at the j-th position, so that

$$\mathfrak{v}_\gamma(f_p) = (f \circ \gamma)'(0) = (f \circ \varphi^{-1})'(\varphi(p)) \circ \left((v_1, \ldots, v_n)\right)$$

$$= \sum_{j=1}^{n} v_j \cdot \frac{\partial(f \circ \varphi^{-1})}{\partial x_j}(\varphi(p)) = \mathfrak{v}(f_p)$$

holds. In particular, the φ-basis for $T_p^{\text{geo}} M$ is thus mapped onto the φ-basis for $T_p^{\text{alg}} M$. $\square$

Exercise 6.6 (Equivalence of Tangent Spaces) Let M be an n-dimensional manifold of type $C^{\mathbb{K},k}$ with $k \in \{\infty, \omega\}$ and $p \in M$. Show that the mapping

$$T_p^{\text{phy}} M \to T_p^{\text{alg}} M, \quad (v_\varphi)_{\varphi \in \Phi_p} \mapsto \sum_{j=1}^{n} (v_\varphi)_j \frac{\partial}{\partial x_j}\bigg|_p$$

is an isomorphism of $\mathbb{K}$-vector spaces and that the resulting diagram

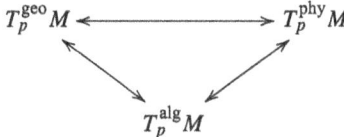

with the linear isomorphisms from Proposition 6.23 and Exercise 6.5 is commutative.

The construction of the tangent spaces shows that for an open subset U of a $C^{\mathbb{K},k}$-manifold M, the tangent spaces $T_p U$ and $T_p M$ can be identified for each $p \in U$.

Derivatives

With the preliminary work on tangent spaces, one can assign a linear mapping $f'(p) := df_p : T_p M \to T_{f(p)} N$ as a "derivative" to a differentiable mapping $f : M \to N$ between two manifolds (see Figure 6.8).

We do this for each of the three types of tangent spaces and show that all three variants are compatible with each other. The geometric and algebraic tangent spaces allow surprisingly simple definitions of the derivative. This is because the actual operation of taking derivatives has already been incorporated into the definitions (equivalence of curves via the derivative, respectively derivation).

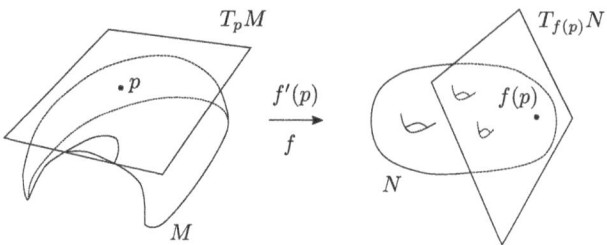

Fig. 6.8 Derivative

Definition 6.24 (Geometric Derivative) Let M and N be differentiable manifolds of type $C^{\mathbb{K},k}$ and $f : M \to N$ a $C^{\mathbb{K},k}$-mapping. Then the *differential* or the *derivative* $f'(p) := df_p : T_p^{\text{geo}} M \to T_{f(p)}^{\text{geo}} N$ of f at $p \in M$ is defined by

$$f'(p)([\gamma]_p) := [f \circ \gamma]_{f(p)}$$

Definition 6.25 (Physicist's Derivative) Let M and N be differentiable manifolds of type $C^{\mathbb{K},k}$ and $f : M \to N$ a $C^{\mathbb{K},k}$-mapping. Then the *differential* or the *derivative* $f'(p) := df_p : T_p^{\text{phy}} M \to T_{f(p)}^{\text{phy}} N$ of f at $p \in M$ is defined by

$$f'(p)\big((v_\varphi)_{\varphi \in \Phi_{M,p}}\big) := \big((\psi \circ f \circ \varphi)'(p) v_\varphi\big)_{\psi \in \Phi_{N,f(p)}}$$

Here, $\Phi_{M,p}$ is the family of charts of M defined in a neighborhood of p, and $\Phi_{N,f(p)}$ is the family of charts of N defined in a neighborhood of $f(p)$. It should be noted that

$$\forall \varphi_1, \varphi_2 \in \Phi_{M,p} : \quad (\psi \circ f \circ \varphi_1)'(p) v_{\varphi_1} = (\psi \circ f \circ \varphi_2)'(p) v_{\varphi_2},$$

because $v_{\varphi_2} = \big((\varphi_2 \circ (\varphi_1)^{-1})'(\varphi_1(p))\big)(v_{\varphi_1})$ holds.

Exercise 6.7 (Equivalence of Derivatives) Show that the geometric derivative and the physicist's derivative transform into each other when combined with the isomorphisms from Exercise 6.5.

Definition 6.26 (Algebraic Derivative) Let M and N be differentiable manifolds of type $C^{\mathbb{K},k}$ with $k \in \{\infty, \omega\}$ and $f : M \to N$ a $C^{\mathbb{K},k}$-mapping. Then the *differential* or the *derivative* $f'(p) := df_p : T_p M \to T_{f(p)} N$ of f at $p \in M$ is defined by

$$\forall \mathfrak{v} \in T_p M, h \in C^\infty\big(f(p)\big): \quad f'(p)(\mathfrak{v})\big(h_{f(p)}\big) := \mathfrak{v}\big((h \circ f)_p\big)$$

6.2 Tangent Spaces and Derivatives

with:

$$
\begin{array}{ccc}
M & \xrightarrow{f} & N \\
{\scriptstyle h \circ f}\downarrow & & \downarrow{\scriptstyle h} \\
\mathbb{R} & \xrightarrow{\text{id}} & \mathbb{R}
\end{array}
$$

Note that it is necessary in this definition to check whether it depends on the choice of the representative $h \in C^{\mathbb{K},k}(U)$ of $h_{f(p)} \in C^{\mathbb{K},k}_{f(p)}$. This is not the case, as one can easily see (exercise).

Exercise 6.8 (Equivalence of Derivatives) Given the conditions of Definition 6.26, let $\mathfrak{v} \in T_p^{\text{alg}} M$ be of the form $\mathfrak{v}_\gamma$ with a $C^{\mathbb{K},k}$-curve $\gamma : I \to M$. Show that $df_p(\mathfrak{v}_\gamma) = \mathfrak{v}_{f \circ \gamma} \in T_{f(p)}^{\text{alg}} N$ holds.

From the previous considerations on tangent spaces and derivatives, it follows that the geometric and the physicist's approach for every type $C^{\mathbb{K},k}$ of manifold with $k \neq 0$ can be directly converted into each other. In the future, we will no longer distinguish between $T_p^{\text{geo}} M$ and $T_p^{\text{phy}} M$ and interpret the different formulas as different descriptions of the same thing. We have to be more careful with the algebraic approach. We only use it when the manifold is of type $C^{\mathbb{K},k}$ with $k \in \{\infty, \omega\}$. In this case, we have also seen that the concept of derivative coincides with that of the other approaches, and therefore we also refrain from a separate notation here.

From now on, we will only work with one kind of tangent space, denoted by $T_p M$, which can be described in different forms. As soon as the description via derivations is used, the condition $k \in \{\infty, \omega\}$ is implicitly assumed.

Exercise 6.9 (Derivative as a Functor) We consider the category $\mathbf{Man}_{\mathbb{K},k}$ of differentiable manifolds of type $C^{\mathbb{K},k}$ and convert it into a category $\mathbf{Man}^*_{\mathbb{K},k}$ of *pointed* spaces by choosing as objects pairs (M, p) with $M \in \text{ob}(\mathbf{Man}_{\mathbb{K},k})$ and $p \in M$. A morphism $f: (M, p) \to (N, q)$ is then a $C^{\mathbb{K},k}$-mapping $f\, M \to N$ with $f(p) = q$. Show (Hint: chain rule) that the assignments

$$(M, p) \mapsto T_p M \quad \text{and} \quad f \mapsto f'(p)$$

define a functor $\mathbf{Man}^*_{\mathbb{K},k} \to \mathbf{Vect}_{\mathbb{K}}$.

For given coordinate systems (U, φ) on M and (V, ψ) on N, we have obtained distinguished bases of $T_p M$ and $T_{f(p)} N$ above. We want to describe the derivative of f with respect to these bases. To this end, we denote the function that assigns to an $x = (x_1, \ldots, x_n)$ the i-th component y_i of the vector $\psi \circ f \circ \varphi^{-1}(x) =: y$ by $F_i : \varphi(U \cap f^{-1}(V)) \to \mathbb{R}$. Let now $\mathfrak{v} \in T_p M$, more precisely

$$\mathfrak{v} = \sum_{j=1}^{m} v_j \frac{\partial}{\partial x_j}\Big|_p, \quad m = \dim(M).$$

Then, with $n = \dim(N)$ for $g \in C^{\mathbb{K},k}_{X,f(p)}$

$$df_p(\mathfrak{v})g_{f(p)} = \mathfrak{v}\big((g \circ f)_p\big) = \sum_{j=1}^{m} v_j \frac{\partial}{\partial x_j}\big|_p \big((g \circ f)_p\big)$$

$$= \sum_{j=1}^{m} v_j \frac{\partial(g \circ f \circ \varphi^{-1})}{\partial x_j}(\varphi(p))$$

$$= \sum_{j=1}^{m} v_j \sum_{i=1}^{n} \Big(\frac{\partial(g \circ \psi^{-1})}{\partial y_i}\big(\psi(f(p))\big) \frac{\partial y_i}{\partial x_j}(\varphi(p))\Big)$$

$$= \sum_{i=1}^{n} \Big(\sum_{j=1}^{m} \frac{\partial F_i}{\partial x_j}(\varphi(p))v_j\Big) \frac{\partial g \circ \psi^{-1}}{\partial y_i}\big(\psi(f(p))\big)$$

$$= \sum_{i=1}^{n} \Big(\sum_{j=1}^{m} \frac{\partial F_i}{\partial x_j}(\varphi(p))v_j\Big) \frac{\partial}{\partial y_i}\big|_{f(p)}(g_{f(p)}).$$

So if $Df(p)$ is the Jacobian matrix of the mapping $F = \psi \circ f \circ \varphi^{-1}$ at p, then $Df(p)$ is exactly the matrix of $f'(p)$ with respect to the bases of T_pM and $T_{f(p)}N$ distinguished by φ and ψ. In the case that $k \in \{\infty, \omega\}$, we have thus shown the following proposition.

Proposition 6.27 (Derivative in Coordinates) *Let M and N be differentiable manifolds of type $C^{\mathbb{K},k}$ and $f : M \to N$ a $C^{\mathbb{K},k}$-mapping. Further, let (U, φ) and (V, ψ) be charts on M and N respectively, and $p \in f^{-1}(V) \cap U$. Finally, we denote the mapping $\psi \circ f \circ \varphi^{-1} : \varphi(f^{-1}(V) \cap U) \to \psi(f(U) \cap V)$ by F. Then the linear mapping $df_p : T_pM \to T_{f(p)}N$ with respect to the φ-basis of T_pM and the ψ-basis of $T_{f(p)}N$ is given by the Jacobian matrix of F.*

Exercise 6.10 (Derivative in Coordinates) Prove Proposition 6.27 also for $k \in \mathbb{N}$, by calculating the matrix representation of the derivative in the φ-bases of the geometric or physicist realization of the tangent spaces.

The following example shows that the given definitions for hypersurfaces actually provide the embedding of the tangent spaces in the surrounding vector space, which was the basis of the geometric definition of the tangent space.

Example 6.28 (Hypersurfaces) Let $f : \mathbb{K}^{n+1} \to \mathbb{K}$ be a $C^{\mathbb{K},k}$-mapping with $\mathrm{grad}(f)(p) \neq 0$ for all $p \in M := f^{-1}(0)$. Then M is a differentiable manifold of dimension n of type $C^{\mathbb{K},k}$ according to Example 6.10. We consider the inclusion mapping $j : M \to \mathbb{K}^{n+1}$ and want to show that it is $\mathcal{C}^{\mathbb{K},k}$ and has an injective derivative $j'(p)$ at every point $p \in M$. Then we can identify each tangent space $T_p M$ with an n-dimensional subspace in $\mathbb{K}^{n+1}$, because all tangent spaces of $\mathbb{K}^{n+1}$ can be identified with $\mathbb{K}^{n+1}$.

To see that j is a $\mathcal{C}^{\mathbb{K},k}$-mapping, we consider a chart (U, φ), as constructed in Example 6.10 for M. Then the mapping $j \circ \varphi^{-1} : \varphi(U) \to \mathbb{K}^{n+1}$ is of the form $F^{-1} \circ \widetilde{j}|_{\varphi(U)} : \varphi(U) \to \mathbb{K}^{n+1}$, where $\widetilde{j} : \mathbb{K}^n \to \mathbb{K}^{n+1}$ is the inclusion defined by $\widetilde{j}(x_1, \ldots, x_n) = (0, x_1, \ldots, x_n)$ and $F : U_1 \to V_1$ is the diffeomorphism defined in Example 6.10. This mapping is of type $\mathcal{C}^{\mathbb{K},k}$. To demonstrate the injectivity of the derivative $j'(p)$, we still need to prove that the rank of the Jacobian matrix of $F^{-1} \circ \widetilde{j}$ at $\varphi(p)$ is equal to n. However, since the Jacobian matrix of F at p is regular and that of $\widetilde{j}$ has rank n, this follows immediately from the chain rule. □

6.3 Tangent and Tensor Bundles

So far, we have considered the tangent space $T_p M$ at a fixed point $p \in M$ of a differentiable manifold. Accordingly, for a given differentiable mapping $f : M \to N$ between two manifolds, we have constructed a linear mapping $f'(p) = df_p : T_p M \to T_{f(p)} N$ as the derivative of f at p for each point $p \in M$. This construction can be made for each point $p \in M$. It is therefore natural to consider all tangent spaces "simultaneously".

Definition 6.29 (Tangent Bundle) Let M be an n-dimensional differentiable manifold of type $\mathcal{C}^{\mathbb{K},k}$. Then the set

$$TM := \coprod_{p \in M} T_p M$$

is called the *tangent bundle* of M.

The term "bundle" comes from the fact that we can imagine TM as being formed from M by "attaching" the vector space $T_p M$ to the point p.

With the definition of the tangent bundle, it is now possible to define a derivative

$$f' := Tf : TM \to TN, \quad T_p M \ni \mathfrak{v} \mapsto (Tf)(\mathfrak{v}) := df_p(\mathfrak{v}) \in T_{f(p)} N$$

for $f : M \to N$. If we start with a $\mathcal{C}^{\mathbb{K},k}$ mapping f with $k > 1$, the question arises whether f' is differentiable in any way. For this, of course, one would first need a differentiable structure on TM and TN. We have already established that the dimension of $T_p M$ is the same for all $p \in M$ (namely $\dim(M) = n$). This means that the set TM exhibits a certain homogeneity. The above image suggests considering TM as an entity of *dimension* $2n$. It is indeed the case that one can define a differentiable structure on TM in a natural way, making TM a manifold of dimension $2n$:

Proposition 6.30 (Tangent Bundle as Manifold) *Let M be an n-dimensional differentiable manifold of type $C^{\mathbb{K},k}$ and $TM = \coprod_{p \in M} T_p M$. Let further $\pi_M : TM \to M$ be the mapping that assigns to each $\mathfrak{v} \in T_p M$ the point $p \in M$. Then there is a Hausdorff topology on TM with respect to which the mapping π is continuous and open. The topological space TM carries the structure of a 2n-dimensional $C^{\mathbb{K},k-1}$-manifold, with respect to which π is a $C^{\mathbb{K},k-1}$-mapping.*

Proof We provide a family of sets $\widetilde{U}_\alpha$ and bijections $\widetilde{\varphi}_\alpha : \widetilde{U}_\alpha \to \widetilde{V}_\alpha$ with $\widetilde{V}_\alpha$ open in $\mathbb{R}^{2n}$, which we want to make into homeomorphisms. If we can then show that the sets $\widetilde{U}_\alpha$ cover the set TM and the mappings

$$\widetilde{\varphi}_{\alpha\beta} = \widetilde{\varphi}_\beta \circ \widetilde{\varphi}_\alpha^{-1} : \widetilde{\varphi}_\alpha(\widetilde{U}_\alpha \cap \widetilde{U}_\beta) \to \widetilde{\varphi}_\beta(\widetilde{U}_\alpha \cap \widetilde{U}_\beta)$$

are of class $C^{\mathbb{K},k-1}$, we have shown two things: First, by declaring the $\widetilde{\varphi}_\alpha$ to be homeomorphisms, we define a topology on TM. Second, the family $(\widetilde{U}_\alpha, \widetilde{\varphi}_\alpha)$ fulfills condition (∗) from Example 6.7 and therefore makes TM a 2n-dimensional $C^{\mathbb{K},k-1}$-manifold.

Let $(U_\alpha, \varphi_\alpha)_{\alpha \in A}$ be a family of charts for M with $\bigcup_{\alpha \in A} U_\alpha = M$. We set $\widetilde{U}_\alpha := \pi^{-1}(U_\alpha)$ and define the mappings $\widetilde{\varphi}_\alpha : \widetilde{U}_\alpha \to \mathbb{K}^{2n}$ by

$$\widetilde{\varphi}_\alpha(\mathfrak{v}) = \left(x_1^\alpha(p), \ldots, x_n^\alpha(p), v_1^\alpha, \ldots, v_n^\alpha\right),$$

where $p \in U_\alpha$ and $\mathfrak{v} = \sum_{j=1}^n v_j^\alpha \cdot \frac{\partial}{\partial x_j^\alpha}\big|_p \in T_p M$. (Note that, as before, the functions x_j^α are defined via the projections of $\varphi(U_\alpha)$ onto the j-th component in $\mathbb{K}^n$). Then $\widetilde{V}_\alpha := \widetilde{\varphi}_\alpha(\widetilde{U}_\alpha) = V_\alpha \times \mathbb{K}^n$ is open in $\mathbb{K}^{2n}$, and

$$\widetilde{\varphi}_\alpha(\widetilde{U}_\alpha \cap \widetilde{U}_\beta) = \varphi_\alpha(U_\alpha \cap U_\beta) \times \mathbb{K}^n$$

(similarly for φ_β). It follows that the mapping $\widetilde{\varphi}_{\alpha\beta}$ is given by

$$(x_1, \ldots, x_n, v_1^\alpha, \ldots, v_n^\alpha) \mapsto \left(\varphi_{\alpha\beta}(x_1, \ldots, x_n), v_1^\beta, \ldots, v_n^\beta\right)$$

for $(x_1, \ldots, x_n) \in \varphi_\alpha(U_\alpha \cap U_\beta)$, where the v_j^β are the coefficients of $\mathfrak{v}$ with respect to the φ_β-basis for $T_p M$. That is, we need to check whether the coefficients of $\mathfrak{v} \in T_p M$ change in a $C^{\mathbb{K},k-1}$-way when the basis changes from the φ_α-basis to the φ_β-basis. To do this, we express the φ_α-basis vectors as a linear combination of the φ_β-basis vectors (see Exercise 6.10; we use the algebraic notation, but this can

6.3 Tangent and Tensor Bundles

also be done in the geometric or physicist's notation, so it is not necessary to assume that $k \in \{\infty, \omega\}$):

$$\frac{\partial}{\partial x_j^\alpha}|_p(f_p) = \frac{\partial(f \circ \varphi_\alpha^{-1})}{\partial x_j^\alpha}(\varphi_\alpha(p))$$

$$= \sum_{k=1}^n \frac{\partial(f \circ \varphi_\beta^{-1})}{\partial x_k^\beta}(\varphi_\beta(p)) \cdot (D\varphi_{\alpha\beta})_{kj}(\varphi_\alpha(p))$$

$$= \sum_{k=1}^n \left((D\varphi_{\alpha\beta})_{kj}(\varphi_\alpha(p))\right) \cdot \frac{\partial}{\partial x_k^\beta}|_p(f_p).$$

Here, $D\varphi_{\alpha\beta}$ denotes the Jacobian matrix of $\varphi_{\alpha\beta}$.

Since the $\varphi_{\alpha\beta}$ are of class $\mathcal{C}^{\mathbb{K},k}$ by assumption, it follows that the $\widetilde{\varphi}_{\alpha\beta}$ are of class $\mathcal{C}^{\mathbb{K},k-1}$ (we set $\infty - 1 := \infty$ and $\omega - 1 = \omega$). This calculation is a special case of the calculation for the derivative in local coordinates.

Our proof is not yet complete: We still need to show that the space TM is a Hausdorff space. Moreover, we have not yet shown the differentiability or openness of π. However, the Hausdorff property follows immediately from the definition of the topology and the Hausdorff property of M. Differentiability and openness of π can be checked locally, i.e., in a chart. There, π is just the projection onto a subspace and thus analytic and open. □

We equip the tangent bundle TM of a differentiable manifold M of type $\mathcal{C}^{\mathbb{K},k}$ with its manifold structure given by Proposition 6.30. The chart $(\widetilde{U}, \widetilde{\varphi})$ constructed there for a chart (U, φ) on M we call the *associated chart* to (U, φ).

Proposition 6.31 (Differentiability of the Derivative) *Let M and N be differentiable manifolds of type $\mathcal{C}^{\mathbb{K},k}$ and $f : M \to N$ a $\mathcal{C}^{\mathbb{K},k}$-map. Then the map $f' = Tf : TM \to TN$ is of class $\mathcal{C}^{\mathbb{K},k-1}$.*

Proof For a given $p \in M$, it suffices to consider charts (U, φ) on M with $p \in U$ and (V, ψ) on N with $f(U) \subset V$ and their associated charts $(\widetilde{U}, \widetilde{\varphi})$ and $(\widetilde{V}, \widetilde{\psi})$. Let $F : \widetilde{\varphi}(\widetilde{U}) \to \widetilde{\psi}(\widetilde{V})$ be defined by $F = \widetilde{\psi} \circ Tf \circ \widetilde{\psi}^{-1}$. We denote elements of $\widetilde{\varphi}(\widetilde{U})$ by (x, v), $x, v \in \mathbb{R}^m$ and elements of $\widetilde{\psi}(\widetilde{V})$ by (y, w), $y, w \in \mathbb{R}^n$, where m and n are the dimensions of M and N, respectively. Then the first $\mathbb{K}^n$-component of $F(\varphi(p), v)$ is just $\psi(f(p))$. The representation of df_p with respect to the φ-basis for T_pM and the ψ-basis for $T_{f(p)}N$, as given in Proposition 6.27, then shows that

$$\forall q \in U: \quad F(\varphi(q), v) = (\psi(f(q)), Df(q)v),$$

where $Df(q)$ is the Jacobian matrix of $\psi \circ f \circ \varphi^{-1}$ at $\varphi(q)$. But since f was assumed to be a $C^{\mathbb{K},k}$-map, the component functions of the Jacobian matrix are $C^{\mathbb{K},k-1}$-maps, i.e., Tf is a $C^{\mathbb{K},k-1}$-map. □

The construction of the derivative is compatible with the composition of maps (see Exercise 6.9), i.e., we have commutative diagrams of the form

$$\begin{array}{ccccc} TM & \xrightarrow{Tf} & TN & \xrightarrow{Tg} & TL \\ \pi_M \downarrow & & \pi_N \downarrow & & \pi_L \downarrow \\ M & \xrightarrow{f} & N & \xrightarrow{g} & L \end{array}$$

and recognize that the assignments $M \mapsto TM$ and $f \mapsto Tf$ define a functor $T: \mathbf{Man}_{\mathbb{K},k} \to \mathbf{Man}_{\mathbb{K},k-1}$, which is also called the *tangent functor*.

Cotangent Bundle

A particularly important special case of differentiable mappings $f: M \to N$ is of course the case $N = \mathbb{K}$. Here, the set of $C^{\mathbb{K},k}$-mappings $f: M \to N$ is just $C^{\mathbb{K},k}(M)$, and the tangent bundle of N can be identified in a canonical way with $N \times \mathbb{K}$, since (N, id_N) is a chart of N. Thus, for every point $p \in M$ a linear mapping from $T_p M$ to $\mathbb{K}$ is obtained by $\mathfrak{v} \mapsto \pi_2 \circ f'(\mathfrak{v})$, where $\pi_2: TN = \mathbb{K} \times \mathbb{K} \to \mathbb{K}$ is the projection onto the second component. This mapping is denoted by $df(p)$. Analogous to the procedure for the derivative, one makes from the $df(p)$ a mapping $df: M \to T^*M$, where T^*M is the disjoint union $\coprod_{p \in M} T_p M^*$ of the dual spaces $T_p M^*$ of $T_p M$. One calls $T_p^* M := T_p M^*$ the *cotangent space* of M at p and T^*M the *cotangent bundle* of M.

Note the close connection between $Tf = f': TM \to T\mathbb{K}$ and $df: M \to T^*M$:

$$\forall p \in M, \mathfrak{v} \in T_p M: \quad df(p)(\mathfrak{v}) = \pi_2 \circ df_p(\mathfrak{v}) = \pi_2 \circ f'(\mathfrak{v}) \in \mathbb{K}.$$

The two mappings differ in that $Tf: TM \to T\mathbb{R}$ always carries the "information f" in the first component, while $df: M \to T^*M$ only describes the derivative df_p depending on p. In many texts, therefore, the notation df is also used for the mapping $f' = Tf$. The difference becomes clear in such texts at the latest when an argument is inserted: $df(p) \longleftrightarrow df(\mathfrak{v})$.

Proposition 6.32 (φ-Bases for Cotangent Spaces) *Let M be a differentiable manifold of type $C^{\mathbb{K},k}$ and (U, φ) a chart on M. Denote the functions $\pi_j \circ \varphi: U \to \mathbb{K}$ with x_j, where π_j is the projection onto the j-th component. Then for $p \in U$ the $dx_1(p), \ldots, dx_n(p)$ form the dual basis to the φ-basis for $T_p M$.*

6.3 Tangent and Tensor Bundles

Proof Let $\mathfrak{v}_1, \ldots, \mathfrak{v}_n$ be the φ-basis for $T_p M$. It suffices to show that $dx_i(p)\mathfrak{v}_j = \delta_{ij}$ holds (Kronecker delta). But since the Jacobian matrix of $x_j \circ \varphi^{-1}$ is just the vector $(0, \ldots, 0, 1, 0, \ldots, 0)^\top$ with the one in the j-th row and the number 1 is the basis vector of $T_r \mathbb{K}$ given by the chart $(\mathbb{K}, \mathrm{id})$, the assertion follows from the representation of $T x_i(\mathfrak{v}) = \bigl(dx_i(p)\bigr)(\mathfrak{v})$, for $\mathfrak{v} \in T_p M$, in local coordinates (see Proposition 6.27). □

We call the basis $dx_1(p), \ldots, dx_n(p)$ constructed in Proposition 6.32 the φ-basis for $T_p^* M$.

Proposition 6.33 (Cotangent Bundle as a Manifold) *Let M be a differentiable manifold of type $\mathcal{C}^{\mathbb{K},k}$ and $\pi: T^* M \to M$ the mapping that assigns to each $\omega \in T_p^* M$ the point $p \in M$. Then there is a Hausdorff topology on $T^* M$ with respect to which the mapping π is continuous and open. The topological space $T^* M$ carries the structure of a manifold of type $\mathcal{C}^{\mathbb{K},k-1}$, with respect to which π is a $\mathcal{C}^{\mathbb{K},k-1}$-mapping.*

Proof The proof can be done analogously to that of Proposition 6.30. We only specify the charts: Let $(U_\alpha, \varphi_\alpha)_{\alpha \in A}$ be a family of charts on M with $\bigcup_{\alpha \in A} U_\alpha = M$. We again set $\widetilde{U}_\alpha := \pi^{-1}(U_\alpha)$. The mapping $\widetilde{\varphi}_\alpha : \widetilde{U}_\alpha \to \mathbb{R}^{2n}$ is then given by

$$\widetilde{\varphi}_\alpha(\omega) := \bigl(x_1^\alpha(p), \ldots, x_n^\alpha(p),\ v_1^\alpha, \ldots, v_n^\alpha\bigr) \tag{6.1}$$

where $p \in U_\alpha$ and $\omega = \sum_{j=1}^n v_j^\alpha \cdot dx_j^\alpha(p) \in T_p^* M$. □

As in the case of the tangent bundle, we call the chart $(\widetilde{U}, \widetilde{\varphi})$ on $T^* M$ constructed from a chart (U, φ) on M the chart *associated* to (U, φ).

Exercise 6.11 (Coordinate Change for Cotangent Spaces) Calculate how the coordinates of $\omega \in T_p^* M$ change in a chart transition $\widetilde{\varphi}_\alpha \circ \widetilde{\varphi}_\beta^{-1} : \widetilde{\varphi}_\alpha(\widetilde{U}_\alpha \cap \widetilde{U}_\beta) \to \widetilde{\varphi}_\beta(\widetilde{U}_\alpha \cap \widetilde{U}_\beta)$.

Proposition 6.34 (Derivative of Scalar Functions) *Let M be a differentiable manifold of type $\mathcal{C}^{\mathbb{K},k}$ and $f : M \to \mathbb{K}$ a $\mathcal{C}^{\mathbb{K},k}$-function. Then the mapping $df : M \to T^* M$ is of class $\mathcal{C}^{\mathbb{K},k-1}$. It satisfies $\pi \circ df = \mathrm{id}_M$.*

Proof It suffices to consider the restriction of df to a chart neighborhood. So let (U, φ) be a chart on M and $(\widetilde{U}, \widetilde{\varphi})$ the chart associated to (U, φ) on $T^* M$. If $\mathfrak{v}_1, \ldots, \mathfrak{v}_n$ is the φ-basis for $T_p M$ and $\mathfrak{v} = \sum_{j=1}^n c_j \mathfrak{v}_j \in T_p M$, then we have according to Proposition 6.27

$$df(p)(\mathfrak{v}) = \sum_{j=1}^n \frac{\partial F}{\partial x_j}\bigl(\varphi(p)\bigr) c_j,$$

where $F: \varphi(U) \to \mathbb{K}$ is the function $f \circ \varphi^{-1}$. Thus, we have:

$$df(p) = \sum_{j=1}^n \frac{\partial F}{\partial x_j}(\varphi(p)) dx_j(p).$$

Since the function F is a $C^{\mathbb{K},k}$-mapping by assumption, the derivatives of F are $C^{\mathbb{K},k-1}$-mappings, and thus, according to the definition of the associated chart, df is also a $C^{\mathbb{K},k-1}$-mapping. □

Vector Bundles

Tangent bundles and cotangent bundles are examples of *vector bundles*, a concept of great importance in differential geometry (as well as in topology and analysis). We provide the general definition here because we will construct other vector bundles from the tangent bundle.

Definition 6.35 (Vector Bundle) Let M be an n-dimensional differentiable manifold of type $C^{\mathbb{K},k}$. A $C^{\mathbb{K},k}$-*vector bundle* of *rank r* over M is a differentiable manifold E of dimension $n + r$, together with a $C^{\mathbb{K},k}$-mapping $\pi : E \to M$, which has the following properties:

(i) For each $p \in M$, $E_p := \pi^{-1}(p)$ has the structure of an r-dimensional $\mathbb{K}$-vector space. This vector space is called the *fiber* over p.
(ii) For each $p \in M$, there is a neighborhood U of p in M and a diffeomorphism $\psi_U : \pi^{-1}(U) \to U \times \mathbb{K}^r$ with $\pi_1 \circ \psi_U = \pi$ (π_1 is the projection onto the U-component), and for each $q \in U$, $\pi_2 \circ \psi_U|_{E_q} : E_q \to \mathbb{R}^r$ is a linear isomorphism (π_2 is the projection onto the $\mathbb{R}^r$-component).

Let $U \subseteq M$ be open and $\sigma : U \to E$ a $C^{\mathbb{K},k}$-mapping with $\pi \circ \sigma = \mathrm{id}_U$. Then σ is called a *section* of (E, π, M) over U. We denote the set of sections of (E, π, M) over U with $C^{\mathbb{K},k}(U; E)$.

Note the obvious similarities in notation between the definitions of vector bundles and étalé spaces. However, these are different concepts, as can be seen from the fact that the fibers of étalé spaces are discrete, while those of vector bundles are equipped with the natural $\mathbb{K}$-vector space topology.

Propositions 6.30 and 6.33 show that tangent bundles and cotangent bundles are indeed vector bundles. Furthermore, Proposition 6.34 shows that $df : M \to T^*M$ for a $C^{\mathbb{K},k}$-function $f : M \to \mathbb{K}$ is a section of the cotangent bundle. Such sections are also called 1-*forms* or *Pfaffian forms*. They are the starting point of the theory of differential forms.

6.3 Tangent and Tensor Bundles

Remark 6.36 (Sheaves of Sections of Vector Bundles) Let M be an n-dimensional differentiable manifold of type $\mathcal{C}^{\mathbb{K},k}$ and (E, π, M) a $\mathcal{C}^{\mathbb{K},k}$-vector bundle over M. Then for each open subset $U \subseteq M$, the set $E|_U := \pi^{-1}(U)$ together with $\pi_U : E|_U \to U$ forms a $\mathcal{C}^{\mathbb{K},k}$-vector bundle $(E|_U, \pi_U, U)$ over U. Let $V \subseteq U$ be another open subset of M. Then $\rho_{VU}(\sigma) := \sigma|_V$ for each section $\sigma : U \to E|_U$ is a section of $(E|_V, \pi_V, V)$. In this way, $U \mapsto \mathcal{C}^{\mathbb{K},k}(U; E)$ becomes a sheaf $\mathcal{E}_\pi$ over M, which we call the *sheaf of sections* or simply *section sheaf* of (E, π, M). Note that $\mathcal{E}_\pi$ is even a locally free $\mathcal{C}^{\mathbb{K},k}_M$-module of finite type (see Definition 5.39 and Example 5.40).

The sheaves of sections of (TM, π_M, M) and (T^*M, π_M, M) are denoted by $\mathcal{T}_M$ and $\mathcal{T}_M^*$ and are called the *tangent sheaf* and the *cotangent sheaf* of M. □

The considerations from Section 5.4 show that from the tangential and cotangential sheaves, a series of other locally free $\mathcal{C}^{\mathbb{K},k}_M$-sheaves of finite type can be constructed. In particular, we can first form $\mathcal{T}_M \oplus \mathcal{T}_M^*$, their tensor powers $(\mathcal{T}_M \oplus \mathcal{T}_M^*)^{\otimes \ell}$ for $\ell \in \mathbb{N}$, and finally the tensor algebra sheaf $\bigoplus_{\ell \in \mathbb{N}_0} (\mathcal{T}_M \oplus \mathcal{T}_M^*)^{\otimes \ell}$, where the 0-th tensor power is simply the free module sheaf $\mathcal{O}_X$. The local sections of the tensor algebra sheaf are called *tensors*. In the tensor algebra sheaf, more precisely within $\bigoplus_{\ell \in \mathbb{N}_0} (\mathcal{T}_M^*)^{\otimes \ell}$, one finds the sheaf Ω_M of alternating tensors, whose sections are called *differential forms*. The sections of $\Omega_M^\ell := (\mathcal{T}_M^*)^{\otimes \ell} \cap \Omega_M$ are called ℓ-forms. In particular, 1-forms are sections of the sheaf $\mathcal{T}_M^*$.

One could now invoke a general result (see Theorem 6.38) to see that there is a vector bundle for each sheaf, whose section sheaf is the respective sheaf. However, it is also possible to first apply the corresponding functorial constructions to the individual tangential and cotangential spaces and then construct the corresponding tensor bundles in analogy to the tangential and cotangential bundles.

In the next section "tensor bundles", we follow the second strategy, but show at the end that both paths lead to the same result. The following considerations on the structural similarity of vector bundles and locally free module sheaves will be of crucial importance for this.

Definition 6.37 (Morphisms of Vector Bundles) Let M be an n-dimensional differentiable manifold of type $\mathcal{C}^{\mathbb{K},k}$. A *morphism* $\varphi: (E_1, \pi_1, M) \to (E_2, \pi_2, M)$ between two $\mathcal{C}^{\mathbb{K},k}$-vector bundles is a $\mathcal{C}^{\mathbb{K},k}$-map $\varphi\, E_1 \to E_2$ with $\pi_1 = \pi_2 \circ \varphi$. We denote the resulting category of $\mathcal{C}^{\mathbb{K},k}$-vector bundles over M by $\mathbf{VB}_M^{\mathbb{K},k}$.

We now want to show that $\mathcal{C}^{\mathbb{K},k}$ vector bundles over M and locally free $\mathcal{C}^{\mathbb{K},k}_M$-module of finite type are essentially the same. More precisely, the two categories are equivalent (see Definition 4.58).

Theorem 6.38 (Locally Free Sheaves and Vector Bundles) *Let M be an n-dimensional differentiable manifold of type $\mathcal{C}^{\mathbb{K},k}$. Then the category $\mathbf{VB}_M^{\mathbb{K},k}$ is equivalent to the full subcategory $\mathbf{Mod}_{\mathcal{C}_M^{\mathbb{K},k}}^{\text{lff}}$ of $\mathbf{Mod}_{\mathcal{C}_M^{\mathbb{K},k}}$, whose objects are the locally free $\mathcal{C}_M^{\mathbb{K},k}$-modules of finite type.*

Proof The assignments $(E, \pi, M) \mapsto \mathcal{E}_\pi$ and

$$\left(\varphi: (E_1, \pi_1, M) \to (E_2, \pi_2, M)\right) \quad \mapsto \quad \left(\mathcal{E}_{\pi_1}(U) \to \mathcal{E}_{\pi_2}(U), \; \sigma_1 \mapsto \varphi \circ \sigma_1\right)$$

define a functor $S: \mathbf{VB}_M^{\mathbb{K},k} \to \mathbf{Mod}_{\mathcal{C}_M^{\mathbb{K},k}}^{\text{lff}}$.

Next, we construct a functor $F: \mathbf{Mod}_{\mathcal{C}_M^{\mathbb{K},k}}^{\text{lff}} \to \mathbf{VB}_M^{\mathbb{K},k}$. Let $\mathcal{E}$ be a locally free $\mathcal{C}_M^{\mathbb{K},k}$-module of finite type and $p \in M$. Then there is an open neighborhood U_p of p in M with $\mathcal{E}|_{U_p}$ isomorphic to $(\mathcal{C}_{U_p}^{\mathbb{K},k})^\ell$ for some $\ell \in \mathbb{N}$. Using the constant 0- and 1-sections in $\mathcal{C}_M^{\mathbb{K},k}(U_p)$, we find a $\mathcal{C}_M^{\mathbb{K},k}(U_p)$-basis $e_1, \ldots, e_\ell$ for $\mathcal{E}|_{U_p}$. Then for each open subset $V \subseteq U_p$ the mapping

$$\varphi_V: \mathcal{C}_M^{\mathbb{K},k}(V)^\ell \to \mathcal{E}(V), \quad (f_1, \ldots, f_\ell) \mapsto \sum_{i=1}^\ell f_i \rho_{V,U_p}^{\mathcal{E}}(e_i)$$

is an isomorphism of $\mathcal{C}_M^{\mathbb{K},k}(V)$-modules. Thus, the induced mappings

$$\varphi_q: (\mathcal{C}_{M,q}^{\mathbb{K},k})^\ell \to \mathcal{E}_q, \quad (f_{1,q}, \ldots, f_{\ell,q}) \mapsto \sum_{i=1}^\ell f_{i,q} e_{i,q}$$

on the stalks at $q \in U_p$ are also isomorphisms. According to Proposition 5.32, the rings $\mathcal{C}_{M,q}^{\mathbb{K},k}$ are local with residue field $\mathbb{K}$. Thus, the φ_q induce isomorphisms $\bar{\varphi}_q: \mathbb{K}^\ell \to \mathcal{E}(q) := \mathcal{E}_q/\mathfrak{m}_q \mathcal{E}_q$, where $\mathfrak{m}_q$ is the maximal ideal of $\mathcal{C}_{M,q}^{\mathbb{K},k}$. Together, the $\bar{\varphi}_q$ for $q \in U_p$ thus provide a bijection

$$\bar{\varphi}_{U_p}: U_p \times \mathbb{K}^\ell \to \coprod_{q \in U_p} \mathcal{E}(q).$$

Note that for $q \in U_p \cap U_{p'}$ the space $\mathcal{E}(q)$ does not depend on whether it was determined over p or over p'. So for each $q \in M$ we have a uniquely determined space $\mathcal{E}(q)$ and can set $E := \coprod_{p \in M} \mathcal{E}(p)$. Then we define $\pi \; E \to M$ by

$$\forall q \in M: \quad \pi^{-1}(q) = \mathcal{E}(q).$$

6.3 Tangent and Tensor Bundles

E is covered by the subsets $\pi^{-1}(U_p)$ with $p \in M$, and for each pair (p, p') of points with $U_p \cap U_{p'} \neq \emptyset$ the mapping

$$\bar{\varphi}_{U_{p'}}^{-1} \circ \bar{\varphi}_{U_p}|_{U_p \cap U_{p'}} : (U_p \cap U_{p'}) \times \mathbb{K}^\ell \to (U_p \cap U_{p'}) \times \mathbb{K}^{\ell'}$$

is given by a basis change in the second variable, which is given by an invertible matrix with entries in $C_M^{\mathbb{K},k}(U_p \cap U_{p'})$. In particular, $\ell = \ell'$ and the topology $\mathfrak{T}_{U_p}$ induced by $\bar{\varphi}_{U_p}$ from $U_p \times \mathbb{K}^\ell$ to $\pi^{-1}(U_p)$ contains $\pi^{-1}(U_p \cap U_{p'})$ as an open set. Thus, a topology on E is defined by

$$U \in \mathfrak{T}_E \quad :\Leftrightarrow \quad \forall p \in M : U \cap U_p \in \mathfrak{T}_{U_p}$$

with respect to which E is a $C^{\mathbb{K},k}$-manifold. Together with the vector space structures on the fibers $\pi^{-1}(p) = \mathcal{E}(p)$, it now follows that (E, π, M) is a vector bundle, which we denote by $F(\mathcal{E})$. Let $\psi : \mathcal{E} \to \mathcal{E}'$ now be a morphism of $C_M^{\mathbb{K},k}$-modules. If one goes through the constructions of $F(\mathcal{E})$ and $F(\mathcal{E}')$ in parallel and tracks the effect of $\psi_V : \mathcal{E}(V) \to \mathcal{E}'(V)$ and $\psi_q : \mathcal{E}_q \to \mathcal{E}'_q$, one finds a vector bundle morphism $F(\psi)\, F(\mathcal{E}) \to F(\mathcal{E}')$, which is given by $F(\psi)|_{\mathcal{E}(q)} = \bar{\psi}_q : \mathcal{E}(q) \to \mathcal{E}'(q)$. That $F(\psi)$ is indeed a $C_M^{\mathbb{K},k}$-mapping is calculated in local coordinates, i.e., over the sets U_p (exercise). This finishes construction the sought-after functor $F : \mathbf{Mod}_{C_M^{\mathbb{K},k}}^{\mathrm{lff}} \to \mathbf{VB}_M^{\mathbb{K},k}$.

It remains to show that the functors $F \circ S$ and $S \circ F$ are each naturally isomorphic to the identity functor. This verification is left to the reader as an exercise.

$\square$

Exercise 6.12 (Derivative as Sheaf Morphism) Let M be an n-dimensional differentiable manifold of type $C^{\mathbb{K},k}$. Show that $f \mapsto df$ defines a sheaf morphism $C_M^{\mathbb{K},k} \to \mathcal{T}_M^*$.

Exercise 6.13 (Gluing of Vector Bundles) Let M be a differentiable $\mathbb{K}$-manifold and $(U_\alpha, \varphi_\alpha)_{\alpha \in A}$ a family of charts that cover M. Further, let V be a finite-dimensional $\mathbb{K}$-vector space and $g_{\alpha\beta} : U_\alpha \cap U_\beta \to \mathrm{Aut}_{\mathbb{K}}(V)$ a family of $C^{\mathbb{K},k}$-mappings that satisfy the following conditions:

(a) $\forall \alpha, \beta, \gamma \in A, x \in U_\alpha \cap U_\beta \cap U_\gamma : \quad g_{\alpha\beta}(x) \circ g_{\beta\gamma}(x) = g_{\alpha\gamma}(x)$.
(b) $\forall \alpha \in A, x \in U_\alpha : \quad g_{\alpha\alpha}(x) = \mathrm{id}_V$.

Show that the trivial vector bundles $U_\alpha \times V$ can be glued together using the *transition functions* $g_{\alpha\beta}$ via

$$\forall \alpha, \beta \in A : \quad (x, v) \sim (x, g_{\alpha\beta}(x)v)$$

to form a vector bundle

$$E := \Big(\coprod_{\alpha \in A} U_\alpha \times V \Big)/\sim \;\to\; M, \quad [(x, v)]_\sim \mapsto x.$$

Tensor Bundles

Let M be a differentiable $C^{\mathbb{K},k}$-manifold and (U,φ) a chart on M. For $p \in U$, we consider the respective φ-bases $\left(\frac{\partial}{\partial x_j}\big|_p\right)_{j=1,\ldots,n}$ and $(dx_j)_{j=1,\ldots,n}$ on T_pM and T_p^*M. Even though we use the notation of the algebraic approach to the tangent space here, we do not have to assume that $k \in \{\infty, \omega\}$, because we could also use the geometric or physicist's notation for φ-bases. The reason for our choice of notation is that it makes the connection to tensor calculus, as practiced in physics, engineering mathematics, and parts of differential geometry, more apparent.

For $r, s \in \mathbb{N}_0$, we define

$$\bigotimes_r^s T_pM := (T_pM)^{\otimes s} \otimes (T_p^*M)^{\otimes r},$$

where the tensor products are taken over $\mathbb{K}$. Then the $n^{(r+s)}$ elements

$$\left(dx \otimes \frac{\partial}{\partial x}\right)_{(\varrho,\sigma)}(p) := dx_{\varrho(1)}(p) \otimes \ldots \otimes dx_{\varrho(r)}(p) \otimes \frac{\partial}{\partial x_{\sigma(1)}}\bigg|_p \otimes \ldots \otimes \frac{\partial}{\partial x_{\sigma(s)}}\bigg|_p$$

is a basis for $\bigotimes_r^s T_pM$. Here, ϱ and σ run through the self-mappings of $\{1, \ldots, r\}$ and $\{1, \ldots, s\}$, respectively.

Proposition 6.39 (Tensor Bundle) *Let M be a differentiable $C^{\mathbb{K},k}$-manifold and $\bigotimes_r^s TM := \bigsqcup_{p \in M} \bigotimes_r^s T_pM$. Furthermore, let $\pi : \bigotimes_r^s TM \to M$ be the mapping that assigns to each $\mathfrak{t} \in \bigotimes_r^s T_pM$ the base point $p \in M$. Then there exists a Hausdorff topology on $\bigotimes_r^s TM$, with respect to which π is open and continuous. The triple $(\bigotimes_r^s TM, \pi, M)$ is a $C^{\mathbb{K},k-1}$-vector bundle of rank $n^{(r+s)}$.*

Proof The proof is completely analogous to the proofs of Propositions 6.30 and 6.33. We only define here the chart $(\widetilde{U}, \widetilde{\varphi})$ associated with a chart (U, φ) on M on $\bigotimes_r^s TM$: For this, we set $\widetilde{U} := \pi^{-1}(U)$ and $\widetilde{\varphi} : \widetilde{U} \to \mathbb{K}^{n+n^{r+s}}$, which for $p \in U$ is given by

$$\widetilde{\varphi}(\mathfrak{t}) := \left(x_1(p), \ldots, x_n(p), (v)_{(\varrho,\sigma)_1}, \ldots, (v)_{(\varrho,\sigma)_{n^{r+s}}}\right)$$

where $\mathfrak{t} = \sum_{j=1}^{n^{(r+s)}} (v)_{(\varrho,\sigma)_j} \left(dx \otimes \frac{\partial}{\partial x}\right)_{(\varrho,\sigma)_j}(p)$ and the pairs (ϱ, σ) are numbered in a fixed way. $\square$

Definition 6.40 (Tensor Bundle and Tensor Fields) We call the vector bundle $\bigotimes_r^s TM$ from Proposition 6.39 the *tensor bundle* of level (r, s) of M. The local sections of the tensor bundle $\bigotimes_r^s TM$ are called *tensor fields* of *level* (r, s).

Note that TM is the tensor bundle of level $(0, 1)$ and T^*M is the tensor bundle of level $(1, 0)$. In particular, the tensor fields of level $(1, 0)$ are just the 1-forms. The tensor fields of level $(0, 1)$, i.e., the sections of the tangent bundle, are also called *vector fields*.

Exercise 6.14 (Change of Charts for Tensor Bundles) Calculate how the coordinates of $t \in \otimes_r^s T_p M$ transform under changes of charts of the form

$$\widetilde{\varphi}_\alpha \circ \widetilde{\varphi}_\beta^{-1} : \widetilde{\varphi}_\beta(\widetilde{U}_\alpha \cap \widetilde{U}_\beta) \to \widetilde{\varphi}_\alpha(\widetilde{U}_\alpha \cap \widetilde{U}_\beta).$$

Each of the tensor bundles $\otimes_r^s TM$ provides a section sheaf, which we denote by $\mathcal{T}_M^{(r,s)}$. The restriction $\mathcal{T}_M^{(r,s)}|_U$ to a coordinate neighborhood U is isomorphic to $\mathcal{C}_U^{\mathbb{K},k-1} \otimes (\otimes_r^s \mathbb{K}^n)$, where $m = \dim M$. A comparison with Example 5.40 shows that indeed

$$\bigotimes_r^s \mathcal{T}_M \cong \mathcal{T}_M^{(r,s)} = S(\otimes_r^s TM)$$

holds, as announced after Remark 6.36.

Remark 6.41 (Tensors and Multilinear Mappings) Let M be a differentiable $\mathcal{C}^{\mathbb{K},k}$-manifold, $V_0, V_1, \ldots, V_s$ be $\mathcal{C}^{\mathbb{K},k}$-vector bundles over M, and $\mathcal{V}_0, \ldots, \mathcal{V}_s$ be the associated $\mathcal{C}_M^{\mathbb{K},k}$-modules (i.e., section sheaves).

(i) For $j = 1, \ldots, s$, let $\mathrm{Hom}(V_j, V_0)$ be the vector bundle over M, whose fiber over $p \in M$ consists of the $\mathbb{K}$-vector space $\mathrm{Hom}_\mathbb{K}(V_{j,p}, V_{0,p})$. The associated $\mathcal{C}_M^{\mathbb{K},k}$-module is (see Example 5.38) $\mathcal{H}om_{\mathcal{C}^{\mathbb{K},k}}(\mathcal{V}_j, \mathcal{V}_0)$ with the local section modules $\mathcal{H}om_{\mathcal{C}^{\mathbb{K},k}}(\mathcal{V}_j, \mathcal{V}_0)(U) = \mathrm{Hom}_{\mathcal{C}^{\mathbb{K},k}(U)}(\mathcal{V}_j(U), \mathcal{V}_0(U))$ for open U in M.

(ii) Let $L(V_1, \ldots, V_s; V_0)$ be the vector bundle over M, whose fiber over $p \in M$ consists of the $\mathbb{K}$-vector space of s-linear mappings $V_{1,p} \times \ldots \times V_{s,p} \to V_{0,p}$. The associated $\mathcal{C}_M^{\mathbb{K},k}$-module $\mathcal{L}_{\mathcal{C}_M^{\mathbb{K},k}}(\mathcal{V}_1, \ldots, \mathcal{V}_s; \mathcal{V}_0)$ has the local section modules

$$\mathcal{L}_{\mathcal{C}_M^{\mathbb{K},k}}(\mathcal{V}_1, \ldots, \mathcal{V}_s; \mathcal{V}_0)(U) = \mathrm{L}_{\mathcal{C}_M^{\mathbb{K},k}(U)}(\mathcal{V}_1(U), \ldots, \mathcal{V}_s(U); \mathcal{V}_0(U))$$

for open U in M. They consist of the s-linear mappings $\mathcal{V}_1(U) \times \ldots \times \mathcal{V}_s(U) \to \mathcal{V}_0(U)$ with respect to $\mathcal{C}^{\mathbb{K},k}(U)$. When V_0 is the trivial bundle $M \times \mathbb{K} \to M$, $\mathcal{V}_0 = \mathcal{C}_M^{\mathbb{K},k}$, and we obtain

$$\mathcal{H}om_{\mathcal{C}^{\mathbb{K},k}}(\mathcal{V}_j, \mathcal{V}_0) = \mathcal{V}_j^\vee.$$

(iii) For $V_0 = M \times \mathbb{K}$, the natural mappings (see Exercise 3.2)

$$\varphi_U : \bigotimes_{j=1}^{s} \mathrm{Hom}_{\mathcal{C}^{\mathbb{K},k}(U)}\left(\mathcal{V}_j(U), \mathcal{C}_M^{\mathbb{K},k}(U)\right)$$
$$\to \mathcal{L}_{\mathcal{C}_M^{\mathbb{K},k}(U)}\left(\mathcal{V}_1(U), \ldots, \mathcal{V}_s(U); \mathcal{C}_M^{\mathbb{K},k}(U)\right),$$

which are defined by

$$(\varphi_U(\psi_1 \otimes \ldots \otimes \psi_s))(v_1, \ldots, v_s) := \prod_{j=1}^{s} \psi_j(v_j),$$

together form a morphism

$$\varphi : \bigotimes_{j=1}^{s} \mathcal{V}_j^{\vee} \to \mathcal{L}_{\mathcal{C}_M^{\mathbb{K},k}}(\mathcal{V}_1, \ldots, \mathcal{V}_s; \mathcal{C}_M^{\mathbb{K},k})$$

of $\mathcal{C}_M^{\mathbb{K},k}$-modules. According to Remark 3.18, the corresponding mappings

$$\bigotimes_{j=1}^{s} V_{j,p}^* \to L(V_{1,p}, \ldots, V_{s,p}; \mathbb{K})$$

are isomorphisms of $\mathbb{K}$-vector spaces. Using the local trivializations, it is therefore seen that φ is a sheaf isomorphism.

(iv) Applying (iii) to the tangent sheaf $\mathcal{T}_M$, one finds a sheaf isomorphism

$$(\mathcal{T}_M^*)^{\otimes s} \to \mathcal{L}_{\mathcal{C}_M^{\mathbb{K},k}}(\mathcal{T}_{M,s}; \mathcal{C}_M^{\mathbb{K},k}),$$

where the notation $\mathcal{T}_{M,s}$ means that one considers s-linear mappings $\mathcal{T}_M(U) \times \ldots \times \mathcal{T}_M(U) \to \mathcal{C}_M^{\mathbb{K},k}(U)$.

(v) Since for finite-dimensional vector spaces the natural embedding $V \to (V^*)^*$ is an isomorphism, local trivialization immediately shows that for a $\mathcal{C}^{\mathbb{K},k}$-vector bundle V over M the sheaves $\mathcal{V}$ and $(\mathcal{V}^{\vee})^{\vee}$ are isomorphic. Thus, (iii), applied to $\mathcal{T}_M^*$, yields a sheaf isomorphism

$$(\mathcal{T}_M)^{\otimes r} \to \mathcal{L}_{\mathcal{C}_M^{\mathbb{K},k}}(\mathcal{T}_{M,r}^*; \mathcal{C}_M^{\mathbb{K},k}).$$

(vi) Combining (iv) and (v), one finds a sheaf isomorphism

$$\bigotimes_r^s \mathcal{T}_M \cong \mathcal{T}_M^{(r,s)} \to \mathcal{L}_{\mathcal{C}_M^{\mathbb{K},k}}(\mathcal{T}_{M,s}, \mathcal{T}_{M,r}^*; \mathcal{C}_M^{\mathbb{K},k}),$$

where $\mathcal{L}_{\mathcal{C}_M^{\mathbb{K},k}}(\mathcal{T}_{M,s}, \mathcal{T}_{M,r}^*; \mathcal{C}_M^{\mathbb{K},k})$ stands for $\mathcal{L}_{\mathcal{C}_M^{\mathbb{K},k}}(\mathcal{T}_M, \ldots, \mathcal{T}_M, \mathcal{T}_M^*, \ldots, \mathcal{T}_M^*; \mathcal{C}_M^{\mathbb{K},k})$ with s copies of $\mathcal{T}_M$ and r copies of $\mathcal{T}_M^*$. □

Vector Fields

The essence of Remark 6.41 is that tensors can be defined via their multilinear effect on vector fields and 1-forms, i.e., local sections of the tangent or cotangent bundle. We will apply this method in the study of differential forms. For this, it will be useful to have another description for the vector fields at hand, which arises from the algebraic point of view and focuses on their effect on functions.

Definition 6.42 (Vector Fields) Let M be a differentiable manifold of type $C^{\mathbb{K},k}$ with tangent bundle TM. A *vector field* on M is a $C^{\mathbb{K},k-1}$-section $\mathfrak{X} : M \to TM$. We denote the set $\mathcal{T}_M(M)$ of all globally defined vector fields on M also with $\mathcal{X}(M)$. The value of a vector field $\mathfrak{X}$ at $x \in M$ is denoted by $\mathfrak{X}(x)$ or $\mathfrak{X}_x$.

Example 6.43 (φ-Basis Fields) Let M be a differentiable manifold of type $C^{\mathbb{K},k}$ and (U, φ) a chart on M. We consider U itself as a differentiable manifold and define mappings $\mathfrak{X}_j : U \to TU$ by $\mathfrak{X}_j(p) := \frac{\partial}{\partial x_j}\big|_p$. We write these mappings as $\mathfrak{X}_j = \frac{\partial}{\partial x_j}$. In the associated chart $(\widetilde{U}, \widetilde{\varphi})$ on TU, these mappings are given by

$$\widetilde{\varphi} \circ \mathfrak{X}_j \circ \varphi^{-1}\big((x_1, \ldots, x_n)\big) = (x_1, \ldots, x_n, 0, \ldots, 0, 1, 0, \ldots, 0),$$

where the 1 is in the $(n + j)$-th column. Thus, $\mathfrak{X}_j$ is a vector field on U for each $j = 1, \ldots, n$. We call these vector fields the φ-basis fields on U. The term "basis field" is doubly justified. On the one hand, the $\mathfrak{X}_1(p), \ldots, \mathfrak{X}_n(p)$ form a basis for $T_pU = T_pM$ for each $p \in U$. On the other hand, it also follows from the above formula that every $\mathfrak{X} \in \mathcal{X}(U)$ is of the form

$$\mathfrak{X}(p) = \sum_{j=1}^n a_j(p)\mathfrak{X}_j(p) \qquad \forall p \in U,$$

where the a_j are mappings of the class $C^{\mathbb{K},k-1}$ on U. □

Every vector field $\mathfrak{X}$ provides (see Proposition 6.19) a mapping

$$C_{X,p}^{\mathbb{K},k} \to \mathbb{K}, \quad a \mapsto \iota(\mathfrak{X}_p)a.$$

If $k \in \{\infty, \omega\}$, we omit ι because we have identified the geometric and the algebraic vector space. If we start with a $C^{\mathbb{K},k}$-function $f: M \to \mathbb{K}$, we obtain a function $\widetilde{\mathfrak{X}}(f): M \to \mathbb{K}$ via $\widetilde{\mathfrak{X}}(f)(p) := \mathfrak{X}(p)(f_p)$. The derivation property of $\mathfrak{X}(p)$ provides a similar property for $\widetilde{\mathfrak{X}}$:

$$\begin{aligned}\widetilde{\mathfrak{X}}(f \cdot g)(p) &= \mathfrak{X}(p)\big((f \cdot g)_p\big) \\ &= \mathfrak{X}(p)(f_p) \cdot g(p) + \mathfrak{X}(p)(g_p) \cdot f(p) \\ &= \widetilde{\mathfrak{X}}(f)(p) \cdot g(p) + f(p) \cdot \widetilde{\mathfrak{X}}(g)(p).\end{aligned}$$

This means, we have the following equality of functions:

$$\widetilde{\mathfrak{X}}(f \cdot g) = \widetilde{\mathfrak{X}}(f) \cdot g + f \cdot \widetilde{\mathfrak{X}}(g).$$

In general, we can only expect that $\widetilde{\mathfrak{X}}(f)$ is a $C^{\mathbb{K},k-1}$-function. This motivates the following generalization of the concept of derivation from Definition 6.18.

Definition 6.44 (Φ-Derivations) Let A and B be two $\mathbb{K}$-algebras and $\Phi: A \to B$ a homomorphism of $\mathbb{K}$-algebras. A $\mathbb{K}$-linear mapping $D: A \to B$ is called a Φ-*derivation*, if it has the following property:

$$\forall f, g \in A: \quad D(f \cdot g) = D(f) \cdot \Phi(g) + \Phi(f) \cdot D(g).$$

We denote the set of all such derivations by $\mathrm{Der}_\Phi(A, B)$. If $A \subseteq B$ and Φ is the inclusion, then we denote the set of all such derivations by $\mathrm{Der}(A, B)$. If even $A = B$ and $\Phi = \mathrm{id}_A$, then we simply speak of derivations of A and denote the set of all such derivations by $\mathrm{Der}(A)$.

Proposition 6.45 (Vector Fields as Φ-Derivations) *Let M be a differentiable manifold of type $C^{\mathbb{K},k}$ and $U \subseteq M$ open. For each vector field $\mathfrak{X} \in \mathcal{T}_M(U)$,*

$$\widetilde{\mathfrak{X}}: C_M^{\mathbb{K},k}(U) \to C_M^{\mathbb{K},k-1}(U)$$

a Φ-derivation, where $\Phi: C_M^{\mathbb{K},k}(U) \to C_M^{\mathbb{K},k-1}(U)$ is the inclusion. In particular, $\widetilde{\mathfrak{X}} \in \mathrm{Der}\big(C_M^{\mathbb{K},k}(U)\big)$, if $k \in \{\infty, \omega\}$.

Proof We have already calculated the derivation property. The differentiability properties remain to be checked. Since they are of local nature, we can assume that

6.3 Tangent and Tensor Bundles

U is a chart neighborhood. So let (U, φ) be a chart on M and $(\widetilde{U}, \widetilde{\varphi})$ the associated chart on TM. Then $F = \widetilde{\varphi} \circ \mathfrak{X} \circ \varphi^{-1} : \varphi(U) \to \widetilde{\varphi}(\widetilde{U})$ has the form

$$F(x_1, \ldots, x_n) = \big(x_1, \ldots, x_n, A_1(x), \ldots, A_n(x)\big) \quad \text{for } x = (x_1, \ldots, x_n) \in \varphi(U).$$

Since $\mathfrak{X}$ is of class $C^{\mathbb{K},k}$, the functions $A_j : \varphi(U) \to \mathbb{K}$ are also of class $C^{\mathbb{K},k}$. For $f \in C^\infty(M)$ and $p \in U$ we calculate

$$\widetilde{\mathfrak{X}}(f)(p) = \mathfrak{X}(p)(f_p) = \sum_{j=1}^n A_j\big(\varphi(p)\big) \cdot \tfrac{\partial}{\partial x_j}\big|_p(f_p)$$

$$= \sum_{j=1}^n A_j\big(\varphi(p)\big) \cdot \tfrac{\partial (f \circ \varphi^{-1})}{\partial x_j}\big(\varphi(p)\big) = \Big(\sum_{j=1}^n (A_j \circ \varphi) \cdot \big(\tfrac{\partial (f \circ \varphi^{-1})}{\partial x_j} \circ \varphi\big)\Big)(p).$$

Since the $\tfrac{\partial (f \circ \varphi^{-1})}{\partial x_j}$ are of class $C^{\mathbb{K},k-1}$, this also applies to $\widetilde{\mathfrak{X}}(f)$. □

Exercise 6.15 (Derivations as Sheaf) In the situation of Proposition 6.45, show that $\operatorname{Der}\big(C_M^{\mathbb{K},k}(U), C_M^{\mathbb{K},k-1}(U)\big)$ is a $C_M^{\mathbb{K},k}(U)$-module with respect to the operations

$$\forall D_1, D_2 \in \operatorname{Der}\big(C_M^{\mathbb{K},k}(U), C_M^{\mathbb{K},k-1}(U)\big), f \in C^{\mathbb{K},k}(U)$$

$$: \quad (D_1 + D_1)(f) := D_1(f) + D_2(f)$$

and

$$\forall D \in \operatorname{Der}\big(C^{\mathbb{K},k}(U), C^{\mathbb{K},k-1}(U)\big), f, h \in C^{\mathbb{K},k}(U): \quad (h \cdot D)(f) := h \cdot \big(D(f)\big).$$

It follows directly from the definitions that the mappings and operations from Proposition 6.45 and Exercise 6.15 are compatible with restrictions to smaller open sets. Therefore, the assignment $U \mapsto \operatorname{Der}\big(C_M^{\mathbb{K},k}(U), C_M^{\mathbb{K},k-1}(U)\big)$ defines a sheaf, more precisely even a $C_M^{\mathbb{K},k}$-module. We denote it by $\mathcal{D}er(C_M^{\mathbb{K},k}, C_M^{\mathbb{K},k-1})$. When $k \in \{\infty, \omega\}$, i.e., when $k-1 = k$, we write instead $\mathcal{D}er(C_M^{\mathbb{K},k})$.

Theorem 6.46 (Characterization of the Tangential Sheaf by Derivations) *Let M be a differentiable manifold of type $C^{\mathbb{K},k}$ with $k \in \{\infty, \omega\}$. Then the assignment $U \mapsto \tau_U$, for U open in M, with*

$$\tau_U : \mathcal{T}_M(U) \to \mathcal{D}er(C_M^{\mathbb{K},k})(U), \quad \mathfrak{X} \mapsto \widetilde{\mathfrak{X}}$$

an isomorphism $\tau : \mathcal{T}_M \to \mathcal{D}er(C_M^{\mathbb{K},k})$ of $C_M^{\mathbb{K},k}$-modules.

Proof After our preliminary remarks, it is clear that τ is a sheaf morphism. We first construct the inverse morphism. Let $U \subseteq M$ be open and $D \in \mathcal{D}er(C_M^{\mathbb{K},k})(U)$. We can assume that U is a coordinate neighborhood, i.e., we have a chart (U, φ). For $p \in U$ and $f \in C_M^{\mathbb{K},k}(V)$ with $p \in V \subseteq U$ open, $[f]_p \mapsto \big(D|_V(f)\big)_p$

defines a derivation $\mathfrak{X}_p$, i.e., an algebraic tangent vector. Since $k \in \{\infty, \omega\}$, Proposition 6.23 shows that $\mathfrak{X}_p$ is a geometric tangent vector, thus an element of $T_p M$. Proposition 6.21 describes $\mathfrak{X}_p$ in coordinates:

$$\mathfrak{X}_p = \sum_{j=1}^n \left(\mathfrak{X}_p((x_j)_p)\right) \cdot \frac{\partial}{\partial x_j}\Big|_p = \sum_{j=1}^n \left(D(x_j)\right)(p) \cdot \frac{\partial}{\partial x_j}\Big|_p.$$

Thus, the mapping $\mathfrak{X} : M \to TM$, $p \mapsto \mathfrak{X}_p$ is given in local coordinates with respect to the chart of TM associated with (U, φ) by the $C^{\mathbb{K},k}$-functions $D(x_j)$ and is therefore itself of class $C^{\mathbb{K},k}$. This makes $\mathfrak{X} \in \mathcal{T}_M(U)$. It is easy to verify (exercise) that $D \mapsto \mathfrak{X}$ is inverse to τ_U. The constructions are also compatible with restrictions, i.e., the τ_U^{-1} define a sheaf morphism.

It remains only to show that τ is a morphism of $C_M^{\mathbb{K},k}$-modules, i.e., the τ_U are $C_M^{\mathbb{K},k}(U)$-linear mappings $\mathcal{T}_M(U) \to \mathcal{D}er(C_M^{\mathbb{K},k})(U)$. The additivity is immediately clear from the definitions. For scalar multiplication, we calculate with $h, f \in C_M^{\mathbb{K},k}(U)$, $\mathfrak{X} \in \mathcal{T}_M(U)$ and $p \in U$:

$$\begin{aligned}\left((\tau_U(h\mathfrak{X}))(f)\right)(p) &= \left((h\mathfrak{X})(p)\right)(f_p) = \left(h(p)\mathfrak{X}(p)\right)(f_p) \\ &= h(p)\left(\mathfrak{X}(p)\right)(f_p) = h(p)\left((\tau_U(\mathfrak{X}))(f)\right)(p) \\ &= \left(h(\tau_U(\mathfrak{X}))(f)\right)(p).\end{aligned}$$

It follows that $\left(\tau_U(h\mathfrak{X})\right)(f) = h\left(\tau_U(\mathfrak{X})\right)(f)$, so $\tau_U(h\mathfrak{X}) = h\tau_U(\mathfrak{X})$. □

Theorem 6.46 is also useful when one does not want to or cannot assume $k \in \{\infty, \omega\}$. Often, the algebraic objects obtained via the derivation property allow an explicit description in local coordinates, from which one can then read off how much differentiability one needs to assume in order to define a corresponding object. We illustrate this principle using the example of the Lie bracket of vector fields.

Remark 6.47 (Lie Bracket of Vector Fields) In the case of $k \in \{\infty, \omega\}$, if one considers the vector fields as derivations, one can perform the operation of vector fields on functions multiple times in succession. In this way, one finds a Lie algebra structure on each $\mathcal{T}_M(U)$. The *Lie bracket*, i.e., the algebra multiplication, is given by

$$\forall \mathfrak{X}, \mathfrak{Y} \in \mathcal{T}_M(U),\, f \in C_M^{\mathbb{K},k}(U): \quad [\mathfrak{X}, \mathfrak{Y}]f := \mathfrak{X}(\mathfrak{Y}f) - \mathfrak{Y}(\mathfrak{X}f)$$

A simple calculation shows that $[\mathfrak{X}, \mathfrak{Y}]$ is itself a derivation again.

In local coordinates $x_1, \ldots, x_n$ with respect to a chart (U, φ), the vector fields

$$\mathfrak{X}(p) = \sum_{j=1}^{n} a_j(p) \left.\frac{\partial}{\partial x_j}\right|_p \quad \text{and} \quad \mathfrak{Y}(p) = \sum_{k=1}^{n} b_k(p) \left.\frac{\partial}{\partial x_k}\right|_p,$$

yield the Lie bracket as

$$[\mathfrak{X}, \mathfrak{Y}](p) = \sum_{j,k=1}^{n} \left(a_j \frac{\partial b_k}{\partial x_j} - b_j \frac{\partial a_k}{\partial x_j} \right)(p) \left.\frac{\partial}{\partial x_k}\right|_p \qquad (*)$$

Equation $(*)$ allows us to define a Lie bracket of $C^{\mathbb{K},k-1}$-vector fields for any $k \geq 2$ (note that for a $C^{\mathbb{K},k}$-manifold, the tangent bundle TM is only a $C^{\mathbb{K},k-1}$-manifold). However, the result is then only a $C^{\mathbb{K},k-2}$-vector field, i.e., a section of $\mathcal{T}_M(U)$, where TM is considered as a $C^{\mathbb{K},k-1}$-manifold. □

6.4 Differential Forms

Differential forms are special tensor fields. They play a prominent role in the theory of manifolds for several reasons. First, one can take meaningful derivatives of differential forms (and as a result, again obtain differential forms), which is not possible for other tensors without additional structures. Second, certain differential forms can also be integrated, leading to a generalization of the fundamental theorem of calculus (the Stokes theorem), which not only generalizes but also simplifies the integral theorems from vector analysis. The third reason is that differential forms can be shaped into a tool (the De Rham cohomology), with which one can investigate in a prototypical way whether problems that can be solved locally can also be solved globally. We will only address the first two aspects in this section.

For our definition of differential forms as special tensor fields, we use the following definition of vector subbundles.

Definition 6.48 (Vector Subbundle) Let M be a differentiable $C^{\mathbb{K},k}$-manifold and (E, π, M) a $C^{\mathbb{K},k}$-vector bundle over M. A subset $E' \subseteq E$ is called a *vector subbundle* of E, if the sets $E'_p := E' \cap E_p$ are vector subspaces of the fibers $E_p := \pi^{-1}(p)$ and the restriction $\pi|_{E'} : E' \to M$ makes the triple $(E', \pi|_{E'}, M)$ a $C^{\mathbb{K},k}$-vector bundle over M.

Let M now be a differentiable $C^{\mathbb{K},k}$-manifold and $p \in M$. We consider the subspace $\Lambda_r(T_p^*M)$ of skew-symmetric tensors in $\bigotimes_k^0 T_pM = (T_p^*M)^{\otimes r} = T^r(T_p^*M)$ (see Definition 3.44). If we identify $(T_p^*M)^{\otimes r}$ with the r-linear mappings $L_{\mathbb{K}}(T_pM, \ldots, T_pM; \mathbb{K})$ via

$$\lambda_1 \otimes \ldots \otimes \lambda_r \mapsto \Big((v_1, \ldots, v_r) \mapsto \lambda_1(v_1) \cdots \lambda_r(v_r) \Big)$$

(see Remark 6.41), then $\Lambda_r(T_p^*M)$ consists of the alternating r-linear mappings $T_pM \times \ldots \times T_pM \to \mathbb{K}$. Using Corollary 3.47, we can also identify $\Lambda_r(T_p^*M)$ with the subspace $\Lambda^r(T_p^*M)$ of the exterior algebra $\Lambda(T_p^*M)$. We set

$$\bigwedge\nolimits^r T^*M := \coprod_{p \in M} \Lambda_r(T_p^*M) \tag{6.2}$$

and define a projection $\pi : \bigwedge^r T^*M \to M$ by $\pi^{-1}(p) = \Lambda_r(T_p^*M)$ for all $p \in M$. By constructing associated maps, one shows (exercise) that $(\bigwedge^r TM^*, \pi, M)$ is a vector subbundle of $\bigotimes_r^0 TM$.

Definition 6.49 (Differential Forms) Let M be an n-dimensional differentiable $C^{\mathbb{K},k}$-manifold and $r \in \{0, \ldots, n\}$. Then the vector bundle $(\bigwedge^r T^*M, \pi, M)$ defined by (6.2) is called the *r-form bundle*. The sections of this bundle are called *alternating r-forms* or simply *r-forms* on M. One also speaks of *differential forms* of *degree r*. The space of differential forms of degree r on M is also denoted by $\Omega^r(M)$.

Since the differential forms are defined as sections of a vector bundle over M, it is immediately clear that the assignment $U \mapsto \Omega^r(U)$ for open subsets of M is a sheaf Ω_M^r.

Remark 6.50 (Differential Forms in Local Coordinates) Let (U, φ) be a chart on M and let $x_1, \ldots, x_n$ be the corresponding coordinates. Then every differential form $\omega \in \Omega_M^r(U)$ can be written in the form

$$\omega = \sum_{i_1 < \ldots < i_r} a_{i_1, \ldots, i_r} dx_{i_1} \wedge \ldots \wedge dx_{i_r}$$

where the $a_{i_1, \ldots, i_r}$ are functions of class $C^{\mathbb{K},k-1}$. If one combines the indices $i_1, \ldots, i_r$ into a multi-index $I \in \mathbb{N}^r$, this is abbreviated to

$$\omega =: \sum_{|I|=r} a_I dx_I$$

Now let (W, ψ) be another chart with local coordinates $y_1, \ldots, y_n$ and ω in these coordinates is given by

$$\omega = \sum_{i_1 < \ldots < i_r} b_{i_1, \ldots, i_r} dy_{i_1} \wedge \ldots \wedge dy_{i_r} = \sum_{|I|=r} b_I dy_I$$

6.4 Differential Forms

On $U \cap W$ one then has

$$dy_J = dy_{j_1} \wedge \ldots \wedge dy_{j_r} = \sum_{i_1=1}^{n} \frac{\partial y_{j_1}}{\partial x_{i_1}} dx_{i_1} \wedge \ldots \wedge \sum_{i_r=1}^{n} \frac{\partial y_{j_r}}{\partial x_{i_r}} dx_{i_r}$$

$$= \sum_{|I|=r} \det\left(\frac{\partial y_J}{\partial x_I}\right) dx_I,$$

where

$$\left(\frac{\partial y_J}{\partial x_I}\right) = \begin{pmatrix} \frac{\partial y_{j_1}}{\partial x_{i_1}} & \cdots & \frac{\partial y_{j_1}}{\partial x_{i_r}} \\ \vdots & & \vdots \\ \frac{\partial y_{j_r}}{\partial x_{i_1}} & \cdots & \frac{\partial y_{j_r}}{\partial x_{i_r}} \end{pmatrix}.$$

The last equality follows by expanding from the skew symmetry of $\wedge$ and the Laplace expansion formula for the determinant. It thus holds

$$\omega = \sum_{|I|=r} \left(\sum_{|J|=r} b_J \det\left(\frac{\partial y_J}{\partial x_I}\right) \right) dx_I.$$

When $r = n = \dim M$, we obtain

$$\omega = a\, dx_1 \wedge \ldots \wedge dx_n = b\, dy_1 \wedge \ldots \wedge dy_n$$

with $a, b \in C_M^{\mathbb{K},k}(U \cap W)$ and

$$a = b \det\left(\frac{\partial y_J}{\partial x_I}\right).$$

□

The $\wedge$-product on the fibers of the differential form bundle can be transferred to the differential forms.

Remark 6.51 (Exterior Product of Differential Forms) Let M be an n-dimensional differentiable $C^{\mathbb{K},k}$-manifold. According to Remark 3.38, $\Lambda_r(T_p^*M) = 0$ for $r > n$. Thus

$$\bigwedge T^*M := \bigoplus_{r=0}^{n} \left(\bigwedge^r T^*M\right) := \coprod_{p \in M} \left(\bigoplus_{r=0}^{n} \Lambda_r(T_p^*M) \right)$$

is a vector bundle, whose section sheaf $\Omega_M := \bigoplus_{r=0}^{n} \Omega_M^r$ is. We denote the multiplication described in Corollary 3.47 on the fiber $\left(\bigwedge T^*M\right)_p = \bigoplus_{r=0}^{n} \Lambda_r(T_p^*M)$ with $\wedge$ and accordingly define a pointwise multiplication $\wedge$ on the local sections:

$$\forall U \in \mathfrak{T}, \alpha, \beta \in \Omega_M(U) \, p \in U: \quad (\alpha \wedge \beta)(p) := \alpha(p) \wedge \beta(p).$$

With this, an associative $C^{\mathbb{K},k}(U)$-algebra structure is defined on each $\Omega_M(U)$. The multiplications are obviously compatible with restrictions, i.e., $\wedge \colon \Omega_M \times \Omega_M \to \Omega_M$ is a $\mathbf{S}(M, \mathbf{Set})$-morphism. □

The Exterior Derivative

The exterior derivative is a specific operation on differential forms that increases the degree of a differential form by one, but decreases the order of differentiability by one. For zero forms, i.e., functions, the exterior derivative is just the operation

$$d \colon C_M^{\mathbb{K},k}(U) \to \Omega_M^1(U) = \mathcal{T}_M^*(U), \quad f \mapsto df$$

from Proposition 6.34. For the definition of the exterior derivative, we revert to the method developed in Section 6.3 to define tensors via their algebraic effect on vector fields, even when k is finite (see Remarks 6.41 and 6.47 as well as Theorem 6.46). For simplicity, we only describe the definitions and results for the case $k \in \{\infty, \omega\}$ and leave it to the reader to determine the minimal requirements for the differentiability degree of the involved objects via the descriptions in local coordinates.

Definition 6.52 (Exterior Derivative) Let M be a differentiable $C^{\mathbb{K},k}$-manifold and $k \in \{\infty, \omega\}$. For a differential form $\alpha \in \Omega_M^r(U)$ of degree r defined on an open subset $U \subseteq M$, the *exterior derivative* $d\alpha$ of α is given by the formula

$$(d\alpha)(\mathfrak{X}_1, \ldots, \mathfrak{X}_{r+1})$$

$$:= \sum_{i=1}^{r+1} (-1)^{i+1} \mathfrak{X}_i \cdot \alpha(\mathfrak{X}_1, \ldots, \widehat{\mathfrak{X}}_i, \ldots, \mathfrak{X}_{r+1})$$

$$\times \sum_{1 \leq i < j \leq r+1} (-1)^{i+j} \alpha([\mathfrak{X}_i, \mathfrak{X}_j], \mathfrak{X}_1, \ldots, \widehat{\mathfrak{X}}_i, \ldots, \widehat{\mathfrak{X}}_j, \ldots, \mathfrak{X}_{r+1}),$$

defined differential form of degree $r+1$, where the hat over a symbol means that this symbol is removed. Note that the formula for $r = 0$ is just the usual derivative $df \in \Omega_M^1(U) = \mathcal{T}_M^*(U)$ of $f \in \Omega_M^0(U) = C_M^{\mathbb{K},k}(U)$.

To see that the given formula actually leads to a differential form, we need to verify that $d\alpha$ is alternating and $C_M^{\mathbb{K},k}(U)$-linear. The former is demonstrated by

6.4 Differential Forms

a simple algebraic calculation using the skew symmetry of α and the Lie bracket (exercise). For the $C_M^{\mathbb{K},k}(U)$-linearity, the identity

$$\forall \mathfrak{X}, \mathfrak{Y} \in \mathcal{T}_M(U), \ f \in C_M^{\mathbb{K},k}(U): \quad [f\mathfrak{X}, Y] = f[\mathfrak{X}, \mathfrak{Y}] + (\mathfrak{Y}(f))\,\mathfrak{X},$$

which is immediately verified from the definition, is used.

The exterior derivatives are compatible with restrictions and obviously $\mathbb{K}$-linear. Therefore, they define a morphism

$$d: \Omega_M \to \Omega_M$$

of sheaves of $\mathbb{K}$-vector spaces. The forms whose exterior derivative is zero are called *closed*.

Remark 6.53 (Exterior Derivative in Local Coordinates) Let (U, φ) be a chart on M and $x_1, \ldots, x_n$ be the corresponding local coordinates. The form $\omega \in \Omega_M^r(U)$ is as in Remark 6.50 by

$$\omega = \sum_{i_1 < \ldots < i_r} a_{i_1,\ldots,i_r} dx_{i_1} \wedge \ldots \wedge dx_{i_r} = \sum_{|I|=r} a_I dx_I$$

given. Then

$$d\omega = \sum_{|I|=r} da_I \wedge dx_I.$$

It suffices to prove this for $\omega = f dx_I$, and by renumbering we can also assume that $I = (1, \ldots, r)$. Note that

$$d\omega = \sum_{|J|=r+1} d\omega\left(\frac{\partial}{\partial x_{j_1}}, \ldots, \frac{\partial}{\partial x_{j_{r+1}}}\right) dx_J.$$

But the defining formula for $d\omega$ shows $d\omega\left(\frac{\partial}{\partial x_{j_1}}, \ldots, \frac{\partial}{\partial x_{j_{r+1}}}\right) = 0$, unless for at least one $s \in \{1, \ldots, r+1\}$ the tuple $(j_1, \ldots, \widehat{j_s}, \ldots, j_{r+1})$ is a permutation of $(1, \ldots, r)$. Because $j_1 < \ldots < j_{r+1}$, this can only happen if $(j_1, \ldots, j_{r+1}) = (1, \ldots, r, \ell)$ for an $\ell > r$. In this case,

$$d\omega\left(\frac{\partial}{\partial x_{j_1}}, \ldots, \frac{\partial}{\partial x_{j_{r+1}}}\right) = (-1)^r \frac{\partial f}{\partial x_\ell},$$

because the Lie brackets of the base fields $\frac{\partial}{\partial x_j}$ vanish according to the Schwarz theorem. This results in

$$d\omega = \sum_{\ell > r} (-1)^r \frac{\partial f}{\partial x_\ell} dx_1 \wedge \ldots \wedge dx_r \wedge dx_\ell$$

$$= \Big(\sum_{\ell=1}^n \frac{\partial f}{\partial x_\ell} dx_\ell\Big) \wedge dx_1 \wedge \ldots \wedge dx_r = df \wedge dx_I.$$

□

Since every differential form can be written locally as an exterior product of 1-forms, the action of d on the 1-forms, together with the following calculation rules, already determines d.

Proposition 6.54 (Calculation Rules for the Exterior Derivative) *The exterior derivative d on a manifold has the following properties:*

(i) *For $\alpha \in \Omega_M^r(U)$ and $\beta \in \Omega_M^s(U)$ it holds*

$$d(\alpha \wedge \beta) = d\alpha \wedge \beta + (-1)^r \alpha \wedge d\beta.$$

(ii) $d \circ d = 0$.

Proof

(i) We prove this in local coordinates (see Remark 6.53 and Exercise 6.16) and can therefore assume $\alpha = f dx_I$ and $\beta = h dx_J$. Then $\alpha \wedge \beta = fh dx_I \wedge dx_J$, and the representation of the exterior derivative in local coordinates shows $d(\alpha \wedge \beta) = d(fh) \wedge dx_I \wedge dx_J$. With $d\alpha = df \wedge dx_I$, $d\beta = dh \wedge dx_J$ and $d(fh) = fdh + hdf$ we get

$$d(\alpha \wedge \beta) = f\,dh \wedge dx_I \wedge dx_J + h\,df \wedge dx_I \wedge dx_J$$
$$= (-1)^{|I|} f\,dx_I \wedge dh \wedge dx_J + d\alpha \wedge \beta$$
$$= (-1)^r \alpha \wedge d\beta + d\alpha \wedge \beta.$$

(ii) Because of (i), $\tilde{d} := d \circ d$ satisfies

$$\tilde{d}(\alpha \wedge \beta) = \tilde{d}\alpha \wedge \beta + \alpha \wedge \tilde{d}\beta,$$

6.4 Differential Forms

so it is a derivation on the algebra of differential forms. To show that it is equal to 0, it is sufficient to show this for the generators of the algebra, i.e., functions and 1-forms. For a function we calculate

$$(d^2 f)(\mathfrak{X}, \mathfrak{Y}) = \mathfrak{X} \cdot \big((df)(\mathfrak{Y})\big) - \mathfrak{Y} \cdot \big((df)(\mathfrak{X})\big) - df([\mathfrak{X}, \mathfrak{Y}])$$
$$= \mathfrak{X}\mathfrak{Y} \cdot f - \mathfrak{Y}\mathfrak{X} \cdot f - [\mathfrak{X}, \mathfrak{Y}] \cdot f = 0,$$

where we have written $\mathfrak{X} \cdot f$ for the action of $\mathfrak{X}$ on f. One can perform a similar, but more complicated calculation also for the 1-forms. Alternatively, one can also notice that locally the 1-forms are of the form $f\, dx_I$ and $d(f\, dx_I) = df \wedge dx_I$ vanishes according to the above calculation.

□

Exercise 6.16 (Cartan Identity) Show: For every vector field $\mathfrak{X}$ the *Cartan identity*

$$d \circ \iota_\mathfrak{X} + \iota_\mathfrak{X} \circ d = L_\mathfrak{X},$$

holds, where for $\alpha \in \Omega_M^r(U)$ and $\mathfrak{X} \in \mathcal{T}_M(U)$ the *Lie derivative* $L_\mathfrak{X} \alpha$ is given by

$$(L_\mathfrak{X} \alpha)(\mathfrak{X}_1, \ldots, \mathfrak{X}_r) = \mathfrak{X} \cdot \alpha(\mathfrak{X}_1, \ldots, \mathfrak{X}_r) - \sum_{i=1}^r \alpha(\mathfrak{X}_1, \ldots, [\mathfrak{X}, \mathfrak{X}_i], \ldots, \mathfrak{X}_r)$$

and the *contraction* $\iota_\mathfrak{X} \alpha$ is given by

$$(\iota_\mathfrak{X} \alpha)(\mathfrak{X}_1, \ldots, \mathfrak{X}_{k-1}) = \alpha(\mathfrak{X}, \mathfrak{X}_1, \ldots, \mathfrak{X}_{k-1}).$$

Use the Cartan identity to give a coordinate-free proof for Proposition 6.54(i) (Hint: Induction).

Exercise 6.17 (Lie Derivative of Differential Forms) Let $\mathfrak{X}$ be a vector field on M. Show that the Lie derivative $L_\mathfrak{X}$ defines a derivation of $\Omega_M(U)$. That is, differential forms $\alpha, \beta \in \Omega_M(U)$ satisfy

$$L_\mathfrak{X}(\alpha \wedge \beta) = (L_\mathfrak{X} \alpha) \wedge \beta + \alpha \wedge (L_\mathfrak{X} \beta).$$

Further, show the formula

$$L_{f\mathfrak{X}}(\alpha) = f L_\mathfrak{X}(\alpha) + df \wedge \iota_\mathfrak{X}(\alpha)$$

for $f \in \mathcal{C}_M^{\mathbb{K},k}(U)$.

Remark 6.55 (Pullback of Differential Forms) Let M and N be differentiable $\mathcal{C}^{\mathbb{K},k}$-manifolds and $f \colon M \to N$ a $\mathcal{C}^{\mathbb{K},k}$-mapping. Choose open subsets $W \subseteq N$ and $U \subseteq f^{-1}(W) \subseteq M$. Then for each differential form $\omega \in \Omega_N^k(W)$, a differential form $f^*\omega \in \Omega_M^k(U)$ can be defined by

$$f^*\omega(\mathfrak{X}_{1,p}, \ldots, \mathfrak{X}_{k,p}) = \omega\big(f'(p)\mathfrak{X}_{1,p}, \ldots, f'(p)\mathfrak{X}_{k,p}\big)$$

where $\mathfrak{X}_{j,p} \in T_p(M)$ for $j = 1, \ldots, k$. The form $f^*\omega$ is called the *pullback* of ω under f or also the *pulled back form*.

The pulling back of forms is compatible with restrictions and exterior products, but also with the exterior derivative: If $\varphi \in C_M^{\mathbb{K},k}(W)$, then

$$f^*(d\varphi) = d\varphi \circ f' = \varphi' \circ f' = (\varphi \circ f)' = d(f^*\varphi).$$

For $\omega = \varphi\, dy_{i_1} \wedge \ldots \wedge dy_{i_k}$, one then calculates

$$\begin{aligned}
d(f^*\omega) &= d\left((\varphi \circ f) f^* dy_{i_1} \wedge \ldots \wedge f^* dy_{i_k}\right) \\
&= d\left((\varphi \circ f) d(f^* y_{i_1}) \wedge \ldots \wedge d(f^* y_{i_k})\right) \\
&= d(\varphi \circ f) \wedge d(f^* y_{i_1}) \wedge \ldots \wedge d(f^* y_{i_k}) \\
&= f^* d\varphi \wedge d(f^* y_{i_1}) \wedge \ldots \wedge d(f^* y_{i_k}) \\
&= f^* d\varphi \wedge f^* dy_{i_1} \wedge \ldots \wedge f^* dy_{i_k} \\
&= f^*(d\varphi \wedge dy_{i_1} \wedge \ldots \wedge dy_{i_k}) \\
&= f^*(d\omega).
\end{aligned}$$

$\square$

6.5 Integration on Real Manifolds

Let M be an n-dimensional real manifold and $f: M \to \mathbb{R}$ a continuous function. We try to define an integral $\int_M f$. For this, we first assume that the support supp f of the function lies in a coordinate neighborhood U of M, and try $\int_M f = \int_U f$. To define $\int_U f$, we consider a chart (U, φ). Then one could set

$$\int_U f := \int_{\varphi(U)} (f \circ \varphi^{-1})(x)\, dx$$

where the integration on the right side can be defined as the Riemann or the Lebesgue integral, because $\varphi(U) \subseteq \mathbb{R}^n$ is open and $f \circ \varphi^{-1}: \varphi(U) \to \mathbb{R}$ is continuous.

What happens if we change the chart used in the definition? So let (U, ψ) be another chart on M. Then we are led to the integral

$$\int_U' f := \int_{\psi(U)} (f \circ \psi^{-1})(y)\, dy$$

6.5 Integration on Real Manifolds

and the transformation formula provides with the mapping

$$F := \psi \circ \varphi^{-1} : \varphi(U) \to \psi(U),$$

that

$$\int_U' f = \int_{\psi(U)} (f \circ \psi^{-1})(y) \, dy = \int_{\varphi(U)} (f \circ \psi^{-1} \circ F)(x) \left| \det \left(F'(x) \right) \right| dx$$

$$= \int_{\varphi(U)} (f \circ \varphi^{-1})(x) \left| \det \left(F'(x) \right) \right| dx.$$

If $x \mapsto \left| \det \left(F'(x) \right) \right|$ is not constantly equal to 1, this integral will not be equal to $\int_U f$ for arbitrary f. This means that our attempted definition of an integral depends on the choice of the chart, so it has no intrinsic meaning for functions on M.

The key idea to solve this problem is not to want to integrate functions, but other objects that generate the same transformation factor during a change of charts as the transformation formula and thus deliver the same result for each chart. A look at Remark 6.50 shows that the differential forms of degree n almost achieve this: Again, we consider the charts (U, φ) and (U, ψ) and denote the associated local coordinates with $(x_1, \ldots, x_n)$ and $(y_1, \ldots, y_n)$ respectively. Then we consider the forms $\omega \in \Omega_M^n(U)$, which in the y-coordinates are given by $\omega(y) = h(y) \, dy_1 \wedge \ldots \wedge dy_n$. In the x-coordinates, the form according to Remark 6.50 is then given by

$$\omega(x) = g(x) \, dx_1 \wedge \ldots \wedge dx_n = (h \circ F)(x) \det \left(F'(x) \right) dx_1 \wedge \ldots \wedge dx_n.$$

This means that

$$\int_{\varphi(U)} g(x) \, dx = \int_{\varphi(U)} (h \circ F)(x) \det \left(F'(x) \right) dx = \int_{\psi(U)} h(y) \, dy.$$

So we can define an integral $\int_U \omega$ independent of the choice of the chart by

$$\int_U \omega := \int_{\varphi(U)} g(x) \, dx \tag{6.3}$$

if we can ensure that $\det \left(F'(x) \right) > 0$ for all $x \in \varphi(U)$. This condition can be met if M can be covered by chart neighborhoods whose charts only yield positive Jacobian determinants for all chart changes. We call such a set of charts an *oriented atlas* of M. The integral defined in (6.3) depends on the choice of the oriented atlas. If, e.g., you were to multiply the first coordinate by -1 in all charts, you would get another oriented atlas, but the sign of the integral of each n-form would reverse. This behavior gives us the opportunity to replicate the situation of the one-dimensional integral of functions, which is also provided with a sign depending on the direction of integration.

Not all manifolds have an oriented atlas. A prominent example is the *Möbius strip* $(]-1, 1[\times [0, 1])_\sim$ for the equivalence relation that identifies $(x, 0)$ with $(-x, 1)$. For non-orientable manifolds, one considers sections of the so-called *density bundle*, which transform with the *absolute value* of the Jacobian determinant. In this section, however, we restrict ourselves to *oriented manifolds*, i.e., manifolds with an oriented atlas, and only consider families of charts whose changes have positive Jacobian determinants.

So far, we have only tried to integrate first functions and then n-forms whose supports were in a fixed coordinate environment. If you want to integrate general n-forms, you have to break them down into forms with supports in coordinate neighborhoods. In ordinary integration theory, functions are broken down into those with supports in prescribed sets by multiplying them with the characteristic functions of these sets. This of course loses continuity, not to mention higher orders of differentiability, as we find them in the n-forms. Under relatively mild topological conditions, the problem can be solved by the concept of the "partition of unity", which represents the constant function 1 as a sum of differentiable functions with compact support and thus localizes many problems. It is important to note that the regularity of these compactly supported functions cannot be better than C^∞. Analytic functions vanish everywhere if they vanish on an open subset.

Partition of Unity

We begin with the formal definition of a (continuous) partition of unity.

Definition 6.56 (Partition of Unity) Let $(M, \mathfrak{T})$ be a topological space. A *partition of unity* is a family $(\rho_i)_{i \in I}$ of continuous functions $\rho_i : M \to [0, 1]$ with compact support, for which each $p \in M$ only has finitely many $\rho_i(p)$ different from zero and which satisfy

$$\sum_{i \in I} \rho_i(p) = 1$$

If $(U_i)_{i \in I}$ is an open cover of M with $\mathrm{supp}(\rho_i) \subseteq U_i$, then the partition of unity is called *subordinate* to the cover $(U_i)_{i \in I}$.

For the purposes of integration theory, it would be sufficient to have a continuous partition of unity. However, this is not good enough for other situations. Therefore, we show that smooth partitions of unity can also be found. The key to this is the following lemma.

6.5 Integration on Real Manifolds

Fig. 6.9 Cut-off function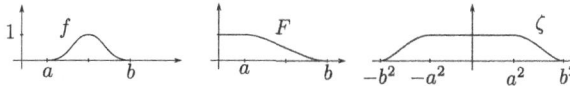

Lemma 6.57 (Cut-Off Functions) *Let M be a differentiable $C^{\mathbb{K},k}$-manifold with $k \neq \omega$. Further, let C be a compact subset of M and V an open neighborhood of C in M. Then there is a function $\chi \in C_M^{\mathbb{R},k}(M)$, which has the value 1 on C and vanishes outside of V.*

Proof We first prove the following claim: For two disjoint subsets A and B of $\mathbb{R}^n$ with A compact and B closed, there is a smooth function $\xi : \mathbb{R}^n \to \mathbb{R}$ that vanishes on B and is constantly equal to 1 on A.

We first show this claim for the special case that A is a (Euclidean) ball and B is the complement of a concentric (open) ball that contains A. We use the rotational symmetry of the situation and first consider the following one-dimensional problem: For $0 < a < b$ consider the function $f : \mathbb{R} \to \mathbb{R}$, which is defined by (see Figure 6.9)

$$f(x) = \begin{cases} \exp\left(\frac{1}{x-b} - \frac{1}{x-a}\right) & \text{for, } a < x < b \\ 0 & \text{otherwise.} \end{cases}$$

Exercise 6.18 Show that f and the function $F : \mathbb{R} \to \mathbb{R}$ defined by

$$F(x) = \frac{\int_x^b f(t)dt}{\int_a^b f(t)dt}$$

are infinitely differentiable (see Figure 6.9).

With F we define the sought function for the specially chosen A and B:

$$\zeta\big((x_1, \ldots, x_n)\big) := F(x_1^2 + \ldots + x_n^2)$$

One finds that $\zeta|_A \equiv 1$ and $\zeta|_B \equiv 0$ for $A = \{x \in \mathbb{R}^n \mid |x|^2 \leq a\}$ and $B = \{x \in \mathbb{R}^n \mid |x|^2 \geq b\}$.

To show the claim for general A and B, we cover A with finitely many open balls K_i (with center $p^{(i)}$ and radius b_i), none of which intersect B. This is possible because A is compact and B is closed.

Exercise 6.19 Show that the numbers $0 < a_i < b_i$ can be chosen so that the closed balls K_i' with center $p^{(i)}$ and radius a_i still cover all of A.

With the already proven special case, we find smooth functions $\zeta_i : \mathbb{R} \to \mathbb{R}$ with $\zeta_i(x) = 1$ for $x \in K_i'$ and $\zeta_i(x) = 0$ for $x \in \mathbb{R}^n \setminus K_i$. With this, we can define the desired function ξ through $\xi := 1 - (1 - \zeta_1)(1 - \zeta_2) \cdots (1 - \zeta_r)$, where r is the number of balls in the cover. This completes the proof of the claim.

Now we can prove the lemma: For every point $p \in C$ there is a chart (U, φ) on M as well as compact neighborhoods K and U' of p with $K \subset \operatorname{int}(U') \subset U \subset V$. Since C is compact, we find m such charts (U_j, φ_j), for which the K_j cover all of C. For each j, the claim provides a function $\zeta_j : \mathbb{R} \to \mathbb{R}$, which is equal to 1 on $\varphi_j(K_j)$ and vanishes outside of $\varphi_j(U_j')$. We define functions $\xi_j : M \to \mathbb{R}$ by

$$\xi_j(p) := \begin{cases} 0 & \text{for } p \in M \setminus U_j, \\ \zeta_j \circ \varphi_j(p) & \text{for } p \in U_j. \end{cases}$$

The ξ_j are smooth, identical to 1 on K_j and equal to 0 on the complement of U_j'. Finally, we define $\xi : M \to \mathbb{R}$ by

$$\xi := 1 - (1 - \xi_1)(1 - \xi_2) \cdots (1 - \xi_m).$$

□

We come to the announced mild assumption on the topology of the manifold that we need to construct a (smooth) partition of unity.

Definition 6.58 (Paracompactness) Let $(M, \mathfrak{T})$ be a topological space and $(U_i)_{i \in I}$ an open cover of M. A second open cover $(V_j)_{j \in J}$ of M is called a *refinement* of $(U_i)_{i \in I}$, if for each $j \in J$ there is an $i \in I$ with $V_j \subseteq U_i$. An open cover $(U_i)_{i \in I}$ is called *locally finite*, if for each $p \in M$ there is an open neighborhood U in M, for which $U \cap U_i \neq \emptyset$ only holds for finitely many $i \in I$. The space $(M, \mathfrak{T})$ is called *paracompact*, if it is a Hausdorff space and every open cover has a locally finite refinement.

It turns out that the condition from Definition 6.1, to have a countable basis of the topology, is already sufficient to guarantee paracompactness.

Proposition 6.59 (Paracompactness of Manifolds) *Every $C^{\mathbb{R},k}$-manifold M is paracompact. More precisely, if $(V_j)_{j \in J}$ is an open cover of M, then there is a family $(U_i, \varphi_i)_{i \in I}$ of charts on M that cover M and have the following properties:*

(i) $(U_i)_{i \in I}$ is a locally finite refinement of $(V_j)_{j \in J}$.
(ii) $\varphi_i : U_i \to B(0, 3) \subseteq \mathbb{R}^3$, where $B(0, r)$ is the open Euclidean ball around zero with radius r, is a homeomorphism.
(iii) The $\widetilde{U}_i := \varphi_i^{-1}(B(0, 1))$ cover M.

6.5 Integration on Real Manifolds

Proof We cover M with a countable family $(W_k)_{k \in \mathbb{N}}$ of open, relatively compact subsets. Then we inductively choose a growing family $(A_k)_{k \in \mathbb{N}}$ of compact sets according to the following scheme: A_1 is the closure of W_1. If A_k is already defined, we set

$$j_k := \min\left\{ j \geq k+1 \,\Big|\, A_k \subseteq \bigcup_{m=1}^{j} W_m \right\}$$

and define A_{k+1} as the closure of $\bigcup_{m=1}^{j_k} W_m$. Then $\bigcup_{k \in \mathbb{N}} A_k = M$, and A_k is contained in the interior A_{k+1}° of A_{k+1}. If now $p \in A_{k+1} \setminus A_k^\circ$ holds, then p is contained in one of the V_j, and there is a coordinate neighborhood $U_{p,j} \subseteq V_j$ of p and a map $\varphi_{p,j} \colon U_{p,j} \to B(0,3)$, which satisfies $\varphi_{p,j}(p) = 0$. We set $\widetilde{U}_{p,j} := \varphi_{p,j}^{-1}(B(0,1))$ and note that the $\widetilde{U}_{p,j}$ cover the compact set $A_{k+1} \setminus A_k^\circ$. So we can find a finite subcover. Let $\widetilde{U}_i$ be the elements of this subcover and U_i the corresponding $U_{p,j}$. The collection of maps that we obtain from all k's has the desired properties. □

Proposition 6.60 (Smooth Partition of Unity) *Let M be a smooth manifold and $(V_j)_{j \in J}$ an open cover of M. Then there is a smooth partition of unity subordinate to this cover.*

Proof We choose a cover by coordinate neighborhoods $(U_i)_{i \in I}$, as constructed in Proposition 6.59. Then Lemma 6.57 provides smooth functions $f_i \colon M \to [0,1]$ with support in U_i, which are constantly equal to 1 on $\widetilde{U}_i$. Since the cover $(U_i)_{i \in I}$ is locally finite, the functions defined by

$$\rho_i(p) := \frac{f_i(p)}{\sum_{i \in I} f_i(p)}$$

for $p \in M$, $\rho_i \colon M \to [0,1]$ indeed form a smooth partition of unity subordinate to $(U_i)_{i \in I}$, and thus also to $(V_j)_{j \in J}$. □

Integration of n-Forms and Stokes' Theorem

Using a partition of unity, we can define integrals of arbitrary n-forms on a paracompact $C^{\mathbb{R},\infty}$-manifold of dimension n.

Definition 6.61 (Integrals of n-Forms) Let M be a n-dimensional $C^{\mathbb{R},\infty}$-manifold and $(U_\alpha, \varphi_\alpha)_\alpha$ an oriented atlas of M. Note that this excludes the case $n = 0$, because the tangent spaces of 0-dimensional manifolds each contain only the zero, so coordinate changes always have 0 as the Jacobian matrix. Further let $\omega \in \Omega^n(M)$.

We can assume that $(U_\alpha)_\alpha$ is a locally finite cover. Let $(\rho_\alpha)_\alpha$ be a smooth partition of unity subordinate to $(U_\alpha)_\alpha$. We define the *integral* of ω by

$$\int_M \omega := \sum_\alpha \int_{U_\alpha} \rho_\alpha \omega,$$

where the integrals on the right side for a given chart are calculated by the formula (6.3). However, we must show that the right side of the equation does not depend on the choice of the cover or the partition of unity. We already know that the formula does not depend on the chart within the oriented atlas when the cover and partition of unity are chosen. So let $(V_\lambda)_\lambda$ and $(\eta_\lambda)_\lambda$ be another choice of cover and partition of unity. Then the following calculation shows the well-definedness of the integral:

$$\sum_\alpha \int_{U_\alpha} \rho_\alpha \omega = \sum_\alpha \int_{U_\alpha} \left(\sum_\lambda \eta_\lambda\right) \rho_\alpha \omega = \sum_\alpha \sum_\lambda \int_{U_\alpha} \eta_\lambda \rho_\alpha \omega$$

$$= \sum_\lambda \sum_\alpha \int_{U_\alpha \cap V_\lambda} \rho_\alpha \eta_\lambda \omega = \sum_\lambda \int_{V_\lambda} \left(\sum_\alpha \rho_\alpha\right) \eta_\lambda \omega = \sum_\lambda \int_{V_\lambda} \eta_\lambda \omega.$$

In our preliminary considerations at the beginning of the section, we already established that the integral of an n-form depends on the choice of the oriented atlas. However, if one replaces the oriented atlas $(U_\alpha, \varphi_\alpha)$ with another oriented atlas (V_β, ψ_β), for which all changes of charts $\psi \circ \beta \circ \varphi_\alpha^{-1}$ have a positive Jacobi determinant, the integrals remain unchanged. We call two such oriented atlases equivalent and note that this indeed defines an equivalence relation on the set of oriented atlases. The equivalence classes of oriented atlases are called *orientations* of M. It can be shown that connected manifolds allow at most two orientations (exercise). The independence of the integral from the choice of the representative of an orientation is the key for both the calculation of concrete integrals and the provability of general theorems about integrals of differential forms, as it allows us to modify the coordinates appropriately to the question and thus apply known results from the integral calculus of one and several variables.

For zero-dimensional M, one could simply define the integral $\int_M \omega$ of a compactly supported 0-form ω by $\sum_{x \in M} \omega(x)$. However, this would mean foregoing the option to orient the different points of M differently, as is done e.g., in the fundamental theorem of calculus with the two boundary points of an interval. Therefore, we also introduce the concept of an *orientation* for zero-dimensional M. This simply refers to a function $o \colon M \to \{\pm 1\}$ and we define the integral

$$\int_M \omega := \sum_{x \in M} o(x) \omega(x)$$

depending on the chosen orientation.

6.5 Integration on Real Manifolds

The aim of this section is the general Stokes theorem, which generalizes the fundamental theorem of calculus and will justify the ad hoc definition of the oriented integral on 0-dimensional manifolds. In Stokes' theorem, a manifold *with boundary* plays the role of the closed interval, with the boundary corresponding to the boundary points of the interval. In order to be able to correctly formulate the prerequisites for differentiability on these sets, we introduce a concept of differentiability on general subsets of $\mathbb{R}^n$.

Definition 6.62 (Differentiability on Non-Open Sets) Let $D \subset \mathbb{R}^n$ be a (arbitrary) subset. A function $f: D \to \mathbb{R}$ is called *smooth*, if for every $x \in D$ there is an open neighborhood $U \subset \mathbb{R}^n$ of x and a smooth function $g: U \to R$ with $g|_{D \cap U} = f|_{D \cap U}$.

Now we can also precisely define what we mean by a manifold with boundary.

Definition 6.63 (Manifold with Boundary) A paracompact Hausdorff space $(M, \mathfrak{T})$, whose topology $\mathfrak{T}$ has a countable basis, is called an *n-dimensional smooth manifold with boundary*, if the following properties are fulfilled:

(a) For every $p \in M$ there is an open neighborhood U and a homeomorphism $\sigma: U \longrightarrow \widetilde{U} \subseteq \mathbb{R}^n$, where $\widetilde{U}$ is either open in $\mathbb{R}^n$ or in $\mathbb{H}_\lambda := \{x \in \mathbb{R}^n \mid \lambda(x) \leq 0\}$, where $\lambda: \mathbb{R}^n \to \mathbb{R}$ is a $\mathbb{R}$-linear mapping.
(b) The coordinate changes $\sigma_j \circ \sigma_i^{-1}: \sigma_i(U_i \cap U_j) \to \sigma_j(U_i \cap U_j)$ are smooth in the sense of Definition 6.62.

In this case, the charts (U, σ) are called charts of M and the U are chart neighborhoods. The *boundary ∂M of M* is the set of $p \in M$ for which there are no maps with an open image in $\mathbb{R}^n$.

Boundary points are mapped to boundary points for changes charts according to the inverse function theorem. Here, as in the definition of a manifold, we could also work with k-times differentiable changes of charts and even for $k = 0$ the concept of the boundary would be well-defined. However, we would then also have to refer back to [MT97, Cor. 7.14] (see Remark 6.6).

Remark 6.64 (The Boundary as a Manifold) Let $(M, \mathfrak{T})$ be a smooth manifold with boundary ∂M. The boundary ∂M is an $(n-1)$-dimensional smooth manifold, because one can define from the maps $\sigma_\alpha: U_\alpha \to \widetilde{U}_\alpha \subseteq \mathbb{H}_\lambda$ the neighborhoods $V_\alpha := U_\alpha \cap \partial M$ and corresponding charts $\psi_\alpha: V_\alpha \to \ker(\lambda_\alpha) \cong \mathbb{R}^{n-1}$. The charts (V_α, ψ_α) define the manifold structure on ∂M by gluing (see Example 6.7). □

Example 6.65 (Manifolds with Boundary)

(i) For $a < b$ in $\mathbb{R}$, the closed interval $[a, b]$ is a manifold with boundary $\{a, b\}$.
(ii) The sphere $M := \overline{B(0; r)} \subset \mathbb{R}^n$ of radius r is a manifold with boundary $\partial M = \mathbb{S}^{n-1}(0; r)$ (the sphere of radius r).
(iii) Suppose the function $f \in C^\infty(\mathbb{R}^n)$ satisfies for $c \in \mathbb{R}$ the condition $f'(x) \neq 0$ for every x with $f(x) = c$. Then $M := \{x \in \mathbb{R}^n \mid f(x) \leq c\}$ is an n-dimensional manifold with boundary $\partial M = \{x \in \mathbb{R}^n \mid f(x) = c\}$.
(iv) Let $T = \mathbb{R}^2/\mathbb{Z}^2$ be a two-dimensional torus. Then $M := \{[x] \in T \mid x_2 \in [0, \frac{1}{2}]$ mod $\mathbb{Z}\}$ is a manifold with boundary.
(v) The set $M := \{x \in \mathbb{S}^n \mid x_{n+1} \geq 0\}$ is a manifold with boundary $\partial M = \{x \in \mathbb{S}^n \mid x_{n+1} = 0\}$. □

Stokes' theorem deals with integrals of differential forms, which we could only define if the underlying manifold was oriented. Therefore, we also need orientations on manifolds with boundary, especially on the boundary.

Remark 6.66 (Orientation on the Boundary) Let M be an n-dimensional smooth manifold with boundary $\partial M \neq \emptyset$. Then $n > 0$. We have a family of charts $(U_\alpha, \varphi_\alpha)_\alpha$, whose chart neighborhoods cover M, called an *oriented atlas* of M, if the Jacobian matrices of the coordinate changes all have positive determinants. We use the same definition also for the manifold with boundary, which is possible according to Definition 6.62, because also at the boundary points the coordinate changes have well-defined Jacobian matrices. It should be noted that this definition would not yield the desired result if we had not allowed arbitrary half-spaces in Definition 6.63, but only lower half-spaces of the form $\{x \in \mathbb{R}^n \mid x_n \leq 0\}$. With such a definition, e.g., the closed interval $[0, 1]$ would not have an orientable atlas.

We can assume without loss of generality that the U_α are all connected. In the case $n = 1$, the images of the charts are then each open or half-open intervals. A point $x \in U_\alpha \cap \partial M$ is mapped by φ_α either to the minimum or the maximum of $\varphi_\alpha(U_\alpha)$. Accordingly, we set $o(x) = -1$ or $o(x) = 1$. This setting is independent of the choice of U_α, because the $(U_\alpha, \varphi_\alpha)$ form an oriented atlas. The orientation of ∂M also does not change if $(U_\alpha, \varphi_\alpha)_\alpha$ is replaced by an equivalent oriented atlas. The orientation of ∂M thus depends only on the given orientation of M.

For $n > 1$, we consider the mappings $\psi_\alpha \colon V_\alpha = U_\alpha \cap \partial M \to \ker(\lambda_\alpha)$ from Remark 6.64. Because $\varphi_\alpha(U_\alpha) \subseteq \mathbb{H}_{\lambda_\alpha}$, $\varphi'_\alpha(x) \colon \mathbb{H}_{\lambda_\alpha} \to \mathbb{H}_{\lambda_\alpha}$ holds for each $x \in V_\alpha$. We choose a basis $v_1^\alpha, \ldots, v_{n-1}^\alpha$ for $\ker(\lambda_\alpha)$ and supplement it with a vector $v_n^\alpha \in \mathbb{H}_{\lambda_\alpha}$ to form a basis for $\mathbb{R}^n$. We can assume that the matrix $(v_1^\alpha, \ldots, v_n^\alpha)$ with the v_j^α as column matrices has a positive determinant (replace v_1 with $-v_1$ if necessary). Let $\widetilde{\psi}_\alpha \colon V_\alpha \to \mathbb{R}^{n-1}$ be the composition of ψ_α with the linear isomorphism given by the basis $v_1^\alpha, \ldots, v_{n-1}^\alpha$ for $\ker(\lambda_\alpha)$. Then the family $(V_\alpha, \widetilde{\psi}_\alpha)_\alpha$ is an orientation of ∂M. To see this, we denote the corresponding coordinates with y_i^α for $i = 1, \ldots, n-1$. The derivative $f'_{\alpha\beta}(\varphi_\alpha(x))$ of the coordinate change from φ_α to φ_β at $x \in U_\alpha \cap U_\beta \cap \partial M$ maps $\mathbb{H}_{\lambda_\alpha}$ bijectively onto $\mathbb{H}_{\lambda_\beta}$. With respect to the bases $v_1^\alpha, \ldots, v_n^\alpha$ and $v_1^\beta, \ldots, v_n^\beta$, the associated matrix has the form

$$A = \begin{pmatrix} A' & * \\ 0 & a \end{pmatrix},$$

where A' is the Jacobian matrix of the coordinate change from $\tilde\psi_\alpha$ to $\tilde\psi_\beta$ at x and $a > 0$. Since the $(U_\alpha, \varphi_\alpha)$ form an orientation, $\det(A) = \det(A')a > 0$. Thus, $\det(A') > 0$, i.e., the $(V_\alpha, \tilde\psi_\alpha)_\alpha$ are an oriented atlas of ∂M. If one replaces $(U_\alpha, \varphi_\alpha)_\alpha$ with an equivalent oriented atlas in this construction, the result is an oriented atlas of ∂M equivalent to $(V_\alpha, \tilde\psi_\alpha)_\alpha$. Thus, here too, the resulting orientation on ∂M depends only on the chosen orientation of M. We call the orientation of ∂M defined by $(V_\alpha, \tilde\psi_\alpha)_\alpha$ the *induced orientation* of the boundary. □

Finally, we extend the concepts of a differential form and the exterior derivative to manifolds with boundary.

Definition 6.67 (Exterior Derivative) Let M be a smooth manifold with boundary. A *smooth differential form* ω of order r on M is a smooth differential form of order r on the manifold (without boundary) $M \setminus \partial M$, for which in each local chart (U_α, x^α) the mappings $\omega^\alpha_{j_1,\ldots,j_k}$, which occur in the coordinate representation

$$\omega = \sum \omega^\alpha_{j_1,\ldots,j_k} dx^\alpha_{j_1} \wedge \ldots \wedge dx^\alpha_{j_k},$$

can be extended to smooth functions on M. For such a form, we define the *exterior derivative* $d\omega$ using the local formulas

$$d\omega := \sum d\omega^\alpha_{j_1,\ldots,j_k} \wedge dx^\alpha_{j_1} \wedge \ldots \wedge dx^\alpha_{j_k}.$$

With this, we can prove the central result of this section. It is not only a generalization of the fundamental theorem of calculus. Once we have set the conceptual framework, the fundamental theorem of calculus is also the essential proof tool.

Theorem 6.68 (Stokes) *Let M be an oriented n-dimensional manifold with boundary ∂M, which is endowed with the induced orientation. Let $i : \partial M \hookrightarrow M$ denote the inclusion. Then for a smooth $n-1$-form $\omega \in \Omega^{n-1}(M)$ on M, it holds that $d\omega \in \Omega^{n-1}(M)$ and*

$$\int_{\partial M} i^*\omega = (-1)^{n-1} \int_M d\omega.$$

The sign $(-1)^{n-1}$ in this theorem is a consequence of the definition of the induced orientation.

Proof We denote the orientation of M with $(U_\alpha, \varphi_\alpha)_\alpha$ and the induced orientation of ∂M with $(V_\alpha, \tilde{\psi}_\alpha)_\alpha$. We can assume that the cover $(U_\alpha)_\alpha$ is locally finite. Then we find a smooth partition of unity $(\rho_\alpha)_\alpha$, which is subordinate to the cover $(U_\alpha)_\alpha$. Then $(\rho_\alpha|_{\partial M})_\alpha$ is a smooth partition of unity on ∂M, which is subordinate to the cover $(V_\alpha)_\alpha$. For $\omega \in \Omega^{n-1}(M)$ we calculate

$$\int_M d\omega = \int_M d\left(\sum_\alpha \rho_\alpha \omega\right) = \int_M \sum_\alpha d(\rho_\alpha \omega) = \sum_\alpha \int_M d(\rho_\alpha \omega).$$

On the other hand, we have

$$\int_{\partial M} i^*\omega = \int_{\partial M} i^*\left(\sum_\alpha \rho_\alpha \omega\right) = \sum_\alpha \int_{\partial M} i^*(\rho_\alpha \omega).$$

So it is enough to prove the theorem for the case that ω is supported in a coordinate neighborhood: $\operatorname{supp}\omega \subset U$. But then we can also assume that $M = U$ is a rectangle in $\mathbb{R}^n$.

Case 1: U contains no boundary point.

In this case, we can assume that $U =]-1, 1[^n$ and it automatically holds $\int_{\partial M} i^*\omega = 0$. For $x = (x_1, \ldots, x_n) \in U$ we find with

$$\omega = \sum_{j=1}^n (-1)^{j-1} \omega_j \, dx_1 \wedge \ldots \wedge \widehat{dx_j} \wedge \ldots \wedge dx_n,$$

that

$$d\omega = \left(\sum_{j=1}^n \frac{\partial \omega_j}{\partial x_j}\right) dx_1 \wedge \ldots \wedge dx_n.$$

Since the orientation of $M = U$ is given by the coordinates, we get

$$\int_U d\omega = \int_U \left(\sum_j \frac{\partial \omega_j}{\partial x_j}\right) dx = \sum_j \int_U \frac{\partial \omega_j}{\partial x_j} dx.$$

Now we write $x^{(j)} := (x_1, \ldots, \widehat{x_j}, \ldots, x_n)$ and $U^{(j)} := \{x^{(j)} \mid x \in U\}$. Since U is a rectangle and $(x^{(j)}, 1), (x^{(j)}, -1) \in \partial U$, we can calculate due to $\operatorname{supp}\omega \subset U$:

$$\int_U \frac{\partial \omega_j}{\partial x_j} dx = \int_{U^{(j)}} \left[\int_{-1}^1 \frac{\partial \omega_j}{\partial x_j}(x^{(j)}, x_j) \, dx_j\right] dx^{(j)}$$

$$= \int_{U^{(j)}} \left[\omega_j(x^{(j)}, 1) - \omega_j(x^{(j)}, -1)\right] dx^{(j)} = 0.$$

6.5 Integration on Real Manifolds

Thus, the theorem is proven in this case.

Case 2: U contains boundary points.

In this case, we have to treat the case $n = 1$ separately. But we start with the case $n > 1$. Then, after orientation-preserving coordinate changes, we can assume that $U =]-1, 1[^{n-1} \times]-1, 0]$ and the orientations are given by the inclusions $U \to \mathbb{R}^n$ and $\partial U =]-1, 1[^{n-1} \times \{0\} = \mathbb{R}^{n-1}$. With the arguments from Case 1, applied to the coordinates $j = 1, \ldots, n-1$, we calculate

$$\int_U d\omega = \sum_{j=1}^n \int_U \frac{\partial \omega_j}{\partial x_j} dx = \int_U \frac{\partial \omega_n}{\partial x_n} dx$$

$$= \int_{U^{(n)}} \left[\int_{-1}^0 \frac{\partial \omega_n}{\partial x_n} (x^{(n)}, x_n) dx_n \right] dx^{(n)}$$

$$= \int_{U^{(n)}} \left[\omega_n(x^{(n)}, 0) - \omega_n(x^{(n)}, -1) \right] dx^{(n)}$$

$$= \int_{U^{(n)}} \omega_n \, dx^{(n)} = \int_{U^{(n)}} \omega_n \, dx_1 \wedge \ldots \wedge dx_{n-1}.$$

The inclusion $i: \partial U \hookrightarrow U$, $(x_1, \ldots, x_{n-1}) \mapsto (x_1, \ldots, x_{n-1}, 0)$ fulfills, due to $\omega = \sum_{j=1}^n (-1)^{j-1} \omega_j \, dx_1 \wedge \ldots \wedge \widehat{dx_j} \wedge \ldots \wedge dx_n$,

$$i^* \omega = \omega \circ Ti = (-1)^{n-1} \omega_n \, dx_1 \wedge \ldots \wedge dx_{n-1}.$$

For $\mathfrak{v}_1, \ldots, \mathfrak{v}_{n-1} \in T_p(\partial U)$ we have

$$\big((i^*\omega)(p)\big)(\mathfrak{v}_1, \ldots, \mathfrak{v}_{n-1}) = \big(\omega(p)\big)(\mathfrak{v}_1, \ldots, \mathfrak{v}_{n-1}),$$

which leads to

$$\int_{\partial U} i^* \omega = \int_{U^{(n)}} (-1)^{n-1} \omega_n \, dx_1 \wedge \ldots \wedge dx_{n-1} = (-1)^{n-1} \int_U d\omega.$$

The case $n = 1$ remains to be treated. In this case, ω is a function and we have either $U = [0, 1[$ or $U =]-1, 0]$. Then $d\omega = \frac{\partial \omega}{\partial x} dx$ and we get

$$\int_U d\omega = \int_0^1 \frac{\partial \omega}{\partial x} dx = -\omega(0) = \int_{\partial U} \omega$$

or

$$\int_U d\omega = \int_{-1}^0 \frac{\partial \omega}{\partial x} dx = \omega(0) = \int_{\partial U} \omega,$$

because $\partial U = \{0\}$ in the first case has the induced orientation $o(0) = -1$ and in the second case $o(0) = 1$.

□

The following corollary is a simple example of a very typical result of our approach, to globalize local results (here the fundamental theorem of calculus) and then to draw conclusions under topological conditions that cannot be realized locally (here, also very typical, compactness).

Corollary 6.69 (Manifolds Without Boundary) *Let M be an oriented compact smooth manifold without boundary and $\omega \in \Omega^{n-1}(M)$. Then $\int_M d\omega = 0$.* □

From Stokes' theorem, the classical integral theorems of Gauss, Stokes, and Green can be derived. We only calculate a simple example here.

Example 6.70 (Circular Disks) For $r > 0$, we consider the manifold $M := \{(x_1, x_2) \in \mathbb{R}^2 \mid x_1^2 + x_2^2 \leq r\}$ with boundary $\partial M = \{(x_1, x_2) \in \mathbb{R}^2 \mid x_1^2 + x_2^2 = r\}$. Furthermore, let U be an open neighborhood of M in $\mathbb{R}^2$ and $f_1, f_2 : U \to \mathbb{R}$ be smooth functions. We consider the 1-form $\omega = f_1 \, dx_1 + f_2 \, dx_2 \in \Omega^1(U)$ on U and pull it back with the inclusion map $i : \partial M \to U$ to a 1-form $i^*\omega = i^*(f_1 \, dx_1 + f_2 \, dx_2) \in \Omega^1(\partial M)$. The exterior derivative $d\omega$ of ω on U is given by $df_1 \wedge dx_1 + df_2 \wedge dx_2$. Stokes' theorem (see Theorem 6.68) yields

$$\int_{\partial M} i^*\omega = -\int_M d\omega,$$

and we want to calculate both sides explicitly.

For the right side, due to $df_i = \frac{\partial f_i}{\partial x_1} dx_1 + \frac{\partial f_i}{\partial x_2} dx_2$ from Definition 6.61 we immediately get

$$\int_M \omega = \int_M \left(\frac{\partial f_2}{\partial x_1} - \frac{\partial f_1}{\partial x_2} \right) d(x_1, x_2).$$

To calculate the left side, we use the parametrization $\gamma :]0, 1] \to \partial M$, $t \mapsto r(\cos 2\pi t, \sin 2\pi t)$ of $\partial M \setminus \{(1, 0)\}$. We pull back $i^*\omega$ again with γ to calculate it in the coordinate t with respect to the chart $(\partial M \setminus \{(1, 0)\}, \gamma^{-1})$. From

$$\gamma^*(i^*\omega)\left(\frac{\partial}{\partial t}\right) = (i \circ \gamma)^*\omega\left(\frac{\partial}{\partial t}\right) = \omega\left((i \circ \gamma)'(t)\left(\frac{\partial}{\partial t}\right)\right)$$

$$= \omega\left(-2\pi r \sin(2\pi t)\frac{\partial}{\partial x_1}, 2\pi r \cos(2\pi t)\frac{\partial}{\partial x_2}\right)$$

$$= -2\pi r \sin(2\pi t) f_1(\gamma(t)) + 2\pi r \cos(2\pi t) f_2(\gamma(t))$$

we obtain $\gamma^*(i^*\omega) = -2\pi r \sin(2\pi t) f_1(\gamma(t)) + 2\pi r \cos(2\pi t) f_2(\gamma(t))$, and Definition 6.61 yields

$$\int_{\partial M\setminus\{(1,0)\}} i^*\omega = \int_{]0,1[} \big(-2\pi r \sin(2\pi t) f_1(\gamma(t)) + 2\pi r \cos(2\pi t) f_2(\gamma(t))\big)\,dt.$$

Since the point $\{(1, 0)\}$ contributes nothing to the integral, using the function $f = (f_1, f_2): U \to \mathbb{R}^2$ and the Euclidean scalar product on $\mathbb{R}^2$ we get

$$\int_{\partial M} i^*\omega = \int_0^1 f(\gamma(t)) \cdot \gamma'(t)\, dt.$$

The right side of this equation is also called the *line integral* of f over γ. With the usual substitution rule (i.e., the transformation formula in one variable), it is also directly seen that it does not depend on the choice of parametrization, but only on $\gamma(]0, 1[)$, i.e., on ∂M. □

6.6 Applications to Complex Differentiability

We want to show, using Example 6.70, that every complex continuously differentiable function on an open subset U of $\mathbb{C}$ is automatically complex analytic, i.e., it can be locally developed into absolutely convergent power series. To do this, we identify $\mathbb{C}$ as a two-dimensional $\mathbb{R}$-vector space with $\mathbb{R}^2$ and compare the real and complex differentiability of a function $U \to \mathbb{C} \cong \mathbb{R}^2$.

Remark 6.71 (Complex and Real Differentiability) If we identify $\mathbb{C}$ with $\mathbb{R}^2$ via $z = \mathrm{Re}\, z + i\,\mathrm{Im}\, z = x + iy \leftrightarrow (x, y)$, then an open subset $U \subseteq \mathbb{C}$ is also open in $\mathbb{R}^2$. A function $f : U \to \mathbb{C}$ can then be understood as a vector-valued function $f = (u, v)\, U \to \mathbb{R}^2$, where we set $f(z) = u(x, y) + iv(x, y)$ with $u, v : U \to \mathbb{R}$. Then f can be real differentiable without having to be complex differentiable. A simple example is the function $\mathbb{C} \to \mathbb{C}, z \mapsto \bar{z}$. It turns out that f is complex differentiable at z_0 if and only if f is real differentiable and its derivative $f'(z_0) : \mathbb{R}^2 \to \mathbb{R}^2$ is complex linear when $\mathbb{R}^2$ is identified with $\mathbb{C}$.

Expressed through the Jacobian matrix with respect to the canonical basis $\{1, i\}$ of $\mathbb{C}$ as a two-dimensional $\mathbb{R}$-vector space, this means

$$\begin{pmatrix} \frac{\partial u}{\partial x} & \frac{\partial u}{\partial y} \\ \frac{\partial v}{\partial x} & \frac{\partial v}{\partial y} \end{pmatrix} = \begin{pmatrix} \frac{\partial \mathrm{Re}\, f}{\partial x} & \frac{\partial \mathrm{Re}\, f}{\partial y} \\ \frac{\partial \mathrm{Im}\, f}{\partial x} & \frac{\partial \mathrm{Im}\, f}{\partial y} \end{pmatrix} = \begin{pmatrix} a & b \\ -b & a \end{pmatrix}.$$

The resulting identities for the partial derivatives are called the *Cauchy-Riemann differential equations*.

If we consider $\mathbb{C}$ as the field of real 2×2 matrices of the form $\begin{pmatrix} a & b \\ -b & a \end{pmatrix}$, then the canonical basis $\{1, i\}$ provides an identification of $\mathbb{C}$ with a subset of $\mathrm{Hom}_{\mathbb{R}}(\mathbb{R}^2, \mathbb{R}^2)$, and the notation $f'(z_0) \in \mathbb{C} \subseteq \mathrm{Hom}_{\mathbb{R}}(\mathbb{R}^2, \mathbb{R}^2)$ becomes unambiguous, regardless of whether we consider the real derivative or the complex derivative of f. □

Let $U \subseteq \mathbb{C}$ now be an open neighborhood of 0 and $f = u + iv \; U \to \mathbb{C}$ continuously complex differentiable. Then $\frac{\partial u}{\partial y} = -\frac{\partial v}{\partial x}$. If we apply the considerations from Example 6.70 to the form $\omega = u dx - v dy \in \Omega^1(U)$ using the notation from Remark 6.71, we find

$$0 = \int_M \left(\frac{\partial u}{\partial y} + \frac{\partial v}{\partial x} \right) d(x, y)$$

$$= \int_0^1 \left(-2\pi r \sin(2\pi t) u(\gamma(t)) - 2\pi r \cos(2\pi t) v(\gamma(t)) \right) dt.$$

Under the identification of $\mathbb{C}$ and $\mathbb{R}^2$, we get $\gamma(t) = r\cos(2\pi t) + ir\sin(2\pi t) = re^{2\pi it}$ and $\gamma'(t) = -2\pi r \sin(2\pi t) + i2\pi r \cos(2\pi t) = 2\pi i r e^{2\pi it}$. It follows that

$$f(\gamma(t))\gamma'(t) = \big((u + iv)(\gamma(t))\big)\big(-2\pi r \sin(2\pi t) + 2\pi i r \cos(t)\big)$$
$$= \big(-2\pi r \sin(2\pi t) u(\gamma(t)) - 2\pi r \cos(2\pi t) v(\gamma(t))\big)$$
$$+ i\big(-2\pi r \sin(2\pi t) v(\gamma(t)) + 2\pi r \cos(2\pi t) u(\gamma(t))\big),$$

in particular,

$$0 = \mathrm{Re}\left(\int_0^1 f(\gamma(t))\gamma'(t)\, dt \right).$$

A similar calculation with $\omega' = v dx - u dy \in \Omega^1(U)$ in view of $\frac{\partial u}{\partial x} = \frac{\partial v}{\partial y}$ first yields

$$0 = \int_M \left(\frac{\partial v}{\partial y} - \frac{\partial u}{\partial x} \right) d(x, y)$$

$$= \int_0^1 \left(-2\pi r \sin(2\pi t) v(\gamma(t)) + 2\pi r \cos(2\pi t) u(\gamma(t)) \right) dt$$

and then

$$0 = \mathrm{Im}\left(\int_0^1 f(\gamma(t))\gamma'(t)\, dt \right).$$

6.6 Applications to Complex Differentiability

Together we thus obtain

$$0 = \int_0^1 f(\gamma(t))\gamma'(t)\,dt.$$

This motivates the following definition.

Definition 6.72 (Complex Line Integral) Let $\gamma : [a, b] \to \mathbb{C}$ be a continuous, piecewise differentiable curve and $f : \gamma([a, b]) \to \mathbb{C}$ continuous. We define the *complex line integral* $\int_\gamma f$ of f over γ by

$$\int_\gamma f := \int_\gamma f(z)dz := \int_a^b f(\gamma(t))\,\gamma'(t)\,dt.$$

With this definition, we can formulate the result of the above calculation as a special form of the Cauchy integral theorem.

Proposition 6.73 (Cauchy Integral Theorem) *Let $U \subseteq \mathbb{C}$ be open and $f \in C^{\mathbb{C},1}(U)$. Then*

$$0 = \int_\gamma f$$

for every curve of the form $\gamma(t) := z_0 + re^{2\pi i t}$ with $\{z \in \mathbb{C} \mid |z - z_0| \leq r\} \subseteq U$.

Proof For each $z_0 \in U$ with $\{z \in \mathbb{C} \mid |z - z_0| \leq r\} \subseteq U$, apply the above calculation to the function $f_{z_0} : U - z_0 \to \mathbb{C}$, $z \mapsto f(z + z_0)$ and the curve $\gamma_{z_0}(t) = re^{2\pi i t}$. □

Example 6.74 (Annuli)

(i) Let $0 < r < R$ and $M := \{z \in \mathbb{C} \mid r \leq |z - z_0| \leq R\}$. Then the boundary ∂M of M consists of two circles, and the induced orientation of the circles is such that they run in opposite directions. We apply the above considerations to a function $f \in C^{\mathbb{C},1}(U)$ whose domain is an open neighborhood of M. Then again $0 = \int_\gamma f$, if γ is a parametrization of ∂M. This time it consists of two parts, $t \mapsto \gamma_r(-t) = z_0 + re^{-2\pi i t}$ and $t \mapsto \gamma_R(t) = z_0 + Re^{2\pi i t}$. So we have

$$\int_{\gamma_r} f = \int_{\gamma_R} f.$$

By varying the radii between r and R, one even shows that the mapping $[r, R] \to \mathbb{C}, s \mapsto \int_{\gamma_s} f$ with $\gamma_s(t) = z_0 + se^{2\pi i t}$ is constant.

(ii) We now consider the set $U_R := \{z \in \mathbb{C} \mid 0 < |z - z_0| < R\}$ and assume that $f \in C^{\mathbb{C},1}(U_R)$ is bounded. With (i), applied to $0 < r' < R' < R$, it follows that

$$\int_{\gamma_{r'}} f = \int_{\gamma_{R'}} f.$$

If $|f(z)| < c$ for all $z \in U_R$, then it follows

$$\left| \int_{\gamma_{r'}} f \right| \leq c \int_0^1 |\gamma'_{r'}(t)| = 2\pi c r'.$$

Now let r' go to 0, it follows

$$0 = \int_{\gamma_{R'}} f.$$

$\square$

Lemma 6.75 (Analyticity of Line Integrals) *Let $\gamma: [a, b] \to \mathbb{C}$ be a piecewise differentiable curve. If $\varphi: \gamma([a, b]) \to \mathbb{C}$ is continuous, then the function defined by*

$$f(z) := \frac{1}{2\pi i} \int_\gamma \frac{\varphi(w)}{w - z} dw$$

is analytic with derivatives

$$f^{(n)}(z) = \frac{n!}{2\pi i} \int_\gamma \frac{\varphi(w)}{(w - z)^{n+1}} dw.$$

Proof Choose $z \in \mathbb{C}$ with $|z - z_0| < d(z_0, \gamma([a, b])) = \mathrm{dist}(z_0, \gamma([a, b])) = \inf_{w \in \gamma([a,b])} |z_0 - w|$. Then

$$\exists c \in \mathbb{R} \, \forall w \in \gamma([a, b]): \quad \left| \frac{z - z_0}{w - z_0} \right| < c < 1.$$

Therefore, the geometric series

$$\sum_{n=0}^\infty \left(\frac{z - z_0}{w - z_0} \right)^n$$

converges uniformly in $w \in \gamma([a,b])$ to

$$\frac{1}{1 - \frac{z-z_0}{w-z_0}} = \frac{w - z_0}{w - z}.$$

With this, one calculates

$$2\pi i\, f(z) = \int_\gamma \frac{\varphi(w)}{w-z} dw = \int_\gamma \sum_{n=0}^\infty \left(\frac{z-z_0}{w-z_0}\right)^n \frac{\varphi(w)}{w-z_0} dw$$

$$= \sum_{n=0}^\infty \int_\gamma \left(\frac{z-z_0}{w-z_0}\right)^n \frac{\varphi(w)}{w-z_0} dw$$

$$= \sum_{n=0}^\infty \left(\int_\gamma \frac{\varphi(w)}{(w-z_0)^{n+1}} dw \right)(z-z_0)^n$$

(for the convergence of the series, we use Fubini's theorem) and find that f is analytic with

$$f^{(n)}(z_0) = \frac{1}{2\pi i} n!\, a_n = \frac{1}{2\pi i} n! \int_\gamma \frac{\varphi(w)}{(w-z_0)^{n+1}} dw.$$

But this was exactly the claim. □

If in the situation of Lemma 6.75 the curve γ is given by $\gamma(t) = z_0 + re^{2\pi i t}$ and φ is constantly equal to 1, then we get

$$\frac{2\pi i}{n!} f^{(n)}(z_0) = \int_\gamma \frac{1}{(w-z_0)^{n+1}} dw = \int_0^1 \frac{1}{r^{n+1}} e^{-2(n+1)\pi i t} 2\pi i r\, e^{2\pi i t} dt$$

$$= \frac{2\pi i}{r^n} \int_0^1 e^{-2n\pi i t} dt = \begin{cases} 0 & \text{for } n > 0 \\ 2\pi i & \text{for } n = 0. \end{cases}$$

With this, we obtain the following corollary to the proof of Lemma 6.75.

Corollary 6.76 (Winding Integral for the Circle) *For the closed curve* $\gamma : [0,1] \to \mathbb{C}$, $z_0 + re^{2\pi i t}$ *with* $z_0 \in U$ *it holds*

$$\forall z \in \{z \in \mathbb{C} \mid |z-z_0| < r\}: \quad 1 = \frac{1}{2\pi i} \int_\gamma \frac{1}{w-z} dw.$$

Theorem 6.77 (Cauchy's Integral Formula) *Let $U \in \mathbb{C}$ be open and $f \in \mathcal{C}^{\mathbb{C},1}(U)$. Then*

$$f(z) = \frac{1}{2\pi i} \int_\gamma \frac{f(w)}{w-z} dw$$

for every curve of the form $\gamma(t) := z_0 + re^{2\pi it}$ with $z \in \{w \in \mathbb{C} \mid |w - z_0| < r\} \subseteq U$.

Proof Let $z \in U \setminus \gamma([a,b])$ be fixed. We consider the function $g : U \to \mathbb{C}$ defined by

$$w \mapsto \begin{cases} \frac{f(w) - f(z)}{w - z} & \text{for } w \neq z, \\ f'(z) & \text{for } w = z. \end{cases}$$

In $U \setminus \{z\}$, g is of the class $\mathcal{C}^{\mathbb{C},1}$. Since f is complex continuously differentiable in z, g is also continuous in z. The argument in Example 6.74(ii) now shows that $\int_\gamma \frac{f(w)-f(z)}{w-z} dw = \int_\gamma g(w) dw = 0$, so due to Corollary 6.76

$$2\pi i f(z) = f(z) \int_\gamma \frac{dw}{w-z} = \int_\gamma \frac{f(w)}{w-z} dw.$$

□

Corollary 6.78 (Holomorphic Functions on $\mathbb{C}$) *The sheaves $\mathcal{C}_\mathbb{C}^{\mathbb{C},1}$ and $\mathcal{C}_\mathbb{C}^{\mathbb{C},\omega}$ coincide.*

Proof Let $U \subseteq \mathbb{C}$ be open and $f \in \mathcal{C}_\mathbb{C}^{\mathbb{C},1}(U)$. We want to show that f is analytic. For this, we choose $z_0 \in U$ and a $r > 0$ with $\{z \in \mathbb{C} \mid |z - z_0| \leq r\} \subseteq U$. Let $\gamma : [0,1] \to \mathbb{C}$, $t \mapsto z_0 + re^{2\pi it}$ and $f^\sharp : \{w \in \mathbb{C} \mid |w - z_0| < r\} \to \mathbb{C}$ be the function defined by

$$f^\sharp(z) := \frac{1}{2\pi i} \int_\gamma \frac{f(w)}{w-z} dw$$

According to Lemma 6.75, $f^\sharp$ is complex analytic. On the other hand, Theorem 6.77 shows that $f^\sharp$ and f coincide on $\{z \in \mathbb{C} \mid |z - z_0| < r\}$. Therefore, f is also complex analytic in a neighborhood of z_0. □

Finally, we want to generalize the result of Corollary 6.78 to $\mathbb{C}^n$. For this, we need to consider products of circular disks, so-called poly-cylinders.

6.6 Applications to Complex Differentiability

Definition 6.79 (Poly-Cylinder) Let $a \in \mathbb{C}$ and $\varrho = (\varrho_1, \ldots, \varrho_n)$ be an n-tuple of positive real numbers, then we call the set

$$P(a, \varrho) := \{z \in \mathbb{C}^n : |z_j - a_j| < \varrho_j \text{ for } j = 1, \ldots, n\}$$

the *polycylinder* with *center* a and *polyradius* ϱ. With $P(a, \varrho)^-$ we denote the closure of $P(a, \varrho)$ in $\mathbb{C}^n$. The set

$$T(a, \varrho) := \{z \in \mathbb{C}^n \mid |z_j - a_j| = \varrho_j \text{ for } j = 1, \ldots, n\}$$

we call the *determining surface* of $P(a, \varrho)^-$. Topologically, this surface is a torus. The sets $P(a, \varrho)$ (and not the higher-dimensional balls) play the role of convergence disks in the theory of several variables.

Now, by iteration, one can find a higher-dimensional analogue of the Cauchy integral formula.

Proposition 6.80 (Cauchy Integral Formula) *Let Ω be open in $\mathbb{C}^n$ and f a holomorphic function on Ω as well as $a \in \Omega$ and $\varrho = (\varrho_1, \ldots, \varrho_n)$ with $\varrho_j > 0$ such, that $P(a, \varrho)^- \subset \Omega$. Then for all $z \in P(a, \varrho)$*

$$f(z) = \frac{1}{(2\pi i)^n} \int_{|\zeta_1 - a_1| = \varrho_1} \cdots \int_{|\zeta_n - a_n| = \varrho_n} \frac{f(\zeta_1, \ldots, \zeta_n)}{(\zeta_1 - z_1) \cdot \ldots \cdot (\zeta_n - z_n)} \, d\zeta_1 \ldots d\zeta_n. \tag{6.4}$$

Proof Let ϱ' be a polyradius for which $P(a, \varrho)^- \subset P(a, \varrho') \subset \Omega$ holds. Choose an arbitrary, but fixed $(\chi_1, \ldots, \chi_{n-1}) \in \mathbb{C}^{n-1}$ with $|\chi_j - a_j| < \varrho'_j$ and consider the function $F_n : \Omega_n \to \mathbb{C}$ on $\Omega_n := \{z_n \in \mathbb{C} \mid (\chi_1, \ldots, \chi_{n-1}, z_n) \in \Omega\} \subset \mathbb{C}$ defined by

$$F_n(z_n) = f((\chi_1, \ldots, \chi_{n-1}, z_n)).$$

As the preimage of an open set with respect to a continuous projection, Ω_n is an open set in $\mathbb{C}$, and according to the conditions on the $|\chi_j - a_j|$, Ω_n contains the circular disk $\{z_n \in \mathbb{C} \mid |z_n - a_n| < \varrho_n\}$. We note further that the function F_n is holomorphic according to the chain rule since the mapping $j_n : \mathbb{C} \to \mathbb{C}^n$, given by $z \mapsto (\chi_1, \ldots, \chi_{n-1}, z)$, is complex linear, thus complex differentiable. Therefore, the Cauchy integral formula in one variable provides that due to $F_n(z_n) = f(\chi_1, \ldots, \chi_{n-1}, z_n)$ and $F_n(\zeta_n) = f(\chi_1, \ldots, \chi_{n-1}, \zeta_n)$, the formula

$$f(\chi_1, \ldots, \chi_{n-1}, z_n) = \frac{1}{2\pi i} \int_{|\zeta_n - a_n| = \varrho_n} \frac{f(\chi_1, \ldots, \chi_{n-1}, \zeta_n)}{\zeta_n - z_n} \, d\zeta_n$$

holds. Analogously, one proceeds with the penultimate variable and obtains

$$f(\chi_1, \ldots, \chi_{n-2}, z_{n-1}, z_n) =$$
$$= \frac{1}{2\pi i} \int_{|\zeta_{n-1}-a_{n-1}|=\varrho_{n-1}} \frac{f(\chi_1, \ldots, \chi_{n-2}, \zeta_{n-1}, z_n)}{\zeta_{n-1} - z_{n-1}} d\zeta_{n-1}$$
$$= \frac{1}{(2\pi i)^2} \int_{|\zeta_{n-1}-a_{n-1}|=\varrho_{n-1}} \int_{|\zeta_n-a_n|=\varrho_n} \frac{f(\chi_1, \ldots, \chi_{n-2}, \zeta_{n-1}, \zeta_n)}{(\zeta_{n-1} - z_{n-1})(\zeta_n - z_n)} d\zeta_n d\zeta_{n-1}.$$

In this way, after n steps, the desired formula is obtained. Note that the order of integration does not matter—one could have started with z_1. □

Lemma 6.81 (Analyticity of Torus Integrals) *Let $P(a, \varrho)$ be a polycylinder in $\mathbb{C}^n$, and $h : T(a, \varrho) \to \mathbb{C}$ a continuous function. Then an analytic function f on $P(a, \varrho)$ is defined by*

$$f(z) = \frac{1}{(2\pi i)^n} \int_{|\zeta_1-a_1|=\varrho_1} \cdots \int_{|\zeta_n-a_n|=\varrho_n} \frac{h(\zeta_1, \ldots, \zeta_n)}{(\zeta_1 - z_1) \cdot \ldots \cdot (\zeta_n - z_n)} d\zeta_1 \ldots d\zeta_n.$$

Proof It suffices to show that f can be represented on $P(a, \varrho)$ by a power series around a. We set

$$c_\alpha := \frac{1}{(2\pi i)^n} \int_{|\zeta_j-a_j|=\rho_j} \frac{h(\zeta)}{\prod_j (\zeta_j - a_j)^{\alpha_j+1}} d\zeta_1 \ldots d\zeta_n \qquad (6.5)$$

and want to show that f is represented by the power series $\sum_{\alpha \in \mathbb{N}^n} c_\alpha (z - a)^\alpha$. To do this, we first verify (exercise) the formula

$$\forall \zeta \in T(a, \varrho), \, z \in P(a, \varrho) : \quad \frac{1}{\prod_j (\zeta_j - z_j)} = \sum_{\alpha \in \mathbb{N}^n} \left(\prod_j \frac{(z_j-a_j)^{\alpha_j}}{(\zeta_j-a_j)^{\alpha_j+1}} \right). \qquad (6.6)$$

Note that the right-hand side of this formula converges uniformly in ζ for fixed $z \in P(a, \varrho)$. We now insert this series into the defining equation of f. The continuity of h in the ζ_1 variable ensures the uniform convergence of the integrand (of the first integral), so we can interchange summation and integration. Now we must use the continuity (not just the continuity in the individual variables) of h to also infer the uniform convergence of the integrand of the second integral and again interchange integration and summation. The claim now follows from n such interchanges. □

Now, analogous to the proof of Corollary 6.78, Proposition 6.80 can be combined with Lemma 6.81 to obtain the following theorem.

Theorem 6.82 (Holomorphic Functions on $\mathbb{C}^n$) *The sheaves $C_{\mathbb{C}^n}^{\mathbb{C},1}$ and $C_{\mathbb{C}^n}^{\mathbb{C},\omega}$ coincide.*

6.6 Applications to Complex Differentiability

To conclude this chapter, we show an exemplary global result that results from combining global topology and local structure: On connected compact complex manifolds, all complex differentiable functions are constant. For this, we need a lemma that is of great importance by itself, because it says that two analytic functions on a connected manifold must coincide if they coincide on an (arbitrarily small) open set.

Lemma 6.83 (Principle of Analytic Continuation) *Let $(M, \mathcal{C}_M^{\mathbb{K},\omega})$ be an analytic manifold and $f \in \mathcal{C}_M^{\mathbb{K},\omega}(M)$. Then the set $E := \{x \in M \mid \forall k \in \mathbb{N}_0 : f^{(k)}(x) = 0\}$ of points where all derivatives of f vanish is open and closed in M.*

Proof Since all derivatives are continuous, E is closed as the intersection of closed sets. To show that E is also open, it suffices to show that there is an open cover of M by open coordinate neighborhoods U with $E \cap U$ open. So we can assume that M is an open subset of $\mathbb{K}^n$. For a multi-index $\alpha \in \mathbb{N}^n$, we denote the corresponding partial derivative by D^α. Then $E = \{x \in M \mid \forall \alpha \in \mathbb{N}^\alpha : D^\alpha f(x) = 0\}$. Now let $x_0 \in E$ and U be a neighborhood of x_0 where the power series $\sum_\alpha c_\alpha (x - x_0)^\alpha$ converges to $f(x)$. Then $c_\alpha = (\alpha!)^{-1} D^\alpha f(x_0) = 0$ for all α. Thus, U is a subset of E, and E is open. $\square$

Theorem 6.84 (Compact Complex Manifolds) *Let $(M, \mathcal{C}_M^{\mathbb{C},1})$ be a connected compact complex manifold and $f \in \mathcal{C}_M^{\mathbb{C},1}(M)$. Then f is constant.*

Proof The compactness of M together with the continuity of f shows that the function $|f| : M \to \mathbb{R}$ has a maximum. Let $z_0 \in M$ be a point where this maximum is attained. The proof strategy is to first show that f is constant on a neighborhood of z_0. Then Lemma 6.83 shows that f is constant.

To show that f is constant in a neighborhood of z_0, we can assume that M is an open coordinate neighborhood in $\mathbb{C}^n$, and then reduce to the case $n = 1$ by restricting the function to the complex line segments $(z_0 + \mathbb{C}v) \cap M$ with $v \in \mathbb{C}^n$. More precisely, we assume that $\{z \in \mathbb{C}^n \mid |z - z_0| \leq R\} \subseteq M$, and for a fixed $v \in \mathbb{C}^n$ with $|v| = 1$ we consider the set $U_v := \{z_0 + \zeta v \mid \zeta \in \mathbb{C}, |\zeta| < R\}$. If we then show that the function $f|_{U_v}$ is constant, then f is constant on the ball around z_0 with radius R.

So now let $\{z \in \mathbb{C} \mid |z - z_0| \leq R\} \subseteq U \subseteq \mathbb{C}$ be open and $f \in \mathcal{C}_{\mathbb{C}}^{\mathbb{C},1}(U)$ with a maximum of $|f|$ at z_0. After multiplying by a constant, we can assume that $f(z_0) \geq 0$. Then the Cauchy integral formula from Theorem 6.77 shows that for $0 < r \leq R$

$$f(z_0) = |f(z_0)| = \left| \int_0^1 f(z_0 + re^{2\pi it}) \, dt \right| \leq M(r) := \sup_t \left| f(z_0 + re^{2\pi it}) \right| \quad (*)$$

holds. Since $|f|$ has its maximum at z_0, it follows that $f(z_0) = M(r)$. The function $h(z) := \operatorname{Re}\bigl(f(z_0) - f(z)\bigr)$ is non-negative on U and $h(z) = 0$ exactly when Re $f(z) = f(z_0)$. Because $|f(z)| \geq \text{Re } f(z)$, then Im $f(z) = 0$, i.e., $f(z) = f(z_0)$. Again with the Cauchy integral formula from Theorem 6.77 it follows that $0 = \int_0^1 h(z_0 + re^{2\pi it}) \, dt$, so the nonnegative function h must vanish on the circle $z_0 + re^{2\pi i \mathbb{R}}$. Thus, $f(z) = f(z_0)$ for all z on this circle. Since $r \leq R$ was arbitrary, f is constant on the entire disk $\{z \in \mathbb{C} \mid R \geq |z - z_0|\}$. □

Literature: There is a huge number of textbooks on topological and differentiable manifolds. The approach via sheaves is rather rarely chosen. Exceptions are [Ra04] and [We15]. The classic books [KN96] and [Sp99] contain the approach via charts and atlases. A relatively direct approach to differential forms and integration can be found in [Sp71]. Complex manifolds are covered in far fewer books than real ones, but there is still a very large selection of texts (see e.g., [GH94], [Ta02] and [We79]).

Algebraic Varieties 7

Compared to manifolds, algebraic varieties are very diverse even locally. A possible entry point into the theory of algebraic varieties is the comparison with submanifolds in $\mathbb{R}^n$ or $\mathbb{C}^n$. We have obtained examples of such submanifolds as zero sets of differentiable functions, whose derivative on the zero set is everywhere different from zero. If one considers polynomial functions instead of differentiable functions, one can also consider such zero sets for other fields. The "algebraic" in algebraic varieties refers to the fact that the functions considered have a polynomial character. Because these functions are simpler than arbitrary differentiable functions, in this context one immediately attempts to say something about zero sets of functions whose (formal) derivatives do not meet any extra regularity conditions. This allows for local singularities, which can look very different. So while the local theory of differentiable manifolds consists of studying calculus on open pieces of $\mathbb{R}^n$ or $\mathbb{C}^n$, the local theory of algebraic varieties consists of studying zero sets of polynomial functions on $\mathbb{K}^n$ for general fields $\mathbb{K}$ and considering functions between such sets that can be considered polynomial in a reasonable way.

With these remarks, it becomes clear that when introducing algebraic varieties, we cannot build on a local theory discussed in elementary courses, but must also develop the local theory first. We do this in Section 7.1 on *algebraic sets*, which we introduce as zero sets of polynomials. On such algebraic sets, we construct sheaves of functions, which are then called *regular* functions. This paves the way for a definition of *algebraic varieties* as ringed spaces that locally look like algebraic sets with their regular functions. Together with the appropriate mappings composed of regular functions, this results in a category of algebraic varieties.

It turns out that there are various technical problems in the study of algebraic sets. For example, polynomials generally have no zeros at all. To guarantee the existence of zeros, one must assume that the field $\mathbb{K}$ is algebraically closed. Since every field $\mathbb{K}$ is contained as a sub-field in an algebraically closed field $\bar{\mathbb{K}}$, one can initially take the position that one considers the zero structures in $\bar{\mathbb{K}}^n$ and only then asks under what circumstances they contain points whose coordinates all lie in $\mathbb{K}$. To be able to systematically deal with such questions, one needs good categorical properties when passing from one field (polynomial ring) to another. Since locally the points of algebraic varieties are described as maximal ideals in rings of regular functions, the question of good categorical properties is in particular a question of the behavior of maximal ideals under ring homomorphisms. Unfortunately, this does not have good categorical properties in contrast to the *prime ideals*. Therefore, the mathematical framework for the study of algebraic varieties is expanded once again and considers ringed spaces that locally no longer look like spaces of maximal ideals in rings, but like spaces of prime ideals in rings. The associated geometric concept is that of a *scheme*.

7.1 Algebraic Sets

We begin with the definition of an algebraic set. Let $\mathbb{K}$ be an arbitrary field and $\mathbb{K}[X_1, \ldots, X_n]$ the polynomial ring over $\mathbb{K}$ in n variables. For $a = (a_1, \ldots, a_n) \in \mathbb{K}^n$ let

$$\mathrm{ev}_a : \mathbb{K}[X_1, \ldots, X_n] \to \mathbb{K}, \quad f \mapsto f(a_1, \ldots, a_n)$$

be the evaluation mapping. Then ev_a is a homomorphism of $\mathbb{K}$-algebras.

Definition 7.1 (Algebraic Sets) A subset $V \subseteq \mathbb{K}^n$ is called *algebraic* or, if one wants to emphasize the role of the field $\mathbb{K}$, $\mathbb{K}$-*algebraic*, if there is an ideal $I \trianglelefteq \mathbb{K}[X_1, \ldots, X_n]$ with

$$V = \mathrm{V}(I) := \{a \in \mathbb{K}^n \mid \forall f \in I : f(a) = 0\}.$$

V is then called the *vanishing* or *zero set* of I.

Remark 7.2 (Zariski Topology on $\mathbb{K}^n$) From the definition of the mapping

$$\mathrm{V} : \{I \trianglelefteq \mathbb{K}[X_1, \ldots, X_n] \mid \text{ideals}\} \longrightarrow \{X \subseteq \mathbb{K}^n \mid \text{subsets}\}, \quad I \mapsto \mathrm{V}(I)$$

one immediately derives the following properties of V:

(i) $\mathrm{V}(0) = \mathbb{K}^n$ and $\mathrm{V}(\mathbb{K}[X_1, \ldots, X_n]) = \emptyset$.
(ii) From $I \subseteq J$ follows $\mathrm{V}(I) \supseteq \mathrm{V}(J)$.
(iii) $\mathrm{V}(I_1 \cap I_2) = \mathrm{V}(I_1) \cup \mathrm{V}(I_2)$.
(iv) $\mathrm{V}\left(\sum_{\lambda \in \Lambda} I_\lambda\right) = \bigcap_{\lambda \in \Lambda} \mathrm{V}(I_\lambda)$.

As a consequence of these properties, it follows that the $\mathrm{V}(I)$ form the closed sets of a topology. This is called the *Zariski topology* on $\mathbb{K}^n$. □

7.1 Algebraic Sets

We want to call functions on an algebraic set in $\mathbb{K}^n$ *polynomial* if they can be written as a restriction of a polynomial function on $\mathbb{K}^n$.

Definition 7.3 (Polynomial Functions on Algebraic Sets) Let $V \subseteq \mathbb{K}^n$ be an algebraic set. A function $f \colon V \to \mathbb{K}$ is called a *polynomial function*, if there is a polynomial $F \in \mathbb{K}[X_1, \ldots, X_n]$ with $f = F|_V$, where $F|_V$ is the restriction of the polynomial function $\mathbb{K}^n \to \mathbb{K}$, $a \mapsto F(a)$ associated with F. The ring $\mathbb{K}[V] := \mathbb{K}[X_1, \ldots, X_n]/\mathrm{I}(V)$ with the *vanishing ideal*

$$\mathrm{I}(V) := \{f \in \mathbb{K}[X_1, \ldots, X_n] \mid \forall a \in V : f(a) = 0\} \trianglelefteq \mathbb{K}[X_1, \ldots, X_n]$$

of V is called the *coordinate ring* of V.

The definition of the coordinate ring should be treated with caution. Although it is presented in this form in various texts on elementary algebraic geometry, it only provides the desired results for infinite fields. This is due to the difference between polynomials and polynomial functions, which becomes apparent when evaluating polynomials at points of $\mathbb{K}^n$. This is easiest to see in one variable, where e.g., the polynomial $X^p - X$ with coefficients in the p-element field $\mathbb{K} = \mathbb{Z}/p\mathbb{Z}$ yields the zero function. In this case, Definition 7.3 for $V = \mathbb{K}$ gives the coordinate ring $\mathbb{K}[V] = \mathbb{K}[X]/(X^p - X)$, while the ring relevant for arithmetic geometry is actually $\mathbb{K}[X]$. As soon as one assumes that the field is algebraically closed, this problem disappears, because algebraically closed fields automatically have infinitely many elements: If $\mathbb{K}$ is finite, the polynomial $f(X) = 1 + \prod_{a \in \mathbb{K}}(X - a)$ has no zero in $\mathbb{K}$.

Remark 7.4 (Coordinate Rings)
Let $V \subseteq \mathbb{K}^n$ be an algebraic set.

(i) Since the polynomial ring $\mathbb{K}[X_1, \ldots, X_n]$ is a finitely generated $\mathbb{K}$-algebra, its quotient $\mathbb{K}[V] = \mathbb{K}[X_1, \ldots, X_n]/\mathrm{I}(V)$ is also a finitely generated $\mathbb{K}$-algebra.

(ii) For $F, G \in \mathbb{K}[X_1, \ldots, X_n]$ it holds

$$F|_V \equiv G|_V \quad \Leftrightarrow \quad F - G \in \mathrm{I}(V).$$

Thus, $\mathbb{K}[V]$ is a ring of functions on V. More precisely, $\mathbb{K}[V]$ is the smallest ring that contains the constant functions and the coordinate functions $a = (a_1, \ldots, a_n) \xmapsto{x_j} a_j$.

(iii) If $f \in \mathbb{K}[V]$ satisfies $f^n = 0 \in \mathbb{K}[V]$, then $\bigl(f(a)\bigr)^n = 0 \in \mathbb{K}$ for all $a \in V$. But then $f(a) = 0$ for all $a \in V$, so according to (ii) $f = 0$. This means, in $\mathbb{K}[V]$ there are no nilpotent elements except zero. By definition, this means that the ring $\mathbb{K}[V]$ is *reduced*. □

We collect some further properties of the mapping

$$\mathrm{I}: \{X \subseteq \mathbb{K}^n \mid \text{subsets}\} \longrightarrow \{I \trianglelefteq \mathbb{K}[X_1, \ldots, X_n] \mid \text{ideals}\}.$$

Proposition 7.5 (Vanishing Ideals)

(i) $\forall X \subseteq Y \subseteq \mathbb{K}^n : \quad \mathrm{I}(X) \supseteq \mathrm{I}(Y)$.
(ii) $\forall X \subseteq \mathbb{K}^n : \quad X \subseteq \mathrm{V}(\mathrm{I}(X))$.
(iii) $X = \mathrm{V}(\mathrm{I}(X))$ *holds in (ii) exactly when X is algebraic.*
(iv) $\forall J \trianglelefteq \mathbb{K}[X_1, \ldots, X_n] : \quad J \subseteq \mathrm{I}(\mathrm{V}(J))$.

Proof The points (i), (ii) and (iv) follow immediately from the definitions. In (iii), $X = \mathrm{V}(\mathrm{I}(X))$ implies by definition that X is algebraic. Conversely, if X is algebraic, i.e., of the form $X = \mathrm{V}(I)$ with $I \trianglelefteq \mathbb{K}[X_1, \ldots, X_n]$, then (iv) provides that $I \subseteq \mathrm{I}(X)$ and therefore $\mathrm{V}(\mathrm{I}(X)) \subseteq \mathrm{V}(I) = X \subseteq \mathrm{V}(\mathrm{I}(X))$ according to Remark 7.2 and (ii). □

Remark 7.6 (Zariski Topology on an Algebraic Set)

Let V be an algebraic set in $\mathbb{K}^n$. We consider the mappings

$$\{I \trianglelefteq \mathbb{K}[V] \mid \text{ideals}\} \underset{\mathrm{I}}{\overset{\mathrm{V}}{\rightleftarrows}} \{X \subseteq V \mid \text{subsets}\},$$

which are defined by

$$\mathrm{V}(I) := \{a \in V \mid \forall f \in I : f(a) = 0\},$$
$$\mathrm{I}(X) := \{f \in \mathbb{K}[V] \mid \forall a \in X : f(a) = 0\}$$

and note that the statements from Remark 7.2 and Proposition 7.5 can be generalized. They just need to be formulated with V instead of $\mathbb{K}^n$. In particular, the $\mathrm{V}(I)$ are the closed sets of a topology on V. This topology is also called the *Zariski topology*. □

Next, we define polynomial mappings between algebraic sets.

Definition 7.7 (Polynomial Mappings)

Let $V \subseteq \mathbb{K}^n$, $W \subseteq \mathbb{K}^m$ be algebraic. A mapping $f : V \to W$ is called *polynomial*, if there are polynomials $F_1, \ldots, F_m \in \mathbb{K}[X_1, \ldots, X_n]$ with $f(a) = (F_1(a), \ldots, F_m(a)) \in \mathbb{K}^m$ for all $a \in V$.

A polynomial mapping $f : V \to W$ is called an *isomorphism* of algebraic sets, if there is a polynomial mapping $g : W \to V$ with $g \circ f = \mathrm{id}_V$ and $f \circ g = \mathrm{id}_W$.

A mapping $f : V \to W$ is polynomial if and only if for all $j = 1, \ldots, m$ the coordinate functions $f_j = y_j \circ f$ with $y_j(b_1, \ldots, b_m) = b_j$ are polynomial, i.e., if $f_j \in \mathbb{K}[V]$.

7.1 Algebraic Sets

Fig. 7.1 Parameterization of the circle

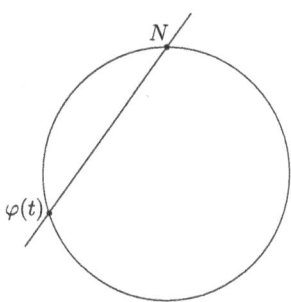

Example 7.8 (Polynomial Mappings)

(i) $\varphi : \mathbb{K}^1 \to C \subseteq \mathbb{K}^3$, $t \mapsto (t^3, t^4, t^5)$ is polynomial.
(ii) $\varphi : \mathbb{R}^1 \to \mathbb{R}^2$, $t \mapsto \frac{1}{t^2+1}(2t, t^2-1)$ is not polynomial (parameterization of the circle; Figure 7.1). □

Theorem 7.9 (Polynomial Mappings and Algebra Homomorphisms) *Let $V \subseteq \mathbb{K}^n$ and $W \subseteq \mathbb{K}^m$ be algebraic sets.*

(i) *A polynomial mapping $f : V \to W$ induces via $f^*(g) := g \circ f$ a $\mathbb{K}$-algebra homomorphism $f^* : \mathbb{K}[W] \to \mathbb{K}[V]$.*
(ii) *Every $\mathbb{K}$-algebra homomorphism $\Phi : \mathbb{K}[W] \to \mathbb{K}[V]$ is of the form $\Phi = f^*$ for a uniquely determined polynomial mapping $f : V \to W$, i.e.,*

$$\{f : V \to W \mid \text{polynomial}\} \to \{\Phi : \mathbb{K}[W] \to \mathbb{K}[V] \mid \mathbb{K}\text{-Algebra-Hom.}\}$$
$$f \mapsto f^*$$

is a bijection.

Proof

(i) Let f be given by $(F_1, \ldots, F_m)$ and g by $G \in \mathbb{K}[Y_1, \ldots, Y_m]$. Then $g \circ f$ is given by $G(F_1, \ldots, F_m)$ (one can substitute polynomials into each other), i.e., $g \circ f$ is polynomial, thus in $\mathbb{K}[V]$. Obviously, $f^*(a) = a$ for $a \in \mathbb{K}$ (constant function). The proof that f^* is indeed an algebra homomorphism is left as an exercise for the reader.
(ii) $y_j : W \to \mathbb{K}$ is the j-th coordinate function for $j = 1, \ldots, m$. It is given by evaluating the polynomials $Y_j \in \mathbb{K}[Y_1, \ldots, Y_m]$, thus it is polynomial. Set $f_j := \Phi(y_j) \in \mathbb{K}[V]$ and $f := (f_1, \ldots, f_m) : V \to \mathbb{K}^m$. First, we want to show that $f(V) \subseteq W$. For this, it is sufficient to show for an arbitrary $G \in I(W) \subseteq \mathbb{K}[Y_1, \ldots, Y_m]$ that for all $Q \in V$ one has $G(f_1(Q), \ldots, f_m(Q)) = 0$. That is, one must show that $G(f_1, \ldots, f_m) = 0 \in \mathbb{K}[V]$. Consider the mapping

$G(y_1, \ldots, y_m) \in \mathbb{K}[W]$, which arises by inserting the mappings $y_j \in \mathbb{K}[W]$ into G. Because $(y_1, \ldots, y_m) \colon W \to \mathbb{K}^m$ is precisely the embedding of W in $\mathbb{K}^m$ and G lies in the vanishing ideal of W, it holds $G(y_1, \ldots, y_m) = 0$. Since Φ is a $\mathbb{K}$-algebra homomorphism, we have

$$G(f_1, \ldots, f_m) = G(\Phi(y_1), \ldots, \Phi(y_m)) = \Phi(G(y_1, \ldots, y_m))$$
$$= \Phi(0) = 0 \in \mathbb{K}[V].$$

Now we know $f(V) \subseteq W$, and together it follows that the mapping $f\colon V \to W$ is polynomial.

Note: $f^*(y_j) = f_j = \Phi(y_j)$, and the y_j generate $\mathbb{K}[W]$, so we have $f^* = \Phi$. The proof of uniqueness is left as an exercise to the reader. □

Corollary 7.10 (Isomorphisms of Algebraic Sets) *A polynomial mapping $f \colon V \to W$ is an isomorphism if and only if $f^* \colon \mathbb{K}[W] \to \mathbb{K}[V]$ is a $\mathbb{K}$-algebra isomorphism.*

Proof If $g\colon W \to V$ with $g \circ f = \mathrm{id}_V$, then $f^* \circ g^* = (g \circ f)^* = \mathrm{id}_V^* = \mathrm{id}_{\mathbb{K}(V)}$. The reverse implication follows from the uniqueness statement in Theorem 7.9. □

Example 7.11 (Polynomial Mappings)
The mapping $\varphi \colon \mathbb{K}^1 \to C := \{(a, b) \in \mathbb{K}^2 \mid b^2 = a^3\} \subseteq \mathbb{K}^2$, $t \mapsto (t^2, t^3)$ is polynomial. If $|\mathbb{K}| = \infty$, then

$$\varphi^* \colon \mathbb{K}[C] = \mathbb{K}[Y_1, Y_2]/(Y_2^2 - Y_1^3) \to \mathbb{K}[\mathbb{K}^1] = \mathbb{K}[X], \quad y_1 \mapsto X^2, \; y_2 \mapsto X^3.$$

In this case, we have $\mathrm{im}\,(\varphi^*) = \mathbb{K}[X^2, X^3] \subsetneq \mathbb{K}[X]$, so φ^* is not an isomorphism.

The condition $|\mathbb{K}| = \infty$ was used for the identification $\mathbb{K}[\mathbb{K}^1] = \mathbb{K}[X]$ and the description of $\mathbb{K}[C]$. In the case of $\mathbb{K} = \{0, 1\}$, the vanishing ideal of $\mathbb{K}^1$ is equal to $(X(X-1)) \trianglelefteq \mathbb{K}[X]$. If x is the image of X in $\mathbb{K}[\mathbb{K}^1]$, then in this case $x^2 = x$, so also $x^3 = x$, and φ^* is surjective. Note that $\mathbb{K}[\mathbb{K}^1] = \mathbb{K} + \mathbb{K}x$ is only two-dimensional. The vanishing ideal $I(C)$ of C is in this case generated by $Y_1 - Y_2$ and $Y_1^2 - Y_1$ (exercise). One also sees that $C = \{(a, a) \in \mathbb{K}^2 \mid a \in \mathbb{K}\}$. The coordinate ring is generated by $y_1 := Y_1 + I(C)$ and is also two-dimensional: $\mathbb{K}[C] = \mathbb{K} + \mathbb{K}y_1$. Because $\varphi^*(y_1) = x$, φ^* is in this case an isomorphism. □

Exercise 7.1 (Algebraic Sets and $\mathbb{K}$-Algebras) Let $\mathrm{Alg}_{\mathbb{K}}^{\mathrm{fgr}}$ be the full subcategory of $\mathrm{Alg}_{\mathbb{K}}$, whose objects are the reduced finitely generated $\mathbb{K}$-algebras ("fg" as in *finitely generated* and "r" as in *reduced*). Show that the algebraic subsets of spaces of the form $\mathbb{K}^n$ with $n \in \mathbb{N}$ together with the polynomial mappings form a category $\mathbf{AS}_\mathbb{K}$ and by the assignments

$$V \mapsto \mathbb{K}[V], \quad f \mapsto f^*$$

a faithful contravariant functor $\mathbf{AS}_\mathbb{K} \to \mathbf{Alg}_\mathbb{K}^{\text{fgr}}$ is defined.

In general, the functor from Exercise 7.1 does not provide an equivalence of the categories $\mathbf{AS}_\mathbb{K}$ and $\mathbf{Alg}_\mathbb{K}^{\text{fgr}}$ (see Proposition 4.59). However, we will show that this is the case for algebraically closed $\mathbb{K}$. In this case, for each $V \in \text{ob}(\mathbf{AS}_\mathbb{K})$ we also find a natural sheaf $\mathcal{O}_V$ of rings on V, which makes $(V, \mathcal{O}_V)$ a ringed space.

Hilbert's Nullstellensatz

Hilbert's Nullstellensatz is the essential tool for proving the equivalence of the categories $\mathbf{AS}_\mathbb{K}$ and $\mathbf{Alg}_\mathbb{K}^{\text{fgr}}$ for algebraically closed $\mathbb{K}$. The following example gives a hint as to why the algebraic closure of $\mathbb{K}$ might play a role in the question of characterizing algebraic sets by properties of their coordinate rings.

Example 7.12 (Zero Sets and Vanishing Ideals)

(i) If $\mathbb{K}$ is not algebraically closed and $f \in \mathbb{K}[X]$ has neither zeros in $\mathbb{K}$ nor is constant, then $J := (f) \neq \mathbb{K}[X]$, but due to $V(J) = \emptyset$ we have $I(V(J)) = \mathbb{K}[X]$, i.e., $J \subsetneq I(V(J))$.
(ii) For $\mathbb{R}^2$ and $f = X_1^2 + X_2^2$ we have $V(f) = \{0\} \subseteq \mathbb{R}^2$, but $I(\{0\}) = (X_1, X_2)$, the ideal generated by X_1 and X_2. In particular, $(f) \subsetneq I(V(f))$.
(iii) For $f \in \mathbb{K}[X_1, \ldots, X_n]$, $2 \leq m \in \mathbb{N}$ we have $V(f^m) = V(f)$ and $f \in I(V(f^m))$, but usually not $f \in (f^m)$. □

Example 7.12 also provides hints on the formulation of the Nullstellensatz. It is about formulating conditions that make V and I a pair of mutually inverse mappings. We will have to restrict ourselves to ideals that contain all roots of a ring element along with the element itself. Such ideals are called *radical ideals*.

Definition 7.13 (Radical Ideal) Let R be a ring and $I \trianglelefteq R$ an ideal. The *radical* of I is

$$\text{rad}(I) := \{f \in R \mid \exists n \in \mathbb{N} : f^n \in I\}.$$

If $\text{rad}(I) = I$, then I is a *radical ideal*.

Remark 7.14 (Radical Ideals) Let R be a factorial ring and $I \trianglelefteq R$ an ideal.

(i) $\mathrm{rad}(I)$ is an ideal: If $f, g \in \mathrm{rad}(I)$, then there exist $n, m \in \mathbb{N}$ with $f^n, g^m \in I$ and

$$(f+g)^r = \sum_{l=0}^{r} \binom{r}{l} f^l \, g^{r-l} \in I \text{ for } r \geq n+m-1.$$

(ii) If I is prime, then I is a radical ideal: From $f^n \in I$ it follows that $f^{n-1} \in I$ or $f \in I$ etc.

(iii) Let $f = \prod_j f_j^{n_j}$ with irreducible (see Exercise 1.8) f_j, none of which can be written as a product of another with a unit from R. Then (exercise) the ideal $I = (f)$ satisfies $\mathrm{rad}(I) = (\prod f_j)$. □

Theorem 7.15 (Hilbert's Nullstellensatz) *Let $\mathbb{K}$ be algebraically closed.*

(i) *Every maximal ideal in $\mathbb{K}[X_1, \ldots, X_n]$ is of the form*

$$\mathfrak{m}_a := (X_1 - a_1, \ldots, X_n - a_n)$$

for an $a = (a_1, \ldots, a_n) \in \mathbb{K}^n$, and $\mathfrak{m}_a$ is the vanishing ideal of a.
(ii) *Let $J \subsetneq \mathbb{K}[X_1, \ldots, X_n]$ be an ideal, then $V(J) \neq \emptyset$.*
(iii) *For every ideal $J \trianglelefteq \mathbb{K}[X_1, \ldots, X_n]$ it holds that $I(V(J)) = \mathrm{rad}(J)$.*

The proof of this theorem requires relatively extensive algebraic preparations. Once we have proven it, the following diagram results for algebraically closed fields:

$$\begin{array}{ccc}
\{\text{ideals } I \subset \mathbb{K}[X_1, \ldots, X_n]\} & \xrightleftharpoons[I]{V} & \{\text{ subsets } X \subseteq \mathbb{K}^n\} \\
\cup & & \cup \\
\{\text{radical ideals}\} & \xleftarrow[\text{bijection}]{V = I^{-1}} & \{\text{ algebraic subsets}\}
\end{array}$$

The essential conceptual tools for the proof of Hilbert's Nullstellensatz are the *finite* and the *finitely generated* R-algebras.

Definition 7.16 (Finite and Finitely Generated Algebras) Let $R \subseteq A$ be commutative rings with identity.

(i) A is called a *finitely generated R-algebra*, if there exist $a_1, \ldots, a_n \in A$ with $A = R[a_1, \ldots, a_n]$, i.e., if A is generated as a ring by R and $\{a_1, \ldots, a_n\}$.
(ii) A is called a *finite R-algebra*, if there exist $a_1, \ldots, a_n \in A$ with

$$A = R\,a_1 + \ldots + R\,a_n$$

i.e., if A is generated as an R-module by $\{a_1, \ldots, a_n\}$.

7.1 Algebraic Sets

The following elementary properties of finite algebras will be used repeatedly in the proof of Hilbert's Nullstellensatz.

Proposition 7.17 (Finite Algebras)

(i) Let $A \subseteq B \subseteq C$ be commutative rings with identity. If B is a finite A-algebra and C is a finite B-algebra, then C is a finite A-algebra.

(ii) Let $A \subseteq B$ and B be a finite A-algebra and $x \in B$. Then there exists a normalized polynomial with coefficients in A that annihilates x:

$$x^n + a_{n-1}x^{n-1} + \ldots + a_0 = 0, \quad a_i \in A.$$

(iii) Conversely, if x is the root of a normalized polynomial with coefficients in A, then $B = A[x]$ is a finite A-algebra.

Proof (i) and (iii) are left as an exercise for the reader. For (ii) let $B = Ab_1 + \ldots + Ab_n$ and $x \in B$. Then $xb_i \in B$, i.e., there exist $a_{ij} \in A$ such that $xb_i = \sum_{j=1}^{n} a_{ij}b_j$ or, written differently, $\sum_{j=1}^{n}(x\delta_{ij} - a_{ij})b_j = 0$. Set $M_{ij} := x\delta_{ij} - a_{ij}$ and $\Delta := \det(M_{ij})_{i,j=1,\ldots,n}$. If M^{adj} is the *adjoint matrix* to M known from Cramer's rule, then (exercise) $M^{\mathrm{adj}} M\, b = \Delta b$, where $b = (b_1, \ldots, b_n)^\top$. With $M\, b = 0$ it follows that $\Delta b = 0$, i.e., $\Delta b_i = 0$ for all $i = 1, \ldots, n$.

Note that $1_B \in B$ is a linear combination of the b_i. Therefore, $\Delta 1_B = 0$ and from this follows $\Delta = 0$, i.e.,

$$\det(x\delta_{ij} - a_{ij}) = 0.$$

This is the sought relation. □

Every finitely generated $\mathbb{K}$-algebra can be written as a finite algebra over a polynomial ring. We will prove this for fields $\mathbb{K}$ with infinitely many elements and derive from it Hilbert's Nullstellensatz. For this, we introduce the concept of algebraically independent elements of a $\mathbb{K}$-algebra.

Definition 7.18 (Algebraically Independent Elements) Let $\mathbb{K}$ be a field with infinitely many elements and A a $\mathbb{K}$-algebra. The elements $a_1, \ldots, a_n \in A$ are called *algebraically independent*, if $f(a_1, \ldots, a_n) = 0$ holds for no non-zero polynomial $f \in \mathbb{K}[X_1, \ldots, X_n]$.

If $a_1, \ldots, a_n \in A$ are algebraically independent, then the evaluation mapping $\mathbb{K}[X_1, \ldots, X_n] \to \mathbb{K}[a_1, \ldots, a_n]$ is an isomorphism of $\mathbb{K}$-algebras.

Lemma 7.19 (Polynomials) *Let $\mathbb{K}$ be a field with infinitely many elements and $0 \neq f \in \mathbb{K}[X_1, \ldots, X_n]$. Then there exists an $a \in \mathbb{K}^n$ with $f(a) \neq 0$.*

Proof We assume without loss of generality that X_n appears in f:

$$f = \sum_{j=0}^{m} g_j X_n^j, \quad g_j \in \mathbb{K}[X_1, \ldots, X_{n-1}], \quad g_m \neq 0.$$

We proceed by induction over n: For $n = 1$, $f \in \mathbb{K}[X_1]$ has only finitely many zeros. For $n > 1$, there exists a $b \in \mathbb{K}^{n-1}$ with $g_m(b) \neq 0$, so $f(b, X_n) = \sum_j g_j(b) X_n^j \neq 0$. Thus, there exists an $a_n \in \mathbb{K}$ with $f(b, a_n) \neq 0$, i.e., $a = (b, a_n)$ has the desired properties. □

Lemma 7.19 shows in particular that for fields with infinitely many elements, polynomials and polynomial functions are essentially the same.

Theorem 7.20 (Noether Normalization) *Let $\mathbb{K}$ be a field with infinitely many elements and $A = \mathbb{K}[a_1, \ldots, a_n]$ a finitely generated $\mathbb{K}$-algebra. Then there exists an $m \leq n$ and $y_1, \ldots, y_m \in A$ such that:*

(a) $y_1, \ldots, y_m$ are algebraically independent over $\mathbb{K}$.
(b) A is a finite $\mathbb{K}[y_1, \ldots, y_m]$-algebra.

Proof Let I be the kernel of the evaluation mapping

$$\mathrm{ev}_a : \mathbb{K}[X_1, \ldots, X_n] \to \mathbb{K}[a_1, \ldots, a_n]$$

for $a = (a_1, \ldots, a_n)$ and $0 \neq f \in I$ (for $I = \{0\}$ there is nothing to show). Set

$$\left.\begin{array}{l} a_1' := a_1 - \alpha_1 a_n \\ \vdots \\ a_{n-1}' := a_{n-1} - \alpha_{n-1} a_n \end{array}\right\} \text{ with } \alpha_j \in \mathbb{K}, \text{ which will be chosen appropriately later.}$$

Then $0 = f(a_1' + \alpha_1 a_n, \ldots, a_{n-1}' + \alpha_{n-1} a_n, a_n)$.

Claim: There exist $\alpha_0, \alpha_1, \ldots, \alpha_{n-1} \in \mathbb{K}$ such that

$$\alpha_0 f(X_1' + \alpha_1 X_n, \ldots, X_{n-1}' + \alpha_{n-1} X_n, X_n) = X_n^k + \text{terms of lower order in } X_n,$$

when f is considered as a polynomial in the variable X_n with coefficients in the ring $\mathbb{K}[X_1', \ldots, X_{n-1}']$ with $X_j' := X_j - \alpha_j X_n$ for $j = 1, \ldots, n-1$.

7.1 Algebraic Sets

To this end, set $d := \deg_{X_1,\ldots,X_n} f$ and $f = F_d + G$ with F_d homogeneous of degree d and $\deg G < d$. We calculate

$$f(X_1, \ldots, X_{n-1}, X_n) = f(X_1' + \alpha_1 X_n, \ldots, X_{n-1}' + \alpha_{n-1} X_n, X_n)$$
$$= F_d(\alpha_1, \ldots, \alpha_{n-1}, 1) X_n^d + \text{terms of lower order in } X_n.$$

If $F_d(\alpha_1, \ldots, \alpha_{n-1}, \alpha_n) = 0$ for all $\alpha = (\alpha_1, \ldots, \alpha_n) \in \mathbb{K}^n$ with $\alpha_n \neq 0$, then $F_d(\alpha_1, \ldots, \alpha_{n-1}, 0) = 0$ as well, because $F_d(\alpha_1, \ldots, \alpha_{n-1}, X)$ is a polynomial. Thus, $F_d(\alpha_1, \ldots, \alpha_{n-1}, \alpha_n) = 0$ for all $\alpha = (\alpha_1, \ldots, \alpha_n) \in \mathbb{K}^n$. However, this is not the case for $F_d \neq 0$ according to Lemma 7.19. Therefore, there exists an $\alpha = (\alpha_1, \ldots, \alpha_n) \in \mathbb{K}^n$ with $\alpha_n \neq 0$ and $F_d(\alpha_1, \ldots, \alpha_{n-1}, \alpha_n) \neq 0$. If we now divide by α_n, we obtain a $(\alpha_1, \ldots, \alpha_{n-1}, 1) \in \mathbb{K}^n$ with $F_d(\alpha_1, \ldots, \alpha_{n-1}, 1) \neq 0$. This proves the claim.

Now let the $\alpha_0, \ldots, \alpha_{n-1}$ be chosen as in the claim. Then

$$0 = a_n^k + \text{terms of lower order in } a_n,$$

where the coefficients lie in $A' := \mathbb{K}[a_1', \ldots, a_{n-1}']$, a normalized equation for a_n over A'. According to Proposition 7.17(iii), A is finite over A', because $A = A'[a_n] = \mathbb{K}[a_1', \ldots, a_{n-1}', a_n]$. The proof of the theorem now proceeds by induction over n: For $n = 1$, $A' = \mathbb{K}$, so A is finite over $\mathbb{K}$, and the claim follows with $m = 0$. For $n > 1$, it follows by induction that A' is a finite $\mathbb{K}[y_1, \ldots, y_m]$-algebra, where the y_j are algebraically independent. According to Proposition 7.17(i), A is then a finite $\mathbb{K}[y_1, \ldots, y_m]$-algebra, and the theorem is proven. $\square$

Lemma 7.21 (Fields that are Finite Algebras) *Let A be a field and $B \subseteq A$ a subring. If A is a finite B-algebra, then B is a field.*

Proof For $b \in B \setminus \{0\}$, $b^{-1} \in A$, and according to Proposition 7.17, there exists a normalized polynomial $f = \sum_{j=0}^{n} b_j X^j$ over B with $b_n = 1$ and $f(b^{-1}) = 0$. But then

$$b^{-1} = -(b_{n-1} + b_{n-2} b + \ldots + b_0 b^{n-1}) \in B.$$

$\square$

Lemma 7.22 (Fields that are Finitely Generated Algebras) *Let $\mathbb{K}$ be a field with infinitely many elements and A a finitely generated $\mathbb{K}$-algebra, i.e., of the form $A = \mathbb{K}[a_1, \ldots, a_n]$. If A is a field, then A is algebraic over $\mathbb{K}$, i.e., every element of A is a root of a polynomial with coefficients in $\mathbb{K}$. In particular, $A = \mathbb{K}$, if $\mathbb{K}$ is algebraically closed.*

Proof We first note that according to Theorem 7.20, algebraically independent elements $y_1, \ldots, y_m \in A$ exist, for which A is a finite $\mathbb{K}[y_1, \ldots, y_m]$-algebra. If A is a field, Lemma 7.21 provides that $\mathbb{K}[y_1, \ldots, y_m] \cong \mathbb{K}[X_1, \ldots, X_m]$ is a field. But then $m = 0$ must hold, i.e., A is a finite $\mathbb{K}$-algebra. The claim then follows from Proposition 7.17(ii). □

Proof of Theorem 7.15

(i) Let $\mathfrak{m} \subseteq \mathbb{K}[X_1, \ldots, X_n]$ be a maximal ideal and $\mathbb{L} = \mathbb{K}[X_1, \ldots, X_n]/\mathfrak{m}$. Then $\mathbb{L}$ is a field and we set $\varphi := \pi \circ \iota \colon \mathbb{K} \to \mathbb{L}$ with the inclusion $\iota \colon \mathbb{K} \to \mathbb{K}[X_1, \ldots, X_n]$ and the canonical quotient mapping $\pi \colon \mathbb{K}[X_1, \ldots, X_n] \to \mathbb{L}$. The mapping φ is automatically a field homomorphism as a ring homomorphism, thus injective. $\mathbb{L}$ is generated as a ring over $\varphi(\mathbb{K})$ by $\pi(X_1), \ldots, \pi(X_n)$, so according to Lemma 7.22, $\varphi(\mathbb{K}) \subseteq \mathbb{L}$ is an algebraic field extension. Because $\mathbb{K}$ is algebraically closed, as is the isomorphic field $\varphi(\mathbb{K})$, Lemma 7.22 even provides $\varphi(\mathbb{K}) = \mathbb{L}$.

Now let $b_j \in \mathbb{L}$ be the image of X_j and $a_j := \varphi^{-1}(b_j)$. Then

$$\pi(X_j - a_j) = \pi(X_j) - \pi \circ \iota(a_j) = \varphi(a_j) - \varphi(a_j) = 0,$$

i.e., $X_j - a_j \in \mathfrak{m}$ for $j = 1, \ldots, n$.

For $a := (a_1, \ldots, a_n) \in \mathbb{K}^n$, the ideal $\mathfrak{m}_a := V(a)$ is the kernel of $\mathrm{ev}_a \colon \mathbb{K}[X_1, \ldots, X_n] \to \mathbb{K}$. Because ev_a is surjective, $\mathbb{K}[X_1, \ldots, X_n]/\mathfrak{m}_a \cong \mathbb{K}$ holds, and $\mathfrak{m}_a$ is a maximal ideal in $\mathbb{K}[X_1, \ldots, X_n]$ according to Proposition 1.20.

By considering the Taylor expansion in a for $f \in \mathbb{K}[X_1, \ldots, X_n]$, it is recognized that $\ker(\mathrm{ev}_a) = (X_1 - a_1, \ldots, X_n - a_n)$ holds. So we have shown above that $\mathfrak{m}_a \subseteq \mathfrak{m}$. The maximality of $\mathfrak{m}_a$ then provides the equality.

(ii) Let $J \subsetneq A = \mathbb{K}[X_1, \ldots, X_n]$ be an ideal. According to Zorn's lemma, J is contained in a maximal ideal $\mathfrak{m}$, which according to (i) is of the form $\mathfrak{m}_a$, so that $a \in V(J)$.

(iii) Let $J \subsetneq \mathbb{K}[X_1, \ldots, X_n]$ be an ideal and $f \in \mathbb{K}[X_1, \ldots, X_n]$. Now consider the ideal

$$J_1 := (J, fY - 1) \subseteq \mathbb{K}[X_1, \ldots, X_n, Y].$$

If $q = (a_1, \ldots, a_n, b) \in V(J_1) \subseteq \mathbb{K}^{n+1}$, then $g(a_1, \ldots, a_n) = 0$ for all $g \in J$ and $a := (a_1, \ldots, a_n) \in V(J)$. Because $0 = (fY - 1)(q) = f(a)b - 1$, it follows that $b = f(a)^{-1}$. But if we have chosen $f \in I(V(J))$, then this yields $V(J_1) = \emptyset$, and (ii) shows $J_1 = \mathbb{K}[X_1, \ldots, X_n, Y]$. It follows

$$1 = g_1 f_1 + g_0 (fY - 1) \in \mathbb{K}[X_1, \ldots, X_n, Y] \qquad (*)$$

with $f_1 \in J$, $g_0, g_1 \in \mathbb{K}[X_1, \ldots, X_n, Y]$.

7.1 Algebraic Sets

Let Y^N be the highest power of Y that occurs in g_0 or g_1. Multiply (∗) by f^N to get

$$f^N = G_1(X_1, \ldots, X_n, fY)f_1 + G_0(X_1, \ldots, X_n, fY)(fY - 1).$$

With

$$(fY)^k = (fY - 1 + 1)^k = \sum_{\ell=0}^{k} \binom{k}{\ell}(fY - 1)^\ell$$

this equality modulo $(fY - 1)$ yields a polynomial $h \in \mathbb{K}[X_1, \ldots, X_n]$ with

$$f^N + (fY - 1) = h(X_1, \ldots, X_n)f_1 + (fY - 1) \in \mathbb{K}[X_1, \ldots, X_n, Y]/(fY - 1).$$

Note that the composition

$$\mathbb{K}[X_1, \ldots, X_n] \hookrightarrow \mathbb{K}[X_1, \ldots, X_n, Y] \to \mathbb{K}[X_1, \ldots, X_n, Y]/(fY - 1)$$

is an injective ring homomorphism, because $fY - 1$ cannot divide any non-zero polynomial in $\mathbb{K}[X_1, \ldots, X_n]$. Because $f, h, f_1 \in \mathbb{K}[X_1, \ldots, X_n]$, we can also read the equation $f^N = hf_1$ in $\mathbb{K}[X_1, \ldots, X_n]$. It follows $f^N = h(X_1, \ldots, X_n)f_1 \in J$ and $f \in \operatorname{rad} J$, so $I(V(J)) \subseteq \operatorname{rad} J$. The converse is clear.

□

Theorem 7.23 (Algebraic Sets and $\mathbb{K}$-Algebras) *Let $\mathbb{K}$ be algebraically closed. Then the categories $\mathbf{Alg}_{\mathbb{K}}^{\text{fgr}}$ and $\mathbf{AS}_{\mathbb{K}}$ are equivalent.*

Proof In Exercise 7.1 we saw that the (contravariant) functor defined by the assignments

$$V \mapsto \mathbb{K}[V], \quad f \mapsto f^*$$

is fully faithful. According to Proposition 4.59, it remains only to show that it is also essentially surjective. For this, we need to find an algebraic set V for a reduced finitely generated $\mathbb{K}$-algebra A such that $\mathbb{K}[V]$ is isomorphic to A as a $\mathbb{K}$-algebra.

Let $a_1, \ldots, a_n \in A$ be generators of A, i.e., let $A = \mathbb{K}[a_1, \ldots, a_n]$. Then the homomorphism

$$\Phi \colon \mathbb{K}[X_1, \ldots, X_n] \to A, \quad F \mapsto F(a_1, \ldots, a_n)$$

of $\mathbb{K}$-algebras is surjective. Set $I := \ker(\Phi)$. Then $A \cong \mathbb{K}[X_1, \ldots, X_n]/I$, and because A is reduced, $I = \operatorname{rad}(I)$. With Hilbert's Nullstellensatz (see Theorem 7.15) it follows that $I = \mathrm{I}(\mathrm{V}(I))$, so

$$\mathbb{K}[\mathrm{V}(I)] = \mathbb{K}[X_1, \ldots, X_n]/\mathrm{I}(\mathrm{V}(I)) = \mathbb{K}[X_1, \ldots, X_n]/I \cong A.$$

□

Construction of the Structure Sheaf

We construct the rings $\mathcal{O}_V(U)$ and their restriction mappings for an algebraic set initially only for the elements U of a basis of the topology and then build a sheaf from them.

Definition 7.24 (Standard Open Sets) Let V be an algebraic set and $f \in \mathbb{K}[V]$. Then $D_V(f) := \{a \in V \mid f(a) \neq 0\}$ is called a *standard open set*.

If $\mathrm{V}(f)$ is the zero set of f, which coincides with the zero set of the principal ideal generated by f, then $D_V(f) = V \setminus \mathrm{V}(f)$. Thus, the standard open sets in V are indeed open with respect to the Zariski topology. According to Remark 7.6, for every ideal $I \trianglelefteq \mathbb{K}[V]$

$$\mathrm{V}(I) = \bigcap_{f \in I} \mathrm{V}(f),$$

so by taking complements, every open set is a union of standard open sets. In other words, the standard open sets form a basis for the Zariski topology.

Construction 7.25 (Structure Sheaf of Algebraic Sets) Let V be an algebraic set and $0 \neq f \in \mathbb{K}[V]$. The mapping $r \colon \mathbb{K}[V] \to \{h \ D_V(f) \to \mathbb{K}\}$, $g \mapsto g|_{D_V(f)}$ is a homomorphism of $\mathbb{K}$-algebras. Consider the set $S := \{f^n \mid n \in \mathbb{N}_0\}$, where $f^0 := 1$ is set. Since $f \neq 0$, $D_V(f) \neq \emptyset$, i.e., there exists an $a \in V$ with $f(a) \neq 0$. But then also $f^n(a) = (f(a))^n \neq 0$. Thus, $0 \notin S$. Because $f^n f^m = f^{n+m} \in S$, S satisfies the prerequisites of Exercise 1.4. Let $\mathbb{K}[V]_f := S^{-1}\mathbb{K}[V]$ be the resulting localization of $\mathbb{K}[V]$ at S and $\varphi_f \colon \mathbb{K}[V] \to \mathbb{K}[V]_f$, $g \mapsto \frac{g}{1}$ the localization mapping, which one immediately sees is not only a ring homomorphism, but even a homomorphism of $\mathbb{K}$-algebras, when introducing the scalar multiplication $c \frac{g}{f^n} := \frac{cg}{f^n}$ on $\mathbb{K}[V]_f$.

Because $f(a) \neq 0$ for all $a \in D_V(f)$, the function $f|_{D_V(f)}$ is invertible in the $\mathbb{K}$-algebra $\{h \ D_V(f) \to \mathbb{K}\}$ of all $\mathbb{K}$-valued functions on $D_V(f)$. By

$$\frac{g}{f^n} \mapsto g|_{D_V(f)} (f|_{D_V(f)})^{-n}$$

7.1 Algebraic Sets

a homomorphism $r_f \colon \mathbb{K}[V]_f \to \{h\ D_V(f) \to \mathbb{K}\}$ of $\mathbb{K}$-algebras is defined, which satisfies $r = r_f \circ \varphi_f$. To see the well-definedness, we assume that $\frac{g_1}{f^{n_1}} = \frac{g_2}{f^{n_2}}$. Then there is an $f^\ell \in S$ with $f^\ell(g_1 f^{n_2} - g_2 f^{n_1}) = 0 \in \mathbb{K}[V]$. Since $f^\ell|_{D_V(f)}$ does not vanish anywhere, the restriction of $g_1 f^{n_2} - g_2 f^{n_1}$ to $D_V(f)$ is zero. This in turn shows $g_1|_{D_V(f)}(f|_{D_V(f)})^{-n_1} = g_2|_{D_V(f)}(f|_{D_V(f)})^{-n_2}$. Thus, r_f is well-defined, the other properties then follow immediately from the definitions. If $r_f\left(\frac{g}{f^n}\right) = 0$, then $g|_{D_V(f)} = 0$ and thus $fg = 0$. But then $g \in \ker(\varphi_f)$ and $\frac{g}{f^n} = 0 \in \mathbb{K}[V]_f$. So r_f is injective, and we can consider $\mathbb{K}[V]_f$ as a ring of functions on $D_V(f)$. We set

$$\mathcal{O}_V(D_V(f)) := \mathbb{K}[V]_f.$$

Next, we construct the restriction mappings. For this purpose, let $f_1, f_2 \in \mathbb{K}[V]$ with $D_V(f_1) \subseteq D_V(f_2)$. We would like to restrict elements of $\mathcal{O}_V(D_V(f_2)) = \mathbb{K}[V]_{f_2}$ to $D_V(f_1)$, but no one guarantees us that the function obtained in this way on $D_V(f_1)$ is an element of $\mathcal{O}_V(D_V(f_1)) = \mathbb{K}[V]_{f_1}$. At this point, Hilbert's Nullstellensatz can help us: We have $V(f_2) \subseteq V(f_1)$, so in particular $f_1 \in I(V(f_2))$. If we now knew that $f_1 \in \mathrm{rad}(f_2)$, we could proceed as follows: There is a $g \in \mathbb{K}[V]$ and an $n \in \mathbb{N}$ with $f_1^n = f_2 g$. In particular, g does not vanish at any point of $D_V(f_1)$. If now $\frac{u}{f_2^m} \in \mathbb{K}[V]_{f_2}$, then on $D_V(f_1)$, $ug^m(f_2^m g^m)^{-1} = ug^m(f_1^{nm})^{-1} = \frac{ug^m}{f_1^{nm}} \in \mathbb{K}[V]_{f_1}$. Considered as a mapping, $\frac{ug^m}{f_1^{nm}}$ is nothing other than the restriction of $\frac{u}{f_2^m}$ to $D_V(f_1)$. Thus, the mapping $\mathbb{K}[V]_{f_2} \to \mathbb{K}[V]_{f_1}$, $\frac{u}{f_2^m} \mapsto \frac{ug^m}{f_1^{nm}}$ is a well-defined homomorphism of $\mathbb{K}$-algebras.

From now on, we assume that $\mathbb{K}$ is *algebraically closed*. With the above argument, we then know that the restriction from $D_V(f_2)$ to $D_V(f_1)$ provides a homomorphism $\rho_{f_1, f_2} \colon \mathbb{K}[V]_{f_2} \to \mathbb{K}[V]_{f_1}$ of $\mathbb{K}$-algebras.

In order to conclude that one can assemble a sheaf $U \mapsto \mathcal{O}_V(U)$ from the $\mathcal{O}_V(D_V(f))$ and their restriction mappings, one needs a gluing property for the $\mathcal{O}_V(D_V(f))$. So let $D_V(f)$ be a standard open set that can be written as a union of standard open sets $D_V(f_\alpha)$ with $\alpha \in A$. We assume that $\frac{g_\alpha}{f_\alpha^{n_\alpha}} \in \mathbb{K}[V]_{f_\alpha}$ are given, which are compatible as functions on subsets of V:

$$\forall \alpha, \beta \in A: \quad \left.\frac{g_\alpha}{f_\alpha^{n_\alpha}}\right|_{D_V(f_\alpha) \cap D_V(f_\beta)} = \left.\frac{g_\beta}{f_\beta^{n_\beta}}\right|_{D_V(f_\alpha) \cap D_V(f_\beta)}. \tag{*}$$

We are looking for a $\frac{g}{f^n} \in \mathbb{K}[V]_f$ with

$$\forall \alpha \in A: \quad \frac{g_\alpha}{f_\alpha^{n_\alpha}} = \left.\frac{g}{f^n}\right|_{D_V(f_\alpha)}. \tag{**}$$

Because $D_V(f) = \bigcup_\alpha D_V(f_\alpha) = \bigcup_\alpha D_V(f_\alpha^2)$ holds, we have

$$V(f) = V\left(\sum_\alpha \mathbb{K}[V] f_\alpha^2\right),$$

and Theorem 7.15 provides that $f \in \mathrm{rad}\left(\sum_\alpha \mathbb{K}[V] f_\alpha^2\right)$. Thus, there exists an $n \in \mathbb{N}$ and $a_\alpha \in \mathbb{K}[V]$, of which only finitely many are different from zero, with $f^n = \sum_\alpha a_\alpha f_\alpha^2$. We set $g := \sum_\alpha a_\alpha g_\alpha f_\alpha \in \mathbb{K}[V]$ and claim that this satisfies $(**)$.

For $\alpha, \beta \in A$, $D_V(f_\alpha) \cap D_V(f_\beta) = D_V(f_\alpha f_\beta)$ holds, and from $(*)$ it follows $f_\alpha f_\beta (g_\alpha f_\beta - g_\beta f_\alpha) = 0 \in \mathbb{K}[V]$. With this, we calculate

$$f_\beta^2 g = f_\beta^2 \sum_\alpha a_\alpha g_\alpha f_\alpha = \sum_\alpha a_\alpha g_\alpha f_\beta^2 f_\alpha = \sum_\alpha a_\alpha g_\beta f_\alpha^2 f_\beta = g_\beta f_\beta f^n.$$

On $D_V(f_\beta) \subseteq D_V(f)$, therefore, $f_\beta g = g_\beta f^n$ and thus $(**)$ holds.

The gluing property just proven for the $\mathcal{O}_V(D_V(f))$ shows that we do not create any ambiguity if we now set for each open subset $U \subseteq V$

$$\mathcal{O}_V(U) := \{h U \to \mathbb{K} \mid \forall a \in U\, \exists f \in \mathbb{K}[V] : a \in D_V(f) \subseteq U, h|_{D_V(f)} \in \mathbb{K}[V]_f\}$$

With this, $U \mapsto \mathcal{O}_V(U)$ together with the restrictions of functions becomes a presheaf of commutative $\mathbb{K}$-algebras. The restriction and gluing properties of the $\mathcal{O}_V(D_V(f))$ then also show that $\mathcal{O}_V$ is a sheaf. Thus, $(V, \mathcal{O}_V)$ is a ringed space.

At this point, we have completed our minimal program for describing the local theory of algebraic varieties over algebraically closed fields. We have constructed the local models along with their structure as ringed spaces. Of course, there would be many properties of these local models to describe, and in texts, even elementary ones, on algebraic geometry, this is done. However, since our main concern here is to explain the concept of an algebraic variety as a local structure described by ringed spaces, we will only provide further information about algebraic sets as needed for doing so. In particular, we refrain from describing tangent spaces, which can be constructed in analogy to the algebraic approach for manifolds.

7.2 Algebraic Varieties

Basically, an algebraic variety over an algebraically closed field $\mathbb{K}$ is a $\mathbb{K}$-ringed space i.e., locally isomorphic to a $\mathbb{K}$-algebraic set with its structure sheaf. Similar to the case of manifolds, however, topological a priori assumptions are also made. Unfortunately, no uniform conventions have been established for this to date. Essentially, it is about two properties:

Quasi-Compactness. This refers to the finite covering property, i.e., it is required that every cover by open sets has a finite subcover.

Separability. This property is a separation property in the sense of topology, a relative of the Hausdorff property. It can be formulated in such a way that for a variety X the closedness of $\{(x, x) \in X \times X \mid x \in X\}$ in the product variety $X \times X$

is required. If the topology on $X \times X$ were the product topology, this would be equivalent to the Hausdorff property. However, the product variety is a categorical product in the category of algebraic varieties, and its topology is *not* the product topology.

The difficulty of describing separability a priori explains why many authors do not include it in the definition of an algebraic variety or introduce the concept of a *prevariety* and develop this until they can say what a separated prevariety is, which they then call a variety. There is agreement, however, on quasi-compactness, which all authors include in one form or another in the definition. To see that this setting is sensible, one has to consider whether the local models, which are also supposed to be algebraic varieties, have this property. This can be derived, e.g., from the Noether property of $\mathbb{K}$-algebraic sets, which states that any descending sequence of closed subsets, and accordingly any ascending sequence of open subsets, becomes stationary after finitely many steps. The Noether property of $\mathbb{K}$-algebraic sets holds without any precondition on $\mathbb{K}$. We discuss it here, although the quasi-compactness of algebraic sets could also be proven differently, because it is a fundamental property of algebraic sets for algebraic geometry.

The Noether Property

Coordinate rings of algebraic sets are not only finitely generated $\mathbb{K}$-algebras, they also have exclusively finitely generated ideals, which makes them so-called *Noetherian rings*.

Definition 7.26 (Finitely Generated Ideals) Let R be a commutative ring with identity and $I \trianglelefteq R$ an ideal. Then I is called *finitely generated*, if there are finitely many elements $x_1, \ldots, x_n \in I$ for which I is the smallest ideal that contains $x_1, \ldots, x_n$.

Proposition 7.27 (Noetherian Rings) *Let R be a commutative ring with identity. Then the following properties are equivalent:*

(1) Every ideal $I \trianglelefteq R$ is finitely generated.
(2) Every ascending sequence $I_1 \subset I_2 \subset \ldots$ of ideals becomes stationary.
(3) Every non-empty subset of ideals in R has a maximal element.

If they are fulfilled, R is called a Noetherian ring.

Proof

(1) $\Rightarrow$ (2): $I := \bigcup_{j=1}^{\infty} I_j$ is an ideal in R. Let I be generated by the elements $f_1, \ldots, f_m$. Then there is an I_k with $f_1, \ldots, f_m \in I_k$, so $I_k = I$ and $I_{k+n} = I_k$ for all n.

(2) $\Rightarrow$ (3): This follows directly from the Zorn's lemma.

(3) ⇒ (1): Let $I \trianglelefteq R$ be an ideal and $\Sigma := \{J \subseteq I \mid J \text{ finitely generated ideal}\}$ the set of finitely generated subideals in I. Due to (3), there is a maximal element J_o of Σ. If $J_o \neq I$, then there is an $f \in I \setminus J_o$. But $J_o + Rf$ is finitely generated with $J_o + Rf \subseteq I$, and this contradiction to the maximality of J_o proves the assertion.

□

The Noether property is inherited by quotients and localizations.

Proposition 7.28 (Noetherian Rings) *Let R be a Noetherian ring.*

(i) *If $I \trianglelefteq R$ is an ideal, then the quotient ring R/I is Noetherian.*
(ii) *Let R be a Noetherian integral domain and K the field of fractions of R. Let $0 \notin S \subseteq R$ be a subset and*

$$R[S^{-1}] := \left\{ \frac{a}{b} \in K \,\middle|\, a \in R, b = 1 \text{ or a product of elements from } S \right\}.$$

Then $R[S^{-1}]$ is a Noetherian ring.

Proof Exercise. Hint for (ii): An ideal in $R[S^{-1}]$ is completely determined by its intersection with R. □

The Noether property is also inherited by polynomial rings. This is a simple consequence of Hilbert's basis theorem.

Theorem 7.29 (Hilbert's Basis Theorem) *Let R be a commutative ring with identity. If R is Noetherian, then the polynomial ring $R[X]$ is also Noetherian.*

Proof Let $J \trianglelefteq R[X]$ be an ideal and set

$$I_n := \{a \in R \mid \exists\, f = aX^n + b_{n-1}X^{n-1} + \ldots + b_0 \in J\}.$$

Then I_n is an ideal for every $n \in \mathbb{N}_0$, and $I_n \subseteq I_{n+1}$ (multiply by X). According to Proposition 7.27, there is an N with $I_N = I_{N+k}$ for all $k \in \mathbb{N}$. Now construct a generating set for J as follows: Let $a_{i1}, \ldots, a_{im(i)}$ be generators of I_i and $f_{ik} = a_{ik}X^i + \ldots \in J$ corresponding polynomials.

Claim: $\mathcal{E} := \{f_{ik} \mid i = 0, \ldots, N,\ k = 1, \ldots, m(i)\}$ generates J.

We show this by induction over $\deg(g)$, that every $g \in J$ lies in the ideal of $R[X]$ generated by $\mathcal{E}$. If $g = 0$, there is nothing to show. So let $0 \neq g \in J$ and

7.2 Algebraic Varieties

$\deg(g) = m$. Then $g = bX^m + \ldots$ and $b \in I_m$. So we can write $b = \sum_k c_{m'k} a_{m'k}$ with $m' = m$ for $m \leq N$ and $m' = N$ otherwise. Set

$$g_1 := g - X^{m-m'} \sum_k c_{m'k} f_{m'k}. \qquad (*)$$

Then $\deg(g_1) \leq \deg(g) - 1$ for $m \geq 1$ and $g_1 = 0$ for $m = 0$. In the case $m = 0$, $(*)$ shows that $g = X^{m-m'} \sum_k c_{m'k} f_{m'k}$ lies in the ideal generated by $\mathcal{E}$, which means the beginning of the induction. For $m > 0$, we get by induction that g_1 lies in the ideal generated by $\mathcal{E}$. So g also lies in this ideal. □

Corollary 7.30 (Finitely Generated K-Algebras are Noetherian) *Let $\mathbb{K}$ be a field, then every finitely generated $\mathbb{K}$-algebra is a Noetherian ring.*

Proof Let A be a finitely generated $\mathbb{K}$-algebra, i.e., there are $a_1, \ldots, a_n \in A$ with $A = \mathbb{K}[a_1, \ldots, a_n]$. Then $A \cong \mathbb{K}[X_1, \ldots, X_n]/I$, where I is the kernel of the evaluation mapping in $a = (a_1, \ldots, a_n) \in \mathbb{K}^n$. With Proposition 7.28(i) and Hilbert's basis theorem, the claim follows. □

Combining Proposition 7.28(i) with Corollary 7.30, yields that coordinate rings of algebraic sets are Noetherian.

Corollary 7.31 (Coordinate Rings are Noetherian) *Let $\mathbb{K}$ be a field and $V \subseteq \mathbb{K}^n$ an algebraic set, then the coordinate ring $\mathbb{K}[V]$ is Noetherian.*

There is also a topological interpretation of this property.

Definition 7.32 (Noetherian Spaces) A topological space $(X, \mathfrak{T})$ is called *Noetherian*, if every descending sequence $K_1 \supseteq K_2 \supseteq \ldots$ of closed subsets becomes stationary.

With Remark 7.6 we see that for a (Zariski-)closed subset K of an algebraic set $V \subseteq \mathbb{K}^n$ we have: $K = V(I(K))$. Since I also reverses inclusions, from a descending sequence $K_1 \supseteq K_2 \supseteq \ldots$ of closed subsets, by applying I, an ascending sequence $I(K_1) \subseteq I(K_2) \subseteq \ldots$ of ideals in $\mathbb{K}[V]$ is obtained. Thus, Corollary 7.31 provides the following proposition.

Proposition 7.33 (Algebraic Sets are Noetherian) *Algebraic sets are noetherian with respect to the Zariski topology.*

Prevarieties over Algebraically Closed Fields

We follow the traditional line of first introducing prevarieties, developing the theory so far that we can give a clean definition of separability, and then defining algebraic varieties as separated prevarieties.

Definition 7.34 (Prevariety) Let $\mathbb{K}$ be an algebraically closed field. A quasi-compact $\mathbb{K}$-ringed space $(X, \mathcal{O}_X)$ is called a *prevariety* or, if one wants to emphasize the role of the field, $\mathbb{K}$-*prevariety*, if it is locally isomorphic to a $\mathbb{K}$-algebraic set with its structure sheaf.

A morphism between two prevarieties is a morphism of $\mathbb{K}$-ringed spaces. This gives the category **PVar**$_\mathbb{K}$ as a full subcategory of the **RSp**$_\mathbb{K}$ of $\mathbb{K}$ ringed spaces.

In the following, $\mathbb{K}$ will always be algebraically closed. Due to Proposition 7.33, every $\mathbb{K}$-algebraic set together with its structure sheaf is a prevariety.

Example 7.35 (Affine $\mathbb{K}$-Varieties) Let A be a reduced finitely generated $\mathbb{K}$-algebra. We know from Theorem 7.23 that the categories **Alg**$_\mathbb{K}^{\text{fgr}}$ and **AS**$_\mathbb{K}$ are equivalent. Therefore, there must be an algebraic set V with $\mathbb{K}[V] \cong A$ for A. We want to give the construction here directly and provide the structure sheaf of V at the same time. We will call the result an *affine variety*.

We denote the set of all maximal ideals $\mathfrak{m} \trianglelefteq A$ with $\mathrm{Spm}(A)$. If one defines for an ideal $I \trianglelefteq A$ the vanishing set

$$V(I) := \{\mathfrak{m} \in \mathrm{Spm}(R) \mid I \subseteq \mathfrak{m}\},$$

one immediately derives the following properties for V (exercise; see also Remark 7.2):

(i) $V(0) = \mathrm{Spm}(A)$ and $V(A) = \emptyset$.
(ii) From $I \subseteq J$ it follows $V(I) \supseteq V(J)$.
(iii) $V(I_1 \cap I_2) = V(I_1) \cup V(I_2)$.
(iv) $V\left(\sum_{\lambda \in \Lambda} I_\lambda\right) = \bigcap_{\lambda \in \Lambda} V(I_\lambda)$.

In particular, the sets $V(I)$ for $I \trianglelefteq A$ form the closed sets of a topology $\mathfrak{Z}$ on $\mathrm{Spm}(A)$, which we again call the *Zariski topology*.

For $\mathfrak{m} \in \mathrm{Spm}(A)$, the mapping $\mathbb{K} \to A/\mathfrak{m}$, $c \mapsto c \cdot 1 + \mathfrak{m}$ is an isomorphism according to Lemma 7.22, which means we can identify $A/\mathfrak{m}$ with $\mathbb{K}$. For $f \in A$, we write $f(\mathfrak{m})$ instead of $f + \mathfrak{m} \in A/\mathfrak{m}$ and set

$$D(f) := \{\mathfrak{m} \in \mathrm{Spm}(A) \mid f(\mathfrak{m}) \neq 0\} = \{\mathfrak{m} \in \mathrm{Spm}(A) \mid f \notin \mathfrak{m}\}.$$

Since maximal ideals are prime, $D(fg) = D(f) \cap D(g)$ holds for $f, g \in A$. This shows that the $D(f)$ with $f \in A$ define the basis of a topology on $\mathrm{Spm}(A)$.

7.2 Algebraic Varieties

However, since $\mathfrak{m} \not\supseteq D(f)$ is equivalent to $f \in \mathfrak{m}$, the complement of $D(f)$ is precisely the vanishing set $V(f)$ of the principal ideal generated by f. Thus, the $D(f)$ are all Zariski-open. Due to (iv), the complement of $V(I)$ is equal to the union of the vanishing sets of a (any) generating system of I, so the $D(f)$ form a basis of the Zariski topology.

Each pair $g, h \in A$ defines a function $D(h) \to \mathbb{K}$, $\mathfrak{m} \mapsto \frac{g(\mathfrak{m})}{h(\mathfrak{m})}$. For an open set $U \in \mathfrak{Z}$ we define $\mathcal{O}_{\mathrm{Spm}(A)}(U)$ as the set of functions $U \to \mathbb{K}$ that are locally of this form, and we call these functions *regular*. This results in a ringed space.

The construction of $(\mathrm{Spm}(A), \mathcal{O}_{\mathrm{Spm}(A)})$ is functorial: If $\psi\colon A \to B$ is a homomorphism of $\mathbb{K}$-algebras (both finitely generated and reduced), then

$$\forall \mathfrak{n} \in \mathrm{Spm}(B): \quad \psi^{-1}(\mathfrak{n}) \in \mathrm{Spm}(A),$$

because $A/\psi^{-1}(\mathfrak{n}) \to B/\mathfrak{n} = \mathbb{K}$ is an injective homomorphism of $\mathbb{K}$-algebras, which implies $A/\psi^{-1}(\mathfrak{n}) = \mathbb{K}$. So we have a mapping $\psi^\sharp\colon \mathrm{Spm}(B) \to \mathrm{Spm}(A)$, $\mathfrak{n} \mapsto \psi^{-1}(\mathfrak{n})$, for which

$$\forall f \in A: \quad (\psi^\sharp)^{-1}(D(f)) = \{\mathfrak{n} \in \mathrm{Spm}(B) \mid \psi^{-1}(\mathfrak{n}) \in D(f)\} = D(\psi(f))$$

holds, because $f \notin \psi^{-1}(\mathfrak{n})$ is true if and if when $\psi(f) \notin \mathfrak{n}$. This shows that $\psi^\sharp$ is continuous. By composing with $\psi^\sharp$, one obtains (exercise) a family of mappings

$$\psi_U \colon \mathcal{O}_{\mathrm{Spm}(A)}(D(f)) \to \mathcal{O}_{\mathrm{Spm}(B)}(D(\psi(f))), \quad h \mapsto h \circ \psi^\sharp,$$

which defines a sheaf morphism $\widetilde{\psi}$. The pair $(\psi^\sharp, \widetilde{\psi})$ is then a morphism of ringed spaces.

Note that the choice of generators $f_1, \ldots, f_n \in A$ of A leads to an injective mapping $\mathrm{Spm}(A) \to \mathbb{K}^n$, $\mathfrak{m} \mapsto (f_1(\mathfrak{m}), \ldots, f_n(\mathfrak{m}))$, whose image is exactly the $\mathbb{K}$-algebraic set that was constructed in the proof of Theorem 7.23. The comparison with Construction 7.25 then also shows that the structure sheaf from there coincides with the one given here (details as an exercise).

While the proof of Theorem 7.23 for the equivalence of $\mathbf{Alg}_\mathbb{K}^{\mathrm{fgr}}$ and $\mathbf{AS}_\mathbb{K}$ relied on the essential surjectivity of the functor $V \mapsto \mathbb{K}[V]$ and Proposition 4.59, here we have given the quasi-inverse functor $A \mapsto \mathrm{Spm}(A)$, which effects the equivalence. An advantage of the description given here is that it does not refer to an embedding into $\mathbb{K}^n$. We denote the $\mathbb{K}$-ringed spaces of the form $(\mathrm{Spm}(A), \mathcal{O}_{\mathrm{Spm}(A)})$ with reduced, finitely generated $\mathbb{K}$-algebra A as *affine variety* and note that it is a prevariety. The prefix "pre" is omitted because it will later be shown (see Example 7.41) that every affine variety is automatically separated and therefore a variety. We denote the full subcategory of $\mathbf{PVar}_\mathbb{K}$, whose objects are the affine $\mathbb{K}$-varieties, with $\mathbf{Var}_\mathbb{K}^{\mathrm{aff}}$.

The equivalence of $\mathbf{Var}_\mathbb{K}^{\mathrm{aff}}$ and $\mathbf{Alg}_\mathbb{K}^{\mathrm{fgr}}$ shows that every morphism

$$(\varphi, \varphi^\flat) \colon (\mathrm{Spm}(A), \mathcal{O}_{\mathrm{Spm}(A)}) \to (\mathrm{Spm}(B), \mathcal{O}_{\mathrm{Spm}(B)})$$

is of the form described above (compare this also with the proof of Theorem 7.9). That is, the sheaf morphism $\varphi^\flat \colon \mathcal{O}_{\mathrm{Spm}(B)} \to \varphi_* \mathcal{O}_{\mathrm{Spm}(A)}$ is given by

$$\varphi^\flat_U \colon \mathcal{O}_{\mathrm{Spm}(B)}(U) \to (\varphi_* \mathcal{O}_{\mathrm{Spm}(A)})(U) = \mathcal{O}_{\mathrm{Spm}(A)})(\varphi^{-1}(U)), \quad h \mapsto h \circ \varphi|_{\varphi^{-1}(U)}$$

for open $U \subseteq \mathrm{Spm}(B)$. Thus, $\varphi^\flat$ is completely determined by φ. In this situation, one simply says, $\varphi \colon \mathrm{Spm}(A) \to \mathrm{Spm}(B)$ is a *regular mapping*, and also writes only $\varphi \colon \mathrm{Spm}(A) \to \mathrm{Spm}(B)$ instead of $(\varphi, \varphi^\flat) \colon (\mathrm{Spm}(A), \mathcal{O}_{\mathrm{Spm}(A)}) \to (\mathrm{Spm}(B), \mathcal{O}_{\mathrm{Spm}(B)})$. □

Remark 7.36 (Affine Open Sets)

(i) Let $(V, \mathcal{O}_V) = (\mathrm{Spm}(A), \mathcal{O}_{\mathrm{Spm}(A)})$ be an affine $\mathbb{K}$-variety. Then for each $f \in \mathbb{K}[V]$ the restriction $(D_V(f), \mathcal{O}_V|_{D_V(f)})$ is also an affine $\mathbb{K}$-variety. Construction 7.25 shows that $(D_V(f), \mathcal{O}_V|_{D_V(f)})$ is isomorphic to $(\mathrm{Spm}(\mathbb{K}[V]_f), \mathcal{O}_{\mathrm{Spm}(\mathbb{K}[V]_f)})$.

(ii) An open subset $U \subseteq X$ of a $\mathbb{K}$-prevariety $(X, \mathcal{O}_X)$ is called an *affine subset* of X, if $(U, \mathcal{O}_X|_U)$ is isomorphic to an affine variety. The affine subsets of X form a basis of the topology of X. To see this, we choose an open subset $U \subseteq X$ and consider an $x \in U$. By definition, x has an open neighborhood U' in X, for which $(U', \mathcal{O}_X|_{U'})$ is isomorphic to an affine variety. Since $U \cap U'$ is open in U', there is according to (i) an affine neighborhood of x, which is contained in $U \cap U'$. Since $x \in U$ was chosen arbitrarily, U is the union of affine sets.

(iii) If $(\varphi, \varphi^\flat) \colon (X, \mathcal{O}_X) \to (Y, \mathcal{O}_Y)$ is a morphism between $\mathbb{K}$-prevarieties, then for an affine open set $V \subseteq Y$, one considers an affine open set $U \subseteq X$ that is contained in $\varphi^{-1}(V)$. One then obtains a restriction $(\varphi|_U, \varphi^\flat|_{\mathcal{O}_Y|_V}) \colon (U, \mathcal{O}_X|_U) \to (V, \mathcal{O}_Y|_V)$, which according to Example 7.35(iii) can simply be considered as a regular mapping $\varphi|_U \colon U \to V$. Since these restrictions completely determine the morphism $(\varphi, \varphi^\flat)$, $\varphi^\flat$ is also completely determined by φ, and we can adopt the convention of speaking only of a *regular mapping* $\varphi \colon X \to Y$, also for prevarieties.

Example 7.37 (Gluing of Prevarieties) Let $(X, \mathfrak{T})$ be a topological space and $\{U_i \in \mathfrak{T} \mid i \in I\}$ an open cover of X. Further, let $\mathcal{O}_{U_i}$ for $i \in I$ be sheaves over U_i, which make $(U_i, \mathcal{O}_{U_i})$ into $\mathbb{K}$-prevarieties. Also, let

$$\{\varphi_{ij} \colon \mathcal{O}_{U_j}|_{U_i \cap U_j} \to \mathcal{O}_{U_i}|_{U_i \cap U_j} \mid i, j \in I\}$$

be a set of gluing data (see Example 5.13) in the category of sheaves of $\mathbb{K}$-algebras. Then the gluing $\mathcal{O}_X$ is a sheaf of $\mathbb{K}$-algebras over X with $\mathcal{O}_X|_{U_i} = \mathcal{O}_{U_i}$. Thus, $(X, \mathcal{O}_X)$ is a $\mathbb{K}$-ringed space and, if I is finite, even a $\mathbb{K}$-prevariety. □

7.2 Algebraic Varieties

Products of affine varieties can be formed. Although no new examples of varieties emerge—namely only affine varieties—by gluing one can also form products of prevarieties and thus obtain a machine that generates new examples.

Example 7.38 (Products of Affine Varieties) Let $(X, \mathcal{O}_X)$, $(Y, \mathcal{O}_Y)$ be two affine algebraic $\mathbb{K}$-varieties. Then there is a categorical product of $(X, \mathcal{O}_X)$ and $(Y, \mathcal{O}_Y)$ in $\mathbf{Var}_{\mathbb{K}}^{\mathrm{aff}}$ (see Definition 4.26). This means, there is an affine variety $(Z, \mathcal{O}_Z)$, unique up to isomorphism, that has the following universal property:

$$\begin{array}{ccc} (Z, \mathcal{O}_Z) & \longrightarrow & (X, \mathcal{O}_X) \\ \downarrow & \searrow{\exists! \Phi_Z} & \uparrow{\forall \Phi_X} \\ (Y, \mathcal{O}_Y) & \underset{\forall \Phi_Y}{\longleftarrow} & (W, \mathcal{O}_W) \end{array}$$

We then write $(X \times Y, \mathcal{O}_{X \times Y})$ for $(Z, \mathcal{O}_Z)$.

Note here that the product topology for algebraic varieties is not suitable, as it does not yield the Zariski topology on the resulting affine space $\mathbb{K}^{n+m}$ for the set-theoretic product of two spaces of the form $\mathbb{K}^n$ and $\mathbb{K}^m$.

For $X = \mathrm{Spm}(A)$ and $Y = \mathrm{Spm}(B)$ we set $Z := \mathrm{Spm}(C)$ with $C = A \otimes_{\mathbb{K}} B$. Applying the contravariant functor Spm to the categorical sum diagram of tensor products of commutative algebras (see Example 4.31), one obtains the categorical product diagram of affine varieties. Since Spm induces an equivalence of categories, $(Z, \mathcal{O}_Z) = (\mathrm{Spm}(C), \mathcal{O}_{\mathrm{Spm}(C)})$ is the categorical product of $(X, \mathcal{O}_X)$ and $(Y, \mathcal{O}_Y)$ in $\mathbf{Var}_{\mathbb{K}}^{\mathrm{aff}}$. □

Proposition 7.39 (Product of Prevarieties) *In the category* $\mathbf{PVar}_{\mathbb{K}}$ *there is a categorical product.*

Proof Let $(X, \mathcal{O}_X)$ and $(Y, \mathcal{O}_Y)$ be two $\mathbb{K}$-prevarieties and $(U_\alpha)_{\alpha \in A}$ and $(V_\beta)_{\beta \in B}$ coverings of X and Y by open affine subsets, respectively. Consider the affine product varieties $(U_\alpha \times V_\beta, \mathcal{O}_{U_\alpha \times V_\beta})$ for $(\alpha, \beta) \in A \times B$, which were constructed in Example 7.38. The universal properties of these products show that

$$\mathcal{O}_{U_\alpha \times V_\beta}|_{(U_\alpha \times V_\beta) \cap (U_{\alpha'} \times V_{\beta'})} = \mathcal{O}_{U_{\alpha'} \times V_{\beta'}}|_{(U_\alpha \times V_\beta) \cap (U_{\alpha'} \times V_{\beta'})}.$$

Thus, one can glue the sheaves $\mathcal{O}_{U_\alpha \times V_\beta}$ for $(\alpha, \beta) \in A \times B$ and obtain a sheaf $\mathcal{O}_{X \times Y}$ over $X \times Y$, which makes $X \times Y$ a prevariety according to Example 7.37.

To show that $(X \times Y, \mathcal{O}_{X \times Y})$ is a categorical product of $(X, \mathcal{O}_X)$ and $Y, \mathcal{O}_Y)$, we consider the diagram

$$\begin{array}{ccc} (X \times Y, \mathcal{O}_{X \times Y}) & \longrightarrow & (X, \mathcal{O}_X) \\ \downarrow \quad \overset{\exists?\Phi}{\nwarrow} & & \uparrow {\scriptstyle \forall \Phi_X} \\ (Y, \mathcal{O}_Y) & \underset{\forall \Phi_Y}{\longleftarrow} & (W, \mathcal{O}_W) \end{array}$$

from which we obtain by restrictions for $(\alpha, \beta) \in A \times B$ due to the universal property of $(U_\alpha \times V_\beta, \mathcal{O}_{U_\alpha \times V_\beta})$ the diagram

$$\begin{array}{ccc} (U_\alpha \times V_\beta, \mathcal{O}_{U_\alpha \times V_\beta}) & \longrightarrow & (U_\alpha, \mathcal{O}_{U_\alpha}) \\ \downarrow \quad \overset{\exists! \Phi_{\alpha,\beta}}{\nwarrow} & & \uparrow {\scriptstyle \forall \Phi_X} \\ (V_\beta, \mathcal{O}_{V_\beta}) & \underset{\forall \Phi_Y}{\longleftarrow} & (W_{\alpha,\beta}, \mathcal{O}_{W_{\alpha,\beta}}) \end{array}$$

The $\Phi_{\alpha,\beta}$ are compatible and can be combined into a morphism $\Phi \colon (W, \mathcal{O}_W) \to (X \times Y, \mathcal{O}_{X \times Y})$. The uniqueness follows from the local uniqueness. □

Varieties over Algebraically Closed Fields

The products of prevarieties not only provide new examples of prevarieties, they also allow us to cleanly define when prevarieties are separated and thus we can finally introduce the concept of a variety.

Definition 7.40 (Algebraic Varieties) A $\mathbb{K}$-prevariety $(X, \mathcal{O}_X)$ is called an *algebraic variety* or simply a *variety* ($\mathbb{K}$-variety, if one wants to emphasize the field), if it is *separated*, i.e., if the diagonal $\Delta_X := \{(x, x) \in X \times X \mid x \in X\}$ in $X \times X$ is closed. We denote the full subcategory of **PVar**$_{\mathbb{K}}$, whose objects are the varieties, with **Var**$_{\mathbb{K}}$.

We can now in particular justify the name "affine varieties".

Example 7.41 (Affine Varieties) Let $(X, \mathcal{O}_X)$ be an affine $\mathbb{K}$-variety. Then we can regard X as a Zariski-closed subset of a $\mathbb{K}^n$. Thus,

$$\Delta_X = \Delta_{\mathbb{K}^n} \cap (X \times X) \subseteq \mathbb{K}^n \times \mathbb{K}^n = \mathbb{K}^{2n},$$

and because $X \times X$ can be considered as a closed subset of $\mathbb{K}^{2n}$ by definition (see Example 7.38), it suffices to show that $\Delta_{\mathbb{K}^n}$ is closed in $\mathbb{K}^{2n}$ to demonstrate that X is a $\mathbb{K}$-variety. However, since $\Delta_{\mathbb{K}^n}$ is the vanishing set of the ideal generated by the polynomials $X_i - X_{i+n}$ for $i = 1, \ldots, n$ in $\mathbb{K}[X_1, \ldots, X_{2n}]$, this is clear. □

7.2 Algebraic Varieties

As a consequence of Example 7.41, it can be seen that for every $\mathbb{K}$-prevariety $(X, \mathcal{O}_X)$, the diagonal Δ_X in $X \times X$ is *locally closed*. A set Y in a topological space X is called locally closed if every point $y \in Y$ has an open neighborhood U_y in X for which $Y \cap U_y$ is closed in U_y. To show the local closedness of Δ_X, one only needs to choose an open affine subset $U_x \subseteq X$ for each $(x, x) \in \Delta_X$ and note that $\Delta_X \cap (U_x \times U_x) = \Delta_{U_x}$, because Δ_{U_x} is closed in $U_x \times U_x$ according to Example 7.41.

Exercise 7.2 (Locally Closed Sets) Let $(X, \mathfrak{T})$ be a topological space and $Y \subseteq X$. Prove the following statements:

(i) Y is locally closed in X if and only if there is a closed set $A \subseteq X$ and an open set $U \subseteq X$ with $Y = A \cap U$.
(ii) If Y is locally closed in X, then Y is open in its closure $\overline{Y}$ in X.

Example 7.42 (Subvarieties) Let $(X, \mathcal{O}_X)$ be a $\mathbb{K}$-prevariety and $Z \subseteq X$ locally closed. Then $(Z, \mathcal{O}_X|_Z)$ is a $\mathbb{K}$-prevariety. Since Z is open in its closure and open subsets of prevarieties are themselves prevarieties, it suffices to show this statement for closed Z. Then the intersections $Z \cap U$ of Z with affine open subsets U of X are closed subsets of affine varieties, thus themselves affine varieties. Therefore, $(Z, \mathcal{O}_X|_Z)$ is locally isomorphic to $(Z \cap U, \mathcal{O}_{Z \cap U})$. Since a finite number of affine open sets suffice to cover X, Z is also quasicompact, thus a prevariety.

If $(X, \mathcal{O}_X)$ is separated, i.e., a variety, then $\Delta_Z = (Z \times Z) \cap \Delta_X$ is closed in $Z \times Z$, because $Z \times Z$ with the relative topology in $X \times X$ is the product of the prevariety Z with itself (see Exercise 7.3). Therefore, $(Z, \mathcal{O}_X|_Z)$ is itself also a variety. □

Exercise 7.3 (Products of Subprevarieties) Let $(X, \mathcal{O}_X)$ and $(Y, \mathcal{O}_Y)$ be prevarieties and $Z \subseteq X$, $W \subseteq Y$ locally closed subsets. Show that $Z \times W$, endowed with the relative topology of the product prevariety $X \times Y$ and the restricted sheaf $\mathcal{O}_{X \times Y}|_{Z \times W}$, is the product of the prevarieties Z and W.
Hint: Consider affine open subsets and reduce the problem to affine varieties, where it is easily solved via the coordinate algebras.

Example 7.43 (Products of Varieties) Let $(X, \mathcal{O}_X)$ and $(Y, \mathcal{O}_Y)$ be varieties. Then the product prevariety $(X \times Y, \mathcal{O}_{X \times Y})$ is also separated, thus a variety. To see this, one only needs to ascertain that $\Delta_{X \times Y} \subseteq (X \times Y) \times (X \times Y)$ can be identified with $\Delta_X \times \Delta_Y \subseteq (X \times X) \times (Y \times Y)$. □

The following lemma is needed for a characterization of separability, which allows us to recognize projective spaces over algebraically closed fields as algebraic varieties.

Lemma 7.44 (Graphs of Regular Mappings) *Let $(X, \mathcal{O}_X)$ and $(Y, \mathcal{O}_Y)$ be two $\mathbb{K}$-prevarieties and $\varphi : X \to Y$ a regular mapping. Then the graph*

$$\Gamma_\varphi := \{(x, \varphi(x)) \in X \times Y \mid x \in X\}$$

is a locally closed subset of $X \times Y$, which is isomorphic to X as a prevariety. If $(Y, \mathcal{O}_Y)$ is a variety, then $\Gamma_\varphi \subseteq X \times Y$ is closed.

Proof Consider the regular mapping $\psi : X \times Y \to Y \times Y$, $(x, y) \mapsto (\varphi(x), y)$. Because $\Gamma_\varphi = \psi^{-1}(\Delta_Y)$ and continuous preimages of (locally) closed sets are (locally) closed, it follows that Γ_φ is locally closed in $X \times Y$ and even closed when $(Y, \mathcal{O}_Y)$ is a variety. In particular, Γ_φ is a prevariety, with respect to which $\gamma_\varphi : X \to \Gamma_\varphi$, $x \mapsto (x, \varphi(x))$ is a regular mapping. Note that the projection $p_1 : \Gamma_\varphi \to X$, $(x, \varphi(x)) \mapsto x$ is also a regular mapping of prevarieties. Thus, X and Γ_φ are isomorphic. □

Proposition 7.45 (Characterization of Separability) *Let $(X, \mathcal{O}_X)$ be a prevariety. Then the following properties are equivalent.*

(1) X is separated.
(2) If U and V are affine open subsets of X, then $U \cap V$ is also affine open, and the following mapping is surjective:

$$\mu_{U,V} : \mathcal{O}_X(U) \otimes_\mathbb{K} \mathcal{O}_X(V) \to \mathcal{O}_X(U \cap V), \quad f \otimes g \mapsto f|_{U \cap V} \cdot g|_{U \cap V}.$$

(3) If $(Z, \mathcal{O}_Z)$ is a prevariety and $\varphi, \psi : Z \to X$ are regular mappings, then $\Delta_{\varphi,\psi} := \{z \in Z \mid \varphi(z) = \psi(z)\}$ is closed in Z.

Proof The implication (3) $\Rightarrow$ (1) follows immediately if one sets $Z = X \times X$ with $\varphi(x_1, x_2) = x_1$ and $\psi(x_1, x_2) = x_2$. Conversely, if X is separated, then in the situation of (3) the set $\Delta_{\varphi,\psi} = (\varphi, \psi)^{-1}(\Delta_X)$ is closed. Thus, (1) and (3) are equivalent.

Since the products $U \times V$ for affine open sets U, V in X form an open cover of $X \times X$, Δ_X is closed in $X \times X$ if and only if all $(U \times V) \cap \Delta_X$ are closed in $U \times V$. According to Lemma 7.44, the mapping $\delta_{U,V} : U \cap V \to (U \times V) \cap \Delta_X$, $x \mapsto (x, x)$ is an isomorphism of prevarieties. But $(U \times V) \cap \Delta_X$ is affine if and only if it is closed in $U \times V$. Thus, $U \cap V$ is affine if and only if $(U \times V) \cap \Delta_X$ is closed in $U \times V$. In this case, the composition of $\mu_{U,V}$ with the isomorphism $\delta_{U,V}$ for $U = \mathrm{Spm}(A)$ and $V = \mathrm{Spm}(B)$ is given by the surjective mapping

$$A \otimes_\mathbb{K} B \to A \otimes_\mathbb{K} B/I,$$

where I is the vanishing ideal of $(U \times V) \cap \Delta_X$ in $\mathbb{K}[U \times V] = A \otimes_\mathbb{K} B$. This shows that (2) follows from (1). Conversely, if (2) holds, then $\mathbb{K}[U \times V] \to \mathbb{K}[U \cap V]$, $f \mapsto f \circ \delta_{U,V}$ is a quotient mapping with kernel $I := \{f \in \mathbb{K}[U \times V] \mid$

7.2 Algebraic Varieties

$f|_{(U \times V) \cap \Delta_X} = 0\}$, and the image of the regular mapping $U \cap V \to U \times V$ coming from $\mathbb{K}[U \times V] \to \mathbb{K}[U \times V]/I \to \mathbb{K}[U \cap V]$ is the (closed) vanishing set $V(I)$ of I in $U \times V$. On the other hand, this image is exactly $(U \times V) \cap \Delta_X$. Thus, this set is closed, i.e., X is separated. This also shows the equivalence of (1) and (2). □

We conclude this section with a family of examples of algebraic varieties, which can be shown to be a genuine generalization of affine varieties.

Example 7.46 (Projective Varieties)

(i) Let $\mathbb{P}^n_{\mathbb{K}}$ be the projective space of lines $[x] = \mathbb{K}x \subseteq \mathbb{K}^{n+1}$ for $0 \neq x \in \mathbb{K}^{n+1}$ (see Example 6.11). For each $i = 0, \ldots, n$ we consider the subsets

$$U_j := \{[x] \in \mathbb{P}^n_{\mathbb{K}} \mid x_j \neq 0\}$$

together with the bijective mappings

$$\varphi_j : U_j \to \mathbb{K}^n, \quad \mathbb{K}(x_0, \ldots, x_n) \mapsto \left(\frac{x_0}{x_j}, \ldots, \frac{x_{j-1}}{x_j}, \frac{x_{j+1}}{x_j}, \ldots, \frac{x_n}{x_j}\right).$$

We use φ_j to define on U_j the structure of an affine $\mathbb{K}$-variety: $\mathcal{O}_{U_j} := \varphi^{-1}(\mathcal{O}_{\mathbb{K}^n})$. For $i, j \in \{0, \ldots, n\}$ we consider the mappings

$$\varphi_i \circ \varphi_j^{-1}|_{\varphi_j(U_j \cap U_i)} : \varphi_j(U_j \cap U_i) \to \varphi_i(U_j \cap U_i).$$

The set $\varphi_j(U_j \cap U_i)$ is affine open in $\mathbb{K}^n$, as it is the complement of the zero set of the projection onto the i-th component. For $i < j$, $\varphi_i \circ \varphi_j^{-1}$ is explicitly given by

$$(z_1, \ldots, z_n) \mapsto (z_1, \ldots, z_j, 1, z_{j+1}, \ldots, z_n)$$

$$\mapsto \left(\frac{z_1}{z_i}, \ldots, \frac{z_{i-1}}{z_i}, \frac{z_{i+1}}{z_i}, \ldots, \frac{z_{j-1}}{z_i}, \frac{1}{z_i}, \frac{z_{j+1}}{z_i}, \ldots, \frac{z_n}{z_i}\right)$$

given. Thus the mappings are regular, i.e., isomorphisms of affine varieties. We can glue the sheaves $(U_j, \mathcal{O}_{U_j})$ and obtain the structure of a $\mathbb{K}$-prevariety on $\mathbb{P}^n_{\mathbb{K}}$. Since the proof of Proposition 7.45 shows that it is sufficient to have condition (2) for a covering family of affine open subsets, we only need to test it here for $U \times V = U_i \times U_j$. We leave this test to the reader as an exercise and only note that $\mathbb{P}^n_{\mathbb{K}}$ is a $\mathbb{K}$-variety.

(ii) If $X \subseteq \mathbb{P}^n_{\mathbb{K}}$ is locally closed, then $(X, \mathcal{O}_{\mathbb{P}^n_{\mathbb{K}}}|_X)$ is a variety according to Example 7.42. Varieties of this kind are called *quasiprojective*. If X is closed in $\mathbb{P}^n_{\mathbb{K}}$, then $(X, \mathcal{O}_{\mathbb{P}^n_{\mathbb{K}}}|_X)$ is called a *projective variety*. □

Projective varieties are generally not affine. One possible proof for this is to introduce the concept of *completeness* for algebraic varieties, and to show that projective varieties are always complete, whereas affine varieties are only complete when they are finite (see [Di74, §3.3]). Since every affine space can be considered as an open subset of a projective space, affine varieties are automatically quasi-projective. Thus, quasi-projective varieties are a true generalization of affine varieties.

7.3 Schemes

Schemes provide an extended conceptual framework for algebraic varieties, in which algebraic sets for fields can be treated that are not algebraically closed. The definition of a scheme is quite simple based on Ch. 5: A scheme is a ringed space (see Definition 5.28), which is locally isomorphic to the spectrum of a ring (see Example 5.30). In the notations, one follows the theory of varieties.

Definition 7.47 (Schemes) An *affine scheme* is a locally ringed space that is isomorphic to the spectrum $(\operatorname{Spec} R, \mathcal{O}_{\operatorname{Spec} R})$ for a commutative ring R with identity. A *scheme* is a locally ringed space $(X, \mathcal{O}_X)$, which is locally isomorphic to an affine scheme. We denote the full subcategory of the category **RSp**$_{\text{loc}}$ of locally ringed spaces, whose objects are the schemes, with **Sch**. The full subcategory of **Sch**, whose objects are affine schemes, we denote with **Sch**^{aff}.

The considerations from Example 5.30 can be extended analogously to the constructions for affine varieties from Example 7.35 and Remark 7.36. To emphasize the analogy with the affine varieties, we now denote prime ideals with $\mathfrak{p}$ instead of P as in Example 5.30.

Remark 7.48 (Affine Schemes) Let R be a commutative ring with identity, $f \in R$ and $V(f) := \{\mathfrak{p} \in \operatorname{Spec}(R) \mid f \in \mathfrak{p}\}$ the vanishing set of the principal ideal $Rf = (f) \trianglelefteq R$. We call

$$D(f) := D_{\operatorname{Spec}(R)}(f) := \operatorname{Spec}(R) \setminus V(f)$$

the *standard open set* associated with f. The $D(f)$ with $f \in R$ form a basis of the topology of $\operatorname{Spec}(R)$. If $U \subseteq \operatorname{Spec}(R)$ is open and $\mathfrak{p} \in U$, then U is of the form $U = \operatorname{Spec}(R) \setminus V(I)$ for an ideal $I \triangleleft R$. Thus, $\mathfrak{p} \notin V(I)$, i.e., $I \not\subseteq \mathfrak{p}$, and there exists a $f \in I$ with $f \notin \mathfrak{p}$. But $(f) \subseteq I$ shows $V(I) \subseteq V(f)$ and thus $\mathfrak{p} \in D(f) \subseteq U$. □

In order to be able to describe the sections on standard open sets as spectra, in analogy to Remark 7.36(i), we need a replacement for Hilbert's Nullstellensatz, which was incorporated via Construction 7.25 in Remark 7.36(i).

7.3 Schemes

Lemma 7.49 (Vanishing Sets and Radicals) *Let R be a commutative ring with identity and $I, J \triangleleft R$. Then*

$$V(I) \subseteq V(J) \iff \operatorname{rad}(J) \subseteq \operatorname{rad}(I).$$

Proof We first show that for each ideal $I \triangleleft R$

$$\operatorname{rad}(I) = \bigcap_{I \subseteq \mathfrak{p} \in \operatorname{Spec}(R)} \mathfrak{p} \qquad (*)$$

holds. Here, the inclusion $\subseteq$ is clear, because prime ideals are radical ideals, so from $I \subseteq \mathfrak{p} \in \operatorname{Spec}(R)$ it follows that $\operatorname{rad}(I) \subseteq \operatorname{rad}(\mathfrak{p}) = \mathfrak{p}$.

To also show the inclusion $\supseteq$, we assume that $f \in R \setminus \operatorname{rad}(I)$, and set $S := \{1, f, f^2, \ldots\}$. Then $S \cap I = \emptyset$ and $S^{-1}I \subsetneq S^{-1}R$, i.e., the localized ideal $S^{-1}I \triangleleft S^{-1}R$ is contained in a maximal ideal $\mathfrak{m} \triangleleft S^{-1}R$. If $\varphi \colon R \to S^{-1}R$ is the canonical homomorphism (see Exercise 1.4), then $\mathfrak{p} := \varphi^{-1}(\mathfrak{m})$ is possibly not maximal, but definitely prime in R (see Exercise 1.7). Because $\varphi(I) \subseteq S^{-1}I \subseteq \mathfrak{m}$, it follows that $I \subseteq \mathfrak{p}$, and because $\varphi(\mathfrak{p}) \subseteq \mathfrak{m} \neq S^{-1}R$, it follows that $\mathfrak{p} \cap S = \emptyset$, so in particular $f \notin \mathfrak{p}$. Together we find $f \in R \setminus \bigcap_{I \subseteq \mathfrak{p} \in \operatorname{Spec}(R)} \mathfrak{p}$. This proves $(*)$.

By definition of V, it holds

$$V(I) \subseteq V(J) \iff \forall \mathfrak{p} \in \operatorname{Spec}(R) : (I \subseteq \mathfrak{p} \Rightarrow J \subseteq \mathfrak{p}).$$

This gives $(*)$, that

$$\operatorname{rad}(I) = \bigcap_{I \subseteq \mathfrak{p} \in \operatorname{Spec}(R)} \mathfrak{p} \supseteq \bigcap_{J \subseteq \mathfrak{p} \in \operatorname{Spec}(R)} \mathfrak{p} = \operatorname{rad}(J).$$

If conversely $\operatorname{rad}(J) \subseteq \operatorname{rad}(I)$ and $I \subseteq \mathfrak{p} \in \operatorname{Spec}(R)$ holds, then it follows $J \subseteq \operatorname{rad}(J) \subseteq \operatorname{rad}(I) \subseteq \operatorname{rad}(\mathfrak{p}) = \mathfrak{p}$. This proves the lemma. □

Now we can show that standard open sets in affine schemes are themselves affine schemes. Since the standard open sets form a basis of the Zariski topology, this implies that the restrictions of affine schemes to open subsets are schemes.

Theorem 7.50 (Standard Open Subschemes) *Let R be a commutative ring with identity and $(\operatorname{Spec}(R), \mathcal{O})$ its spectrum. Then:*

(i) $\forall \mathfrak{p} \in \operatorname{Spec}(R) : \quad \mathcal{O}_{\mathfrak{p}} \cong R_{\mathfrak{p}}$.
(ii) $\forall f \in R : \quad \mathcal{O}(D(f)) \cong R_f$.
  Here, R_f is the localization $S^{-1}R$ with $S := \{1, f, f^2, \ldots\}$.
(iii) $\forall f \in R : \quad (D(f), \mathcal{O}|_{D(f)}) \cong \operatorname{Spec}(R_f)$.

(iv) $\mathcal{O}(\operatorname{Spec}(R)) \cong R$ and the following diagram commutes

$$\begin{array}{ccc} \mathcal{O}(\operatorname{Spec}(R)) & \longrightarrow & \mathcal{O}_{\mathfrak{p}} \\ \cong \downarrow & & \downarrow \cong \\ R & \longrightarrow & R_{\mathfrak{p}} \end{array}$$

Proof

(i) Each $s \in \mathcal{O}_{\mathfrak{p}}$ is of the form $s = [(U,t)]_{\mathfrak{p}} =: t_{\mathfrak{p}}$ with $t \in \mathcal{O}(U)$. We set $\varphi: \mathcal{O}_{\mathfrak{p}} \to R_{\mathfrak{p}}$, $s \mapsto t(\mathfrak{p})$ and show that φ is an isomorphism. The homomorphism property is clear. To show surjectivity, we consider $r \in R_{\mathfrak{p}}$. Then r is of the form $r = \frac{a}{b}$ with $b \notin \mathfrak{p}$. With $V := D(b)$ we find $\mathfrak{p} \in V$, and $t \in \mathcal{O}(V)$ as well as $\varphi(t_{\mathfrak{p}}) = r$ holds for

$$t: V \to \coprod_{\mathfrak{q} \in V} R_{\mathfrak{q}}, \quad \mathfrak{q} \mapsto \frac{a}{b}.$$

For the injectivity of φ, we consider $s, s' \in \mathcal{O}_{\mathfrak{p}}$ with $\varphi(s) = \varphi(s')$. Then there is a neighborhood V of $\mathfrak{p}$ as well as $a, a' \in R$ and $b, b' \in R \setminus \mathfrak{p}$, for which $t = \frac{a}{b}, t' = \frac{a'}{b'} \in \mathcal{O}(V)$ with $s = t_{\mathfrak{p}}$ and $s' = t'_{\mathfrak{p}}$ holds. Because of $\frac{a}{b} = \frac{a'}{b'}$ there is a $h \in R \setminus \mathfrak{p}$ with $ab'h = a'bh$, and that yields

$$\forall \mathfrak{q} \in \operatorname{Spec}(R) \text{ with } b, b', h \notin \mathfrak{q}: \quad \frac{a}{b} = \frac{a'}{b'} \in R_{\mathfrak{q}}.$$

This in turn shows $t|_{D(b) \cap D(b') \cap D(h)} = t'|_{D(b) \cap D(b') \cap D(h)}$ and thus, because $\mathfrak{p} \in D(b) \cap D(b') \cap D(h)$, also $s = t_{\mathfrak{p}} = t'_{\mathfrak{p}} = s'$.

(ii) Consider the homomorphism

$$\psi: R_f \to \mathcal{O}(D(f)), \quad \frac{a}{f^n} \mapsto \left(\mathfrak{q} \mapsto \frac{a}{f^n} \in R_{\mathfrak{q}} \right).$$

We want to show that ψ is an isomorphism. To prove the injectivity of ψ, we assume that $\psi(\frac{a}{f^n}) = \psi(\frac{b}{f^m})$. We choose a $\mathfrak{p} \in D(f)$ and a $h \in R \setminus \mathfrak{p}$ with $haf^m = hbf^n$. Let

$$I := \operatorname{Ann}(af^m - bf^n) := \{ r \in R \mid r(af^m - bf^n) = 0 \}.$$

Then $h \in I$ and $I \not\subseteq \mathfrak{p}$, i.e., $\mathfrak{p} \notin V(I)$. This argument shows $D(f) \cap V(I) = \emptyset$, thus $V(I) \subseteq V(f)$. Now Lemma 7.49 yields that $f \in \operatorname{rad}(I)$. Therefore, there

7.3 Schemes

is an $\ell \in \mathbb{N}$ with $af^{m+\ell} = bf^{n+\ell}$, and the equality

$$\frac{a}{f^m} = \frac{b}{f^m} \in R_f$$

is proven.

To show the surjectivity of ψ, we consider $s \in \mathcal{O}(D(f))$. Locally, s is of the form

$$D(h_i) \to \coprod_{\mathfrak{q} \in D(h_i)} R_{\mathfrak{q}}, \quad \mathfrak{q} \mapsto \frac{a_i}{b_i} \in R_{\mathfrak{q}},$$

where $a_i, h_i \in R$ and $b_i \in R \setminus \mathfrak{q}$ for $\mathfrak{q} \in D(h_i)$ must be chosen appropriately. According to Remark 7.48, we can assume $D(f) = \bigcup_{i \in I} D(h_i)$ and find (see Example 5.30)

$$V(f) \supseteq \bigcap_{i \in I} V(h_i) = V\left(\sum_{i \in I}(h_i)\right).$$

With Lemma 7.49, it follows that $f \in \operatorname{rad}\left(\sum_{i \in I}(h_i)\right)$, i.e., there is a $m \in \mathbb{N}$ and $c_i \in R$ with $f^m = \sum_{i \in I} c_i h_i$, where only finitely many summands are different from zero. Because

$$\mathfrak{p} \in D(f) \iff f \notin \mathfrak{p} \iff (\exists n \in \mathbb{N} : f^n \notin \mathfrak{p})$$

this yields $D(f) \subseteq \bigcup_{i \in I} D(h_i)$, where a finite union is sufficient. Moreover, this characterization of the standard open sets also shows $D(h_i) \subseteq D(b_i)$, thus $h_i \in \operatorname{rad}(Rb_i)$. So there are $d_i \in R$ and $k_i \in \mathbb{N}$ with $h_i^{k_i} = d_i b_i$. Because $D(h_i) = D(h_i^{k_i})$ we can assume without loss of generality that $k_i = 1$. By replacing a_i with $d_i a_i$, we can assume $h_i = b_i$. Now if $\mathfrak{q} \in D(h_i) \cap D(h_j) = D(h_i h_j)$, then

$$s(\mathfrak{q}) = \frac{a_i}{h_i} = \frac{a_j}{h_j} \in R_{\mathfrak{q}},$$

i.e., $\psi(\frac{a_i h_j}{h_i h_j}) = \psi(\frac{a_j h_i}{h_i h_j})$. If we now apply the first part of the proof to $h_i h_j$, it follows $\frac{a_i h_j}{h_i h_j} = \frac{a_j h_i}{h_i h_j}$. Thus, we find an $n \in \mathbb{N}$ with

$$0 = (h_i h_j)^n (a_i h_j^2 h_i - a_j h_j h_i^2) = (h_i h_j)^{n+1}(a_i h_j - a_j h_i)$$
$$= h_j^{n+2}(a_i h_i^{n+1}) - h_i^{n+2}(a_j h_j^{n+1}).$$

Analogous to the above calculations, we now replace h_i with h_i^{n+2} and a_i with $h_i^{n+1}a_i$. This allows us to assume

$$\forall i,j \in I: \qquad h_j a_i = h_i a_j$$

As before, we can write $f^m = \sum_{i \in I} c_i h_i$, where again only finitely many summands are different from zero. Then we set $a := \sum_{i \in I} c_i a_i$ and find $h_j a = \sum_{i \in I} c_i a_i h_j = \sum_{i \in I} c_i a_j h_i = f^m a_j$, thus

$$\forall \mathfrak{q} \in D(h_j) \subseteq D(f): \qquad \frac{a}{f^m} = \frac{a_j}{h_j} \in R_\mathfrak{q}$$

and finally $\psi(\frac{a}{f^m}) = s$.

(iii) For $f \in R$ and $f \notin \mathfrak{p} \in \mathrm{Spec}(R)$, let $\mathfrak{p}_f \trianglelefteq R_f$ be the ideal generated by $j_f(\mathfrak{p})$, where $j_f: R \to R_f, r \mapsto \frac{r}{1}$ is the canonical homomorphism. Then $\mathfrak{p}_f \in \mathrm{Spec}(R_f)$, because for $\frac{a}{f^n} \frac{b}{f^m} \in \mathfrak{p}_f$ there is an $\ell \in \mathbb{N}_0$ and $c \in \mathfrak{p}$ with $f^{n+m}c = f^\ell ab \in \mathfrak{p}$. But this shows $a \in \mathfrak{p}$ or $b \in \mathfrak{p}$, thus $\frac{a}{f^n} \in \mathfrak{p}_f$ or $\frac{b}{f^m} \in \mathfrak{p}_f$.

Note that $j_f^{-1}(\mathfrak{p}_f) = \{r \in R \mid \exists t \in S: tr \in \mathfrak{p}\} = \mathfrak{p}$. Conversely, if $\mathfrak{q} \in \mathrm{Spec}(R_f)$, then $f \notin j_f^{-1}(\mathfrak{q})$ and $\left(j_f^{-1}(\mathfrak{p})\right)_f = \mathfrak{q}$, i.e., $\mathfrak{p} \mapsto \mathfrak{p}_f$ and $\mathfrak{q} \mapsto j_f^{-1}(\mathfrak{q})$ are mutually inverse mappings for the sets

$$\mathrm{Spec}(R_f) = \{\mathfrak{p}_f \mid f \notin \mathfrak{p} \in \mathrm{Spec}(R)\}. \underset{\mathrm{bij.}}{\longleftrightarrow} D(f) = \{\mathfrak{p} \in \mathrm{Spec}(R) \mid f \notin \mathfrak{p}\}.$$

We set $\varphi: D(f) \to \mathrm{Spec}(R_f)$, $\mathfrak{p} \mapsto \mathfrak{p}_f$ and claim that

$$\forall J \trianglelefteq R_f: \qquad \varphi^{-1}(V(J)) = V(j_f^{-1}(J)), \qquad (*)$$

which then shows the continuity of φ. If $\mathfrak{p}_f \in V(J)$, i.e., $J \subseteq \mathfrak{p}_f$, then $j_f^{-1}(J) \subseteq j_f^{-1}(\mathfrak{p}_f) = \mathfrak{p}$, i.e., $\mathfrak{p} \in V(j_f^{-1}(J))$. Conversely, if $\mathfrak{p} \in V(j_f^{-1}(J))$, then for $\frac{b}{f^n} \in J$, that $b \in j_f^{-1}(J) \subseteq \mathfrak{p}$, thus $\frac{b}{f^n} \in \mathfrak{p}_f$. This shows $J \subseteq \mathfrak{p}_f$, thus $\mathfrak{p}_f \in V(J)$ and thus $(*)$.

To also show the continuity of the inverse mapping, we prove

$$\forall I \trianglelefteq R: \qquad \varphi(V(I)) = V(S^{-1}I). \qquad (**)$$

If $\mathfrak{p} \in D(f) \cap V(I)$, then $I \subseteq \mathfrak{p}$, but $f \notin \mathfrak{p}$. This shows $S^{-1}I \subseteq S^{-1}\mathfrak{p} = \mathfrak{p}_f$, thus $\mathfrak{p}_f \in V(S^{-1}I)$. Conversely, if $\mathfrak{p}_f \in V(S^{-1}I)$ and $a \in I$, then $\frac{a}{1} \in S^{-1}I \subseteq \mathfrak{p}_f$, thus $a \in j_f^{-1}(\mathfrak{p}_f) = \mathfrak{p}$. It follows that $\mathfrak{p} \in V(I)$, and $(**)$ is shown. We now know that φ is a homeomorphism.

Now let $U \subseteq \mathrm{Spec}(R_f) = \varphi(D(f))$ be open. A section $s \in \mathcal{O}_{\mathrm{Spec}(R_f)}(U)$ is given as a mapping $s: U \to \coprod_{\mathfrak{p}_f \in U}(R_f)_{\mathfrak{p}_f}$. However, for $\mathfrak{p}_f = \varphi(\mathfrak{p}) \in \mathrm{Spec}(R_f)$ with $\mathfrak{p} \in D(f)$, it holds that $f \notin \mathfrak{p}$ and therefore

$(R_f)_{\mathfrak{p}_f} = R_\mathfrak{p}$. Thus, s is identified by concatenation with φ with a section of $\mathcal{O}_{\mathrm{Spec}(R)}(\varphi^{-1}(U))$, and all sections of $\mathcal{O}_{\mathrm{Spec}(R)}(\varphi^{-1}(U))$ are obtained in this way.

(iv) The first statement follows with $f = 1$ immediately from (ii). We can refine the diagram with the mappings constructed in (i) and (ii):

$$\begin{array}{ccc} \mathcal{O}\big(\mathrm{Spec}(R)\big) & \xrightarrow{\mathrm{ev}_\mathfrak{p}} & \mathcal{O}_\mathfrak{p} \\ \psi \uparrow & & \downarrow \varphi \\ R & \xrightarrow{j_\mathfrak{p}} & R_\mathfrak{p} \end{array}$$

Here, $\mathrm{ev}_\mathfrak{p}$ is the evaluation at $\mathfrak{p}$ and $j_\mathfrak{p}: R \to R_\mathfrak{p}$, $a \mapsto \frac{a}{1}$ is the canonical mapping.

$\square$

The existence of bases for the topology, which consist of affine open sets, allows us to construct new schemes by gluing (see Example 5.13); we illustrate the procedure on a variant of projective space.

Example 7.51 (Projective Spaces over Rings) Let R be a commutative ring with identity and $S := R[X_0, \ldots, X_n, X_0^{-1}, \ldots, X_n^{-1}]$ the ring of all *Laurent polynomials* in $n+1$ variables. Here, S is by definition the quotient of the polynomial ring $R[X_0, \ldots, X_n, Y_0, \ldots, Y_n]$ by the ideal generated by the elements $X_0 Y_0 - 1, \ldots, X_n Y_n - 1$. In S, we consider for $i = 0, \ldots, n$ the subrings

$$R^{(i)} := R\left[\frac{X_0}{X_i}, \ldots, \widehat{\frac{X_i}{X_i}}, \ldots, \frac{X_n}{X_i}\right],$$

where, as usual, the hat over a term means that it is omitted. Thus, the $R^{(i)}$ are all isomorphic to the polynomial ring over R in n variables. We consider for $i, j = 0, \ldots, n$ with $i \neq j$ the affine schemes $U_i := \mathrm{Spec}(R^{(i)})$ and therein the standard open subsets

$$U_{ij} := D_{U_i}\left(\frac{X_j}{X_i}\right),$$

which according to Theorem 7.50 are themselves affine schemes for the rings $R^{(i)}_{\frac{X_j}{X_i}}$. Further, we set $\varphi_{ii} = \mathrm{id}_{U_i}$ and define $\varphi_{ji}: U_{ij} \to U_{ji}$ via the equality

$$R^{(i)}_{\frac{X_j}{X_i}} = R^{(j)}_{\frac{X_i}{X_j}}.$$

This gives gluing data (see Example 5.13) and it is found that the gluing $\mathbb{P}_R^n$ is a ringed space i.e., locally isomorphic to an affine scheme. We call the scheme $\mathbb{P}_R^n$ the *projective space of relative dimension n over R*.

It is clear that R can be embedded as a ring of constant functions in the ring $\mathcal{O}_{\mathbb{P}_R^n}(\mathbb{P}_R^n)$. On the other hand, the sections $\mathcal{O}_{\mathbb{P}_R^n}(U_i)$ are just the polynomials in $\frac{X_0}{X_i}, \ldots, \widehat{\frac{X_i}{X_i}}, \ldots, \frac{X_n}{X_i}$. No non-constant polynomial can be extended to a global section (exercise), so the embedding is even an isomorphism. If $\mathbb{P}_R^n$ were an affine scheme of the form $\mathrm{Spec}(S)$, then according to Theorem 7.50 $\mathcal{O}_{\mathbb{P}_R^n}(\mathbb{P}_R^n) = S$, i.e., $S = R$, would have to hold. This would imply $\mathbb{P}_R^n = \mathrm{Spec}(R)$, which in general is wrong (e.g., when $R = \mathbb{K}$ is a field, so that the right side consists of only one point). □

Geometric Interpretation of Affine Schemes

The constructions for schemes carried out so far bear a striking resemblance to the constructions from Section 7.2 for $\mathbb{K}$-varieties for algebraically closed fields $\mathbb{K}$, but the maximal ideals must be replaced everywhere by prime ideals. This difference has a geometric interpretation. We limit ourselves to affine schemes because they are geometric interpretations of the *points* of a scheme, so the restriction to affine open subsets makes no difference.

According to Hilbert's Nullstellensatz, points in an affine variety correspond precisely to the maximal ideals of the coordinate algebra. General prime ideals, on the other hand, correspond to certain closed subvarieties. Which closed subvarieties these are, can be precisely formulated.

Definition 7.52 (Irreducible Topological Spaces) A topological space $(X, \mathfrak{T})$ is called *irreducible*, if there is no decomposition of X of the form $X = X_1 \cup X_2$ with closed subsets $X_1, X_2 \subsetneq X$.

If X is an algebraic set over a field $\mathbb{K}$ and $\mathfrak{T}$ is the Zariski topology, then irreducibility of X means that X cannot be written as a union of two proper algebraic subsets. Since algebraic subsets are constructed from vanishing sets of ideals, one can expect a characterization of irreducibility via properties of the vanishing ideals.

Proposition 7.53 (Irreducible Components of Algebraic Sets) *Let $\mathbb{K}$ be a field and $X \subseteq \mathbb{K}^n$ algebraic. Then:*

(i) *X is irreducible if and only if $\mathrm{I}(X)$ is prime.*
(ii) *There is a unique decomposition $X = X_1 \cup \ldots \cup X_r$ up to order with X_i irreducible algebraic subsets, for which*

$$\forall i \neq j: \quad X_i \not\subseteq X_j$$

applies. The X_i are then called the irreducible components *of X.*

7.3 Schemes

Proof

(i) If X is not irreducible, then we can write $X = X_1 \cup X_2$ with $X_1, X_2 \subsetneq X$ algebraic. Because $X_i = V(I(X_i))$, $I(X_1)I(X_2) \subseteq I(X) \subsetneq I(X_1), I(X_2)$. So there are $f_i \in I(X_i) \setminus I(X)$ with $f_1 f_2 \in I(X)$, i.e., $I(X)$ is not prime.

Conversely, if $I(X)$ is not prime, there are $f_1, f_2 \notin I(X)$ with $f_1 f_2 \in I(X)$. Let $I_i := (I(X), f_i)$ and $V(I_i) =: X_i$, then $X_i \subsetneq X$, because for $f_j \notin I(X)$ there is an $a_j \in X$ with $f_j(a_j) \neq 0$. This a_j is not in X_j. But from $b \in X$ it follows $f_1 f_2(b) = 0$, i.e., $f_1(b) = 0$ or $f_2(b) = 0$ and thus $b \in X_1$ or $b \in X_2$.

(ii) Let $X_1 \supseteq X_2 \supseteq \ldots \supseteq X_n \supseteq \ldots$ be a chain of algebraic sets. Then: The corresponding chain of vanishing ideals $I(X_1) \subseteq I(X_2) \subseteq \ldots \subseteq I(X_n) \subseteq \ldots$ becomes stationary. Because $X_n = V(I(X_n))$ (see Proposition 7.5), the first chain also becomes stationary. Therefore, every set of algebraic subsets in $\mathbb{K}^n$ has a minimal element (Zorn's lemma).

Now let Σ be the set of all algebraic subsets $X \subseteq \mathbb{K}^n$ for which no decomposition as in (ii) exists. If $\Sigma \neq \emptyset$, then there is a minimal element X in Σ. Then X is not irreducible (that would be then already the decomposition), so we find $X = X_1 \cup X_2$ with $X_1, X_2 \subsetneq X$ algebraic. Due to the minimality, X_1 and X_2 have decompositions, and together one finds a decomposition for X. This provides a contradiction! Therefore, $\Sigma = \emptyset$, and this proves the existence of the decomposition. The uniqueness is left as an exercise for the reader. □

If one applies this proposition to algebraically closed $\mathbb{K}$, the diagram from Hilbert's Nullstellensatz can be supplemented as follows:

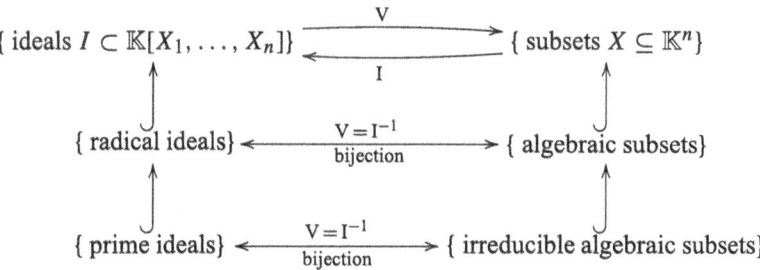

Now if A is a finitely generated reduced $\mathbb{K}$-algebra, then the maximal ideals describe the points of an affine $\mathbb{K}$-variety, namely $(\mathrm{Spm}(A), \mathcal{O}_{\mathrm{Spm}(A)})$. The points of the associated affine scheme $(\mathrm{Spec}(A), \mathcal{O}_{\mathrm{Spec}(A)})$ on the other hand are the irreducible subvarieties of $\mathrm{Spm}(A)$, of which the points are only the smallest. The definition of the Zariski topology on $\mathrm{Spec}(A)$ shows (exercise) that the closure of a point $\mathfrak{p} \in \mathrm{Spec}(A)$ is given by $\{\mathfrak{q} \in \mathrm{Spec}(A) \mid \mathfrak{p} \subseteq \mathfrak{q}\}$. This means, an irreducible algebraic subset $\mathfrak{q}$ of $\mathrm{Spm}(A)$ belongs to the closure of $V(\mathfrak{p})$ if it is an algebraic subset of $V(\mathfrak{p})$. In particular, a point $\mathfrak{m} \in \mathrm{Spm}(A)$ belongs to the closure of $V(\mathfrak{p})$ if

$\mathfrak{m} \in V(\mathfrak{p})$. A point in $\mathfrak{p} \in \mathrm{Spec}(A)$ is therefore only closed if $\mathfrak{p} \trianglelefteq A$ is maximal, i.e., belongs to $\mathrm{Spm}(A)$.

Since A can be recovered both as coordinate algebra $\mathbb{K}[\mathrm{Spm}(A)]$ from $\mathrm{Spm}(A)$ and as $\mathcal{O}_{\mathrm{Spec}(A)}\big(\mathrm{Spec}(A)\big)$ from $\mathrm{Spec}(A)$, one sees that through

$$\mathrm{Spm}(A) \longleftarrow\!\dashv A \vdash\!\longrightarrow \mathrm{Spec}(A)$$

a fully faithful functor $\mathbf{Var}_{\mathbb{K}}^{\mathrm{aff}} \to \mathbf{Sch}^{\mathrm{aff}}$ is defined. By gluing affine varieties and affine schemes, one even obtains a fully faithful functor $\mathbf{Var}_{\mathbb{K}} \to \mathbf{Sch}$. This proves the statement made at the beginning of this section that schemes provide an extended conceptual framework for dealing with algebraic varieties.

Schemes Over (General) Fields

Finally, we want to briefly sketch how to use schemes to talk about "algebraic varieties" over not necessarily closed fields or even rings. For the missing details, we refer to [GW10].

A *scheme over a ring* R is a scheme morphism $(X, \mathcal{O}_X) \to (\mathrm{Spec}(R), \mathcal{O}_{\mathrm{Spec}(R)})$. For a not necessarily algebraically closed field $\mathbb{K}$, a scheme over $\mathbb{K}$ is thus a morphism $(X, \mathcal{O}_X) \to (\star, \mathbb{K})$, where $(\mathrm{Spec}(\mathbb{K}), \mathcal{O}_{\mathrm{Spec}(\mathbb{K})}) = (\star, \mathbb{K})$ is the one-point ringed space with ring $\mathbb{K}$. In particular, $\mathcal{O}_X$ is then a sheaf of $\mathbb{K}$-algebras. We say $(X, \mathcal{O}_X)$ "is of locally finite type" if $(X, \mathcal{O}_X)$ is locally of the form $(\mathrm{Spec}(A), \mathcal{O}_{\mathrm{Spec}(A)})$ with a finitely generated $\mathbb{K}$-algebra A. We fix such a scheme of locally finite type over $\mathbb{K}$. For $x \in X$, the ring $\mathcal{O}_{X,x}$ is local according to Theorem 5.34. Let $\kappa(x)$ be the associated residue field, i.e., the quotient of $\mathcal{O}_{X,x}$ by the maximal ideal. The point x is called $\mathbb{K}$-*rational*, if the restriction $\mathbb{K} \to \kappa(x)$ of the quotient mapping is an isomorphism. We denote the set of $\mathbb{K}$-rational points of X by $X_{\mathbb{K}}(\mathbb{K})$. It can be shown that the mapping

$$X(\mathbb{K}) := \mathbf{Sch}(\mathrm{Spec}(\mathbb{K}), X) \to X_{\mathbb{K}}(\mathbb{K}), \quad (f, f^{\flat}) \mapsto f(\star)$$

is a bijection. The set $X(\mathbb{K})$ of morphisms $\mathbf{Sch}(\mathrm{Spec}(\mathbb{K}), X)$ is called the set of $\mathbb{K}$-*valued points* of X.

If X is the vanishing set $V(f_1, \ldots, f_k)$ for $f_1, \ldots, f_k \in A := \mathbb{K}[X_1, \ldots, X_n]$ in $\mathrm{Spec}(A)$ (see Example 5.30), then (exercise)

$$X_{\mathbb{K}}(\mathbb{K}) = \{x \in \mathbb{K}^n \mid f_1(x) = \ldots = f_k(x) = 0\}.$$

That is, for this example, the setting $X_{\mathbb{K}}(\mathbb{K})$ provides the desired points.

The identification of the $\mathbb{K}$-rational points with the $\mathbb{K}$-valued points makes the $\mathbb{K}$-rational points a representable functor and thus opens up the treatment of rationality questions using the very powerful technique of point functors (see [EH07, Chap. VI]).

7.3 Schemes

An important application of the theory of rational points of schemes is the theory of algebraic groups in the context of number-theoretical questions, in which the field $\mathbb{K}$ is often finite extensions of $\mathbb{Q}$—hence the name "rational points". In order to even begin to build such a theory, one needs the ability to form products of schemes over $\mathbb{K}$ in order to formulate what a compatible products of elements should be. For this, one resorts to the ideas of Section 4.3 and shows that the category **Sch** allows pullbacks in the sense of Example 4.36 (see [GW10, Thm. 4.18]). For two schemes X and Y over $\mathbb{K}$, one considers the diagram

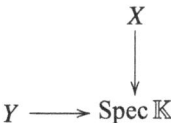

and denotes the associated pullback, which is then also a scheme over $\mathbb{K}$, by $X \times Y$.

Literature: The classic text for an introduction to the scheme-theoretic approach to algebraic geometry is [Ha77]. Meanwhile, there are several more reader-friendly presentations (see e.g., [EH07], [Ho12] and [GW10]). If one wants to understand how this approach generalizes the theory of varieties, it is advisable to also look at texts like [Di74], [Mi05] and [Pe95], which discuss varieties in detail. As an introduction to the local theory, [Re90] and [Re95] are also suitable. The cross-connection to manifolds is explained in [GH94].

Part III

Outlook

From the structures that we have discussed in Parts I and II, a large number of new structures could easily be obtained by combination. For example, one could consider variants of each algebraic structure in which the underlying sets are topological spaces, manifolds, or varieties, and the operations are morphisms in the respective category. However, algebraic and local structures can also be mixed with other structures that are neither algebraic nor local. For example, one could require that there is a measure on a manifold that is defined on the σ-algebra generated by the open subsets. Or one might want to presuppose the existence of an orientation. However, most conceivable combined structures are not considered let alone studied in detail in practice.

There are different motivations to study new structures. One motivation can be that examples of such structures appear in already studied mathematical problems. For example, linear algebra originated as a structural abstraction of the study of (linear) differential equations. Another motivation is that additional structural elements are suggested by the mathematical modeling of physical or other phenomena. For many applications, the existence of certain functions is necessary for modeling the relevant situations. In relativity theory, for example, some kind of distance measurement in space-time is needed. In this section, we describe various classes of such specialized structures. We provide motivations and precise definitions, but largely refrain from detailed results. The structures described are each the subject of their own mature theory, and we limit ourselves to providing standard literature on the respective theory, which may serve the reader as a starting point for a search for more detailed information.

One could call the entire field of functional analysis, whose development is closely linked to the mathematical modeling of quantum physics, an example class for the study of specialized structures. It combines vector spaces with norms or topologies, that is, functional analysis deals with special vector spaces. However, the focus is on the fact that functional analysis generalizes methods and results of linear algebra to certain infinite-dimensional vector spaces. This draws attention to generalizations of structures and theories that are developed to solve given problems that could not be tackled with conventional methods. For example, it happens that a certain combination of properties is needed to be able to use a known technique. The introduction of complex numbers in the study of polynomial equations can be read in this way, as can the introduction of distributions to solve differential equations. In both cases, these are extensions of the mathematical framework to be able to describe solutions to given problems in this extended framework, which can then be further investigated. The extension of $\mathbb{C}$-varieties to schemes to study rationality questions can also be justified in this way.

Modifying or replacing an existing structure can also aim to unify or simplify arguments from separate contexts. Given the rapid development of mathematical knowledge in many different directions, unification is a quite serious task. We will discuss two principles of unification separately here. The transfer of arguments to other areas, illustrated by homological algebra as a technology of structure comparison, and the transfer of structural elements, illustrated by the concept of the group object for the formalization of symmetries. Chapter 8 is dedicated to these illustrations. In Chap. 9, we then discuss specialized structures as well as possible generalizations and unifications.

Since very different contents are only treated very briefly in this part, each section contains its own annotated bibliography.

8 Transfer of Arguments and Structures

In this chapter, we illustrate the transfer of arguments and structural elements to other areas with two examples. Homological algebra as a technology for structural comparison and the concept of a group object for the formalization of symmetries.

8.1 Technologies of Structural Comparison

Chapter 4 was dedicated to the comparison of mathematical structures. There, we discussed the conceptual foundations for the precise description of such comparisons.

However, there are entire areas of mathematics that are concerned with the comparison of mathematical structures from the outset. A typical example is *algebraic topology*, which aims to find algebraic measures for families of topological spaces that can be used to distinguish the members of the family. Technically speaking, one is looking for functors from categories of topological spaces to categories of algebraic structures. Examples of this are various (co)homology or homotopy functors. If the images of the functors are not isomorphic, neither were the pre-images (see e.g. [Ha02]). Thus, one tries to construct the functors coarse enough to be able to say something meaningful about their images, but at the same time fine enough to be able to distinguish interesting topological spaces from each other. Other questions are those of obstructions to the existence of topological spaces with certain additional properties, which often can be rephrased as a question about the vanishing of certain elements of a cohomology group.

A completely different form of structural comparison is the subject of *Galois theory*. Starting from Galois' insights into the role of group theory for the solvability of polynomial equations, Galois theory deals with the interplay between field extensions $\mathbb{K} \subseteq \mathbb{L}$ and the groups $\mathrm{Aut}(\mathbb{L})^{\mathbb{K}}$ of field automorphisms that leave $\mathbb{K}$ pointwise fixed (see [La93, Part 2]). Galois theory is an indispensable tool in algebraic number theory and arithmetic geometry. Moreover, the local class field theory, which deals with field extensions with abelian Galois group, is the starting point for the *Langlands program*, one of the major mathematical research projects of the present. Nevertheless, we refrain from sketching it separately here, but refer to the literature (see [Bu96] and [Go11]).

Homological Algebra

From the constructions of algebraic topology, a mathematical technology has developed over time, which is now used in many areas of mathematics under the heading *homological algebra*. From the geometric-combinatorial beginnings of simplicial homology, a theory of families of mappings between modules was developed. Today, homological algebra is a theory of certain categories and associated functors. It can also be applied to sheaves, especially module sheaves, and then leads to corresponding sheaf cohomologies. These are often the key to investigating global properties of local structures. Such techniques were first used in the theory of functions of several complex variables, but it was soon recognized that they are also of utmost use in algebraic and arithmetic geometry. Nowadays, sheaf cohomology and its further developments play an important role in a multitude of mathematical disciplines. We refrain from sketching these developments, but refer directly to the relevant literature (see [Br97, Go73, Ha77, KV95, Ra04, Ta02, We15, We79, Wa83]). To explain the basic ideas of homological algebra, we consider the surface of a tetrahedron as an example of a topological space given as a simplicial complex. It is composed of four triangles, each having a boundary consisting of three intervals. The intervals, in turn, have their endpoints as a boundary. To be able to calculate with these geometric building blocks, one must form formal (linear) combinations of the geometric objects. For example, one wants to see the entire surface as the sum of the triangles constituting it and the boundary of a triangle as the sum of the intervals forming the boundary. The boundary of a triangle is closed, i.e., a closed line segment. It therefore has no boundary itself. To correctly describe the boundary of the triangle as the sum of the endpoints of its intervals when forming sums, the endpoints must add up to zero. This can be achieved by defining the boundary of an interval as the difference of the endpoints. The surface of the tetrahedron is also closed, i.e., it has no boundary. Therefore, the boundaries of the triangles must also add up to zero, which can be achieved by defining the boundary of a triangle as the alternating sum of its boundary intervals.

Since an alternating sum of simplices of the same dimension (here either points, intervals, or triangles) is just a special $\mathbb{Z}$-valued function on the set of these simplices, these considerations lead to the spaces C_j of $\mathbb{Z}$-valued functions on the set of j-simplices. The combinatorial definition of the boundary then becomes a $\mathbb{Z}$-linear mapping $\partial_j : C_j \to C_{j-1}$ and the geometric fact that a boundary has no boundary translates into the identity $\partial_{j-1} \circ \partial_j = 0$. It is clear that $\text{im}(\partial_j) \subseteq \ker(\partial_{j-1})$ and one can define the homology groups $H_j := \ker(\partial_{j-1})/\text{im}(\partial_j)$.

8.1 Technologies of Structural Comparison

Families of morphisms

$$\cdots \longrightarrow C_j \xrightarrow{\partial_j} C_{j-1} \xrightarrow{\partial_{j-1}} C_{j-2} \longrightarrow \cdots$$

with $\partial_{j-1} \circ \partial_j = 0$ can of course be considered for any category of R-modules. It doesn't matter whether the indices are counted down as here or up as in the case of the exterior derivative (cf. Proposition 6.54)

$$\cdots \longrightarrow \Omega_M^j(U) \xrightarrow{d_j} \Omega_M^{j+1}(U) \xrightarrow{d_{j+1}} \Omega_M^{j+2}(U) \longrightarrow \cdots,$$

for which the de Rham cohomology group $H_M^{j+1}(U) := \ker(d_{j+1})/\operatorname{im}(d_j)$ is defined.

It turns out that the de Rham cohomology groups, which were constructed from local data of the manifold, are of functorial nature (pull back differential forms via Remark 6.55) and contain global information about the manifold. This can be seen as a first hint of the repeatedly confirmed relevance of cohomological information. If one, thus motivated, looks for further applications of these ideas, it is obvious to consider sequences of morphisms in any category **C**. But if one also wants to generalize the identity $d_{j+1} \circ d_j = 0$, one needs a candidate for "0" in the morphism sets $\mathbf{C}(A, B)$. This is obtained if one requires that the $\mathbf{C}(A, B)$ are each abelian groups. If one also assumes that the composition of morphisms with respect to these $\mathbb{Z}$-module structures is bilinear, the category **C** is called *preadditive*. If one then wants to continue talking about images and kernels of morphisms, one needs in **C** also a *zero object* $0 \in \mathrm{Ob}(\mathbf{C})$, for which $\mathbf{C}(0, A)$ and $\mathbf{C}(A, 0)$ each have exactly one element. In addition, for the definitions of kernel and image of a morphism, the existence of finite limits and colimits in the sense of the Definitions 4.34 and 4.35 is needed. The *kernel* $\ker(f)$ of a morphism $f \in \mathbf{C}(A, B)$ is the limit of the diagram.

$$A \underset{0}{\overset{f}{\rightrightarrows}} B,$$

where $0 \in \mathbf{C}(A, B)$ is the uniquely determined morphism that factors through the zero object. Analogously, the *cokernel* $\operatorname{coker}(f)$ of f is obtained as the colimit of this diagram. Thus, we have $\ker(f) \to A$ and $B \to \operatorname{coker}(f)$. The cokernel of $\ker(f) \to A$ is denoted by $\operatorname{coim}(f)$ and the kernel of $B \to \operatorname{coker}(f)$ by $\operatorname{im}(f)$. From the definitions, a natural factorization

$$A \longrightarrow \operatorname{coim}(f) \longrightarrow \operatorname{im}(f) \longrightarrow B$$

of $A \to B$ is obtained. If $\operatorname{coim}(f) \to \operatorname{im}(f)$ is always an isomorphism, **C** is called an *exact* category. An exact preadditive category is called an *abelian* category, see e.g. [We15]. It turned out that the properties of an abelian category are needed to get homological algebra, as it was developed for modules over commutative rings, up and running.

In addition to simplicial homology, there is a second root of algebraic topology, namely homotopy theory. This theory is about deforming mappings between topological spaces continuously along a parameter $t \in [0, 1]$. Without going into the details, we only note here how such deformations affect cohomology functors. If a cohomology theory as above is given by *chain complexes* of the form

$$(C_X^\bullet, d_{X,\bullet}) := (\ldots \longrightarrow C_X^j \xrightarrow{d_{X,j}} C_X^{j+1} \xrightarrow{d_{X,j+1}} C_X^{j+2} \longrightarrow \ldots)$$

then a continuous mapping $f : X \to Y$ between two topological spaces induces morphisms $f^j : C_Y^j \to C_X^j$, which make the diagram

$$\begin{array}{ccccccc}
\ldots \longrightarrow & C_Y^j & \xrightarrow{d_{Y,j}} & C_Y^{j+1} & \xrightarrow{d_{Y,j+1}} & C_Y^{j+2} & \longrightarrow \ldots \\
& \downarrow f^j & & \downarrow f^{j+1} & & \downarrow f^{j+2} & \\
\ldots \longrightarrow & C_X^j & \xrightarrow{d_{X,j}} & C_X^{j+1} & \xrightarrow{d_{X,j+1}} & C_X^{j+2} & \longrightarrow \ldots
\end{array}$$

commutative. This commutativity shows that $(f^\bullet)$ induces morphisms between the cohomology spaces. It turns out that a homotopy between two continuous mappings $f, g : X \to Y$ induces morphisms $H^j : C_Y^j \to C_X^{j-1}$ for which

$$f^j - g^j = d_{X,j-1} \circ H^j + H^{j+1} \circ d_{Y,j} : C_Y^j \to C_X^j$$

holds. With this, one calculates that homotopic mappings induce the same morphism in cohomology.

If one now has an abelian category **C**, then one can introduce a category **C**$^\bullet$ of complexes of the form

$$(C^\bullet, d_\bullet) := (\ldots \longrightarrow C^j \xrightarrow{d_j} C^{j+1} \xrightarrow{d_{j+1}} C^{j+2} \longrightarrow \ldots)$$

with $d_{j+1} \circ d_j = 0$, for which morphisms are given by commutative diagrams of morphisms of the form

$$\begin{array}{ccccccc}
\ldots \longrightarrow & D^j & \xrightarrow{\delta_j} & D^{j+1} & \xrightarrow{\delta_{j+1}} & D^{j+2} & \longrightarrow \ldots \\
& \downarrow f^j & & \downarrow f^{j+1} & & \downarrow f^{j+2} & \\
\ldots \longrightarrow & C^j & \xrightarrow{d_j} & C^{j+1} & \xrightarrow{d_{j+1}} & C_X^{j+2} & \longrightarrow \ldots
\end{array}$$

Then one can define the concept of a *homotopy* between two morphisms $f^\bullet$ and $g^\bullet$ as a family $H^j : D^j \to C^{j-1}$ of morphisms, for which

$$f^j - g^j = d_{j-1} \circ H^j + H^{j+1} \circ \delta_j : D^j \to C^j.$$

Homotopy of morphisms is an equivalence relation. By replacing morphisms between complexes with homotopy classes of complexes, we obtain the *homotopy category* $K(\mathbf{C})$ of complexes over $\mathbf{C}$. With the construction of $K(\mathbf{C})$, we have taken the first step in the construction of the so-called *derived category* $D(\mathbf{C})$ (see e.g. [Iv86, Chap. XI]).

Abelian categories and their derived categories are the conceptual framework for modern homological algebra. Clausen and Scholze say in the introduction to [CS22] the fact that the category **ABtop** of topological abelian groups does not form an abelian category is a central motivation for their novel *condensed mathematics*. In this theory, one considers *condensed sets*, which by definition are **Set**-sheaves over the category **CHaus** of compact Hausdorff spaces with continuous mappings as morphisms. This refers to a functor $\mathbf{CHaus}^{\mathrm{op}} \to \mathbf{Set}$ (**Set**-presheaf, cf. Definition 5.1) that satisfies the appropriate sheaf conditions (see [LS23, Chap. 3 & Def. 4.1.3]). A *condensed abelian group* is then an **Ab**-sheaf over the category **CHaus**. The condensed abelian groups form an abelian category **ABcond** and there is a faithful functor **ABtop** $\to$ **ABcond** (see [LS23, Cor. 6.1.5 & Prop. 6.2.1]). This has once again created an extension of the framework in which one has additional tools to find "generalized solution" for problems, the relevance of which for the original problems can then be investigated.

Literature: [Mu18] is an introduction to topology discussing also elementary homotopy. [Ha02] and [Sp66] are introductions to algebraic topology containing also simplicial homology. In all three books manifolds and de Rham cohomology are only touched upon. Proofs for the de Rham theorem, which establishes the connection between de Rham cohomology and the singular cohomology discussed in the aforementioned books (an advancement of simplicial cohomology), can be found in [Wa83] and [We15]. The book by Wedhorn also includes an appendix on homological algebra in abelian categories. Comprehensive presentations up to the definition of derived categories are provided by [Iv86, We94]. The books [GM03, KS06] go even further. The essay collection [AC21a] contains a sketch of recent conceptual developments in homotopy theory.

8.2 Group Objects and Group Actions

We started this book with algebraic structures and then moved on to local structures that could be described by sheaves. The objects thus obtained, such as manifolds or algebraic varieties, can themselves carry algebraic structures. In this case, we want the algebraic structure mappings to be morphisms of the corresponding category. If the algebraic structure includes operations, this is only possible for categories that allow finite products. This combination of local and algebraic structures may seem arbitrary and unmotivated at first glance, but at least in the case of groups, the examples show that such combinations can provide interesting insights and paradigmatic examples. We therefore describe the procedure exemplarily for the case of groups.

Definition 8.1 (Group Objects) Let **C** be a category that allows finite products and contains a *terminal* object $\star$ (i.e., from every object $A \in \mathbf{C}$ there is exactly one morphism $t_A : A \to \star$). An object $G \in \mathrm{ob}(\mathbf{C})$, together with three morphisms m: $G \times G \to G$, i $G \to G$ and e: $\star \to G$, is called a *group object* in **C**, if the following diagrams commute:

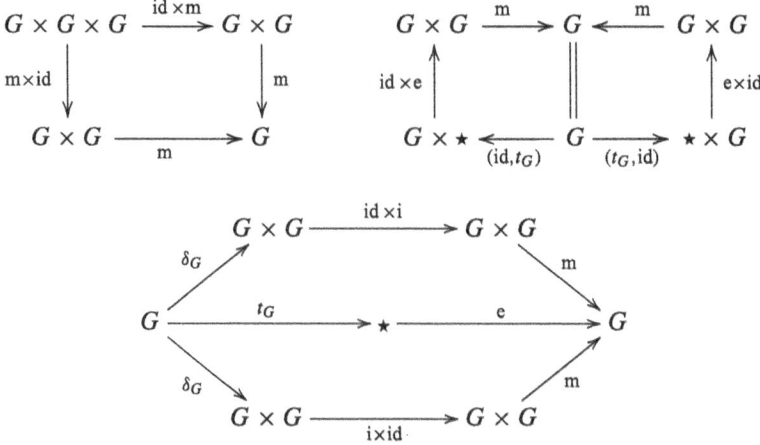

For manifolds and varieties over algebraically closed fields, we have established the existence of finite products (see Exercise 6.2 and Example 7.43). Terminal objects also exist in both categories. They each consist of the one-point ringed space with ring equal to $\mathbb{K}$. The corresponding group objects are called $\mathbb{K}$-*Lie groups* or *algebraic groups* over $\mathbb{K}$. Together with the morphisms, which are also group homomorphisms, they each form a category, which in turn are the subject of rich theories (see e.g., [HN12] or [Mi17]). Motivated by questions especially from number theory, it is of great importance for algebraic groups not to be restricted to algebraically closed fields. In Section 7.3, corresponding rationality questions were already addressed, but one would also like to be able to work with rings, e.g., to be able to deal with groups of matrices whose entries are integers. Conceptually, this can be done using *schemes over a fixed base scheme* $(S, \mathcal{O}_S)$ (simply, S-scheme). An S-scheme is defined as a scheme morphism $(X, \mathcal{O}_X) \to (S, \mathcal{O}_S)$. The existence of pullbacks in **Sch**, which we already used in Section 7.3 for the existence of products of schemes over fields, allows in general products of S-schemes. This then also allows the introduction of group objects in the category **Sch**$_S$ of S-schemes, whose morphisms are commutative triangles

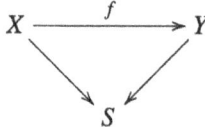

of scheme morphisms.

There are a number of other relevant examples of group objects. For example, the categories of supermanifolds relevant for the physical theories of supersymmetry (see e.g., [CCF11]) also allow products. The corresponding group objects are called *Lie supergroups*. They are examples of the fact that group objects of a category do not have to be groups themselves.

The existence of group objects in categories with products also suggests the introduction of group actions.

Definition 8.2 (Group Actions) Let **C** be a category that allows finite products and contains a terminal object $\star$. Further let $G, X \in \mathrm{ob}(\mathbf{C})$. If G is a group object in **C**, a morphism $\mu : G \times X \to X$ is called a group action if the following diagrams commute:

$$
\begin{array}{ccc}
G \times G \times X & \xrightarrow{\mathrm{id} \times \mu} & G \times X \\
{\scriptstyle m \times \mathrm{id}} \downarrow & & \downarrow {\scriptstyle \mu} \\
G \times X & \xrightarrow{\mu} & X
\end{array}
\qquad
\begin{array}{ccc}
\star \times X & \xrightarrow{e \times \mathrm{id}} & G \times X \\
& {\scriptstyle \cong} \searrow & \downarrow {\scriptstyle \mu} \\
& & X
\end{array}
$$

With this definition, one has a framework in which one can study not only the effects of Lie groups on manifolds and of algebraic groups on varieties, but also more exotic operations such as the effects of Lie supergroups on supermanifolds (see e.g., [AHW18]).

New Structures from Orbits and Quotients

Group actions often give rise to the introduction of new structures. For example, if one lets $(\mathbb{R}, +)$ act on the torus $\mathbb{R}^2/\mathbb{Z}^2$ by

$$(t, (x + \mathbb{Z}, y + \mathbb{Z})) \mapsto (x + t + \mathbb{Z}, y + \sqrt{2}t + \mathbb{Z})$$

then every orbit of the action is dense in $\mathbb{R}^2/\mathbb{Z}^2$ (see [HN12, Bsp. 9.3.12]). In this way the 2-dimensional manifold $\mathbb{R}^2/\mathbb{Z}^2$ is written as the union of 1-dimensional manifolds that are not closed subsets of $\mathbb{R}^2/\mathbb{Z}^2$, even though every point has a neighborhood that is homeomorphic to the product of the orbit of an open interval

$]-\epsilon, \epsilon[\subseteq \mathbb{R}$ through the point and an open interval. This provides an example of a *foliation* of a manifold (see e.g., [Mi08, § 3.23] and [MM03], but also [LS20] for algebraic variants).

More than when considering the orbits of a group action, the need to consider new structures arises when considering quotients, i.e., the set of orbits. For the above $\mathbb{R}$-action on the torus, the space of orbits, like the space of leaves for many other foliations, has no good properties with respect to the quotient topology. Traditional methods cannot analyze such spaces. *Non-commutative geometry* picks up the idea that some spaces can be completely described by associated algebras of functions. This applies, e.g., to affine $\mathbb{C}$-varieties, which can be recovered from their regular functions. It also applies to compact topological spaces, which are completely determined by the $\mathbb{C}$-valued continuous functions defined on them. In non-commutative geometry, one constructs non-commutative C^*-algebras for singular spaces, such as the spaces of leaves, which replace the algebras of continuous functions and can be studied using methods of functional analysis (see e.g. [Co94, GVF01]). An important step in the construction of foliation C^*-algebras is the geometric construction of certain *groupoids* for foliations. Groupoids are algebraic structures that are most easily described as small categories in which every morphism is invertible. This means that the endomorphisms of objects form groups. If there is only one object, the groupoid structure reduces to a group structure.

For some actions, the quotients are singular, but still manageable with geometric methods. An example is the so-called modular surface, which is obtained as a quotient of the upper half-plane in $\mathbb{C}$ with respect to the Möbius transformations given by the group $SL(2, \mathbb{Z})$. The modular surface is an example of an *orbifold*, for which, roughly speaking, in the definition of a manifold by local charts, the open subsets of $\mathbb{R}^n$ are replaced by quotients of such sets by finite groups. For the modular surface, these finite groups appear as stabilizers of the points i (isomorphic to $\mathbb{Z}/2\mathbb{Z}$) and $\pm\frac{1}{2} + i\frac{\sqrt{3}}{2}$ (isomorphic to $\mathbb{Z}/3\mathbb{Z}$). Orbifolds were introduced as objects under the name V-manifolds as early as 1956 by Satake (see [Sa56]), but the definition of morphisms is much more subtle (see [Po17]). Quotients of algebraic varieties and schemes, which are already locally more complex than manifolds, also give objects which are no longer in the starting categories. In this context, the theory of *stacks* is considered, which has proven to be very useful for the study of families of geometric objects. Stacks are generalizations of schemes, defined in the style of functor categories.

Literature: There are many books about Lie groups, often they are written with a view to certain classes (e.g., compact or reductive) or applications (e.g., linear representations) of Lie groups. [HN12] is dedicated to the general structure theory of Lie groups.

Classic texts about algebraic groups, which can be written as groups of matrices, are [Bo91, Sp88]. The book [Mi17] by Milne deals with rationality questions in a systematic way within the framework of schemes over fields. The preface to [Mi17] contains an explanation of the differences in the treatment of rationality questions,

8.2 Group Objects and Group Actions

as [Bo91] presents it. The connections between algebraic groups and number theory are extensively discussed in [PR94].

A systematic presentation of the theory of orbifolds in book form is still pending, but [MM03] contains a discussion of orbifolds and their cross-connections with foliations and groupoids. Groupoids and generalizations thereof not only play a role in non-commutative geometry, but appear in various contexts. Many of these are described in the two volumes [AC21a, AC21b].

There are now various books about supermanifolds, each treating the topic in very different ways (see e.g., [CCF11, DM99, Tu04]), but a generally accepted systematic presentation for this field is also not yet available. [AC21b] also contains an article on this topic.

Stacks are introduced in [AC21a]. There is a special experiment on this topic: The *Stacks Project* is a novel attempt to further develop stacks and their role in algebraic geometry as part of a community effort in the form of an open access online textbook, see [SP].

9 Specialization, Generalization and Unification of Structures

In this chapter, we exemplify specialized structures that can then be generalized and unified. The aim of the chapter is not to provide an exhaustive list of well-motivated structures that have not yet been listed in the first two parts. Otherwise, e.g., the absence of *combinatorial structures* would be inexcusable. Rather, the aim is to convey a sense that the structural considerations of the first two parts allow us to place a multitude of mathematical terms without much effort.

9.1 Special Tensors

In Ch. 6, we saw that from a manifold structure, given by a sheaf, i.e., by local data, can derive a number of other structures. In particular, we found the tangent spaces, which are defined pointwise, but vary continuously or differentiably depending on the class of the manifold. Together they form a vector bundle, which according to Theorem 6.38 is characterized by its section sheaf, and should therefore also be considered a local structure. More generally, we have tensor and form bundles for each differentiable manifold. In various geometric or physical theories in which manifolds play a role, in addition to the manifold structure and the derived structures, there are individual elements of these additional structures, each of which has a geometric or physical interpretation. Here we consider, as examples, pseudo-Riemannian metrics, symplectic forms, and Poisson tensors.

Pseudo-Riemannian Metrics

Riemannian geometry is the generalization of Euclidean geometry to curved spaces. The starting point for this theory is the idea that lengths and angles can no longer be described globally, but initially only in the tangent space, whose elements are interpreted as "velocities" with directions and absolute values.

Definition 9.1 (Pseudo-Riemannian Manifold) Let M be a $\mathcal{C}_{\mathbb{R}}^k$-manifold and $g \in \mathcal{T}_M^{(0,2)}(M)$ a global tensor field, which is pointwise non-degenerate and symmetric. Then (M, g) is called a *pseudo-Riemannian manifold* and g a *pseudo-Riemannian metric*. If g is pointwise positive definite, (M, g) is called a *Riemannian manifold* and g a *Riemannian metric*. If g has the signature $(n - 1, 1)$ pointwise, (M, g) is called a *Lorentz manifold* and g a *Lorentz metric*.

Lorentz manifolds of dimension 4 model space-time in general relativity. The three spatial directions can no longer be clearly separated from the one time direction, but the signature of the form models what remains of "spatiality" and "temporality" in Einstein's physical theory of gravitation.

Example 9.2 (Pseudo-Riemannian Manifolds)

(i) Let $M = \mathbb{R}^n$ and $g_0: \mathbb{R}^n \times \mathbb{R}^n \to \mathbb{R}$ be a non-degenerate symmetric bilinear form. Then $g(x) = g_0$ defines a pseudo-Riemannian metric on the manifold $\mathbb{R}^n$. It is Riemannian if g_0 is positive definite, and Lorentzian if the signature of g_0 is $(n - 1, 1)$. Special cases include the *Euclidean spaces*, for which g_0 is given by the identity matrix $\mathbf{1}_n$ with respect to the standard basis, and the *Minkowski space*, for which $n = 4$ and g_0 is equal to the diagonal matrix $\mathrm{diag}(1, 1, 1, -1)$ with respect to the standard basis.

(ii) Let (M, g) be a Riemannian $\mathcal{C}_{\mathbb{R}}^k$-manifold and N a $\mathcal{C}_{\mathbb{R}}^k$-manifold. If $\varphi: N \to M$ is an injective $\mathcal{C}_{\mathbb{R}}^k$-mapping whose derivative $\varphi'(x): T_x(N) \to T_{\varphi(x)}(M)$ is also injective for every $x \in N$, then $x \mapsto g_N(x) := g(\varphi(x)) \circ (\varphi'(x) \times \varphi'(x))$ defines a Riemannian metric g_N on N. This construction does not generally work for pseudo-Riemannian metrics because the restriction of a non-degenerate bilinear form to a subspace can be degenerate.

(iii) By combining (i) and (ii), one obtains a Riemannian metric for the $\mathbb{R}$-surfaces from Example 6.10. □

Every differentiable manifold carries Riemannian metrics, because Riemannian metrics on coordinate neighborhoods, which obviously exist, can be added up to a Riemannian metric on the entire manifold using a smooth partition of unity subordinate to an atlas.

If (M, g) is a Riemannian manifold and $\gamma: [a, b] \to M$ is a piecewise differentiable curve, one can assign a length $\ell_g(\gamma)$ to γ by integrating over the length of the tangent vectors determined by g:

$$\ell_g(\gamma) := \int_a^b \sqrt{g(\gamma'(t), \gamma'(t))}\, dt.$$

9.1 Special Tensors

With this, one can define a distance function $d_g : M \times M \to \mathbb{R}$ by the infimum of all lengths of piecewise differentiable connecting curves. If any two points in M can be connected by such curves, this distance function is a metric, and if a curve realizes the distance between two points, it is called a *geodesic*.

Theorem 9.3 (Hopf-Rinow) *Let (M, g) be a Riemannian manifold that is complete as a metric space. Then any two points can be connected by a geodesic.*

In Proposition 6.34 we saw that derivatives of scalar functions on manifolds are sections of the cotangent bundle, i.e., 1-forms. In introductory lectures, derivatives of scalar functions on open subsets of $\mathbb{R}^n$ are often introduced as vectors whose components are the partial derivatives. The fact that these two definitions match is due to a tacit use of the natural Euclidean metric on $\mathbb{R}^n$, with the help of which the gradient of the functions is defined. This gradient is the object that is often simply referred to as the derivative of f in vector analysis.

Definition 9.4 (Gradient) Let (M, g) be a pseudo-Riemannian manifold and $f: M \to \mathbb{R}$ a differentiable function. Then the equation

$$\forall x \in M, v \in T_x(M): \quad g(x)\bigl(\mathrm{grad}(f)(x), v\bigr) = df(x)(v)$$

a vector field $\mathrm{grad}(f)$ on M, which is called the *gradient* of f.

From the proof of Proposition 6.34 one can read off (exercise) that for $M = \mathbb{R}^n$ with the Euclidean metric as in Example 9.2(i) the gradient of $f: \mathbb{R}^n \to \mathbb{R}$ with respect to the standard basis is indeed given by the partial derivatives.

Symplectic Forms

Symplectic geometry models classical mechanics in Hamiltonian formulation. The space of possible states, i.e., the "phase space", is a manifold additionally equipped with a non-degenerate 2-form. This 2-form allows formulating the equations of motion for a given energy distribution on the states, i.e., a "Hamilton function". We formulate the central definition here only for real manifolds, but it can also be transferred to complex manifolds.

Definition 9.5 (Symplectic Manifold) Let M be a $\mathcal{C}_{\mathbb{R}}^k$-manifold and $\omega \in \Omega^2(M)$ a closed 2-form, which is pointwise non-degenerate as a bilinear form. Then (M, ω) is called a *symplectic manifold* and ω a *symplectic form*.

A *Hamilton function* is a differentiable function $f\, M \to \mathbb{R}$, and the equations of motion are given by vector fields, which are obtained analogously to the gradients from the symplectic form.

Definition 9.6 (Symplectic Gradient) Let (M, ω) be a symplectic manifold and $f \colon M \to \mathbb{R}$ a differentiable function. Then the equation

$$\forall x \in M, v \in T_x(M) : \quad \omega(x)\big(\mathfrak{X}_f(x), v\big) = df(x)(v)$$

defines a vector field $\mathfrak{X}_f$ on M, which is called the *symplectic gradient* or the *Hamiltonian vector field* of f.

The differential equation obtained from a vector field $\mathfrak{X} \colon M \to TM$ on a differentiable manifold is given by $\gamma' = \mathfrak{X} \circ \gamma$ for a differentiable curve $\gamma \colon I \to M$. In the case of Hamiltonian vector fields, this differential equation is then called the *Hamiltonian equations of motion*.

Example 9.7 (Symplectic Manifolds)

(i) Let $M = \mathbb{R}^{2n}$ and $\omega_0 \colon \mathbb{R}^{2n} \times \mathbb{R}^{2n} \to \mathbb{R}$ be a non-degenerate skew-symmetric bilinear form. Then $\omega(x) := \omega_0$ defines a symplectic form on the manifold $\mathbb{R}^{2n}$, because the exterior derivative of ω is zero (see Remark 6.53).

(ii) Let N be a differentiable $C_{\mathbb{R}}^k$-manifold with $k \geq 2$ and $M := T^*(N)$. If $\pi \colon M \to N$ is the canonical projection, then the derivative $\pi' \colon T(M) \to T(N)$ provides a 1-form

$$M = T^*(N) \supseteq T^*_{\pi(\beta)}(N) \ni \beta \mapsto \Theta_\beta := \beta \circ \pi'|_{T_\beta(M)} \in T_\beta(M)^*$$

on M. The exterior derivative $\omega := d\Theta$ is closed according to Proposition 6.54. If one calculates with respect to a chart (U, φ) with φ-basis (see Proposition 6.32), one finds (exercise) that Θ in local coordinates has the form $\sum_{i=1}^n y_i dx_i$, where $x_1, \ldots, x_n$ are the coordinates on N and $y_1, \ldots, y_n$ are the fiber coordinates. Accordingly, $\omega = \sum_{i=1}^n dy_i \wedge dx_i$ is non-degenerate pointwise. □

Example 9.8 (Co-adjoint Orbits) Let G be a Lie group and $\mathfrak{g}$ the tangent space of G at the identity element e. The group G acts on itself through the conjugations $c_g \colon G \to G$, $h \mapsto ghg^{-1}$ by diffeomorphisms. Since $c_g(e) = e$, the derivative $\mathrm{Ad}(g) := c'_g(e)$ maps the space $\mathfrak{g}$ onto itself. The action of G on $\mathfrak{g}$ defined in this way is called the *adjoint action*. By setting

$$\forall g \in G, v \in \mathfrak{g}^*, X \in \mathfrak{g} : \quad \langle \mathrm{Ad}^*(g)v, X \rangle := \langle v, \mathrm{Ad}(g^{-1})X \rangle$$

one defines on the dual space $\mathfrak{g}^*$ of $\mathfrak{g}$ the *co-adjoint action*. It turns out that the orbits of the co-adjoint action are naturally symplectic manifolds.

To be able to give the definition of the symplectic form ω, we need the natural Lie bracket $[\cdot, \cdot]$ on $\mathfrak{g}$, (cf. Remark 6.47). It is obtained by extending elements of $\mathfrak{g}$ by left translation to translation-invariant vector fields, using the Lie bracket of vector fields from Remark 6.47, and evaluating the result at the identity element. This connection

9.1 Special Tensors

between Lie groups and Lie algebras is functorial and the Lie algebra $\mathfrak{g}$ of a Lie group G is often also denoted by $\mathrm{Lie}(G)$.

If now $v \in \mathfrak{g}^*$, then all tangent vectors to the co-adjoint orbit $\mathrm{Ad}^*(G)v$ at v are of the form $\mathrm{ad}^*(X)v := (\mathrm{Ad}^*)'(e)v$ with $X \in \mathfrak{g}$. By

$$\forall X, Y \in \mathfrak{g}: \quad \omega(v)(\mathrm{ad}^*(X)v, \mathrm{ad}^*(Y)v) := \langle v, [X, Y] \rangle$$

one can define a non-degenerate skew-symmetric bilinear form on $T_v(\mathrm{Ad}^*(G)v)$. With the help of the co-adjoint action, this bilinear form can be extended to a G-invariant symplectic form ω on $\mathrm{Ad}^*(G)v$. □

If V is an $\mathbb{R}$-vector space and ω_0 is a non-degenerate skew-symmetric bilinear form on V, then (V, ω_0) is called a *symplectic vector space*. Every symplectic vector space has even dimension $2n$ and a basis with respect to which the form ω_0 is given by the block matrix

$$J_n := \begin{pmatrix} 0 & \mathbf{1} \\ -\mathbf{1} & 0 \end{pmatrix}$$

where $\mathbf{1}$ denotes the unit matrix.

In contrast to (pseudo-)Riemannian manifolds, which can also differ strongly locally, symplectic manifolds all look the same locally. There are always local charts (U, φ), for which the symplectic form with respect to the φ-basis from Proposition 6.32 is represented by the matrix J_n.

Theorem 9.9 (Darboux) *Let (M, ω) be a symplectic manifold. Then for every point $x \in M$ there is a chart (U, φ) such that the symplectic form ω in coordinates has the form $\omega = \sum_{i=1}^{n} dx_i \wedge dx_{i+n}$.*

Coordinates for (M, ω) as in Theorem 9.9 are called *symplectic coordinates*.

A real vector space V becomes a complex vector space if there is an $\mathbb{R}$-linear mapping $I : V \to V$ with $I^2 = -\mathrm{id}$, because then one can define scalar multiplication with complex numbers by

$$\forall x, y \in \mathbb{R}, v \in V: \quad (x + iy) \cdot v := x \cdot v + y \cdot Iv$$

Such an $I \in \mathrm{Hom}_\mathbb{R}(V, V) = V^* \otimes V$ is called a *complex structure* on V.

For a real differentiable manifold M, a (global) tensor field $I \in \mathcal{T}_M^{(1,1)}(M)$ is called a *almost-complex structure* on M if each $I_x \in \bigoplus_1^1 T_x M = \mathrm{End}_\mathbb{R}(T_x(M))$ is a complex structure on $T_x(M)$. The reason for the almost in "almost-complex structure" is that while every complex manifold considered as a real manifold has an almost-complex structure, conversely an almost-complex structure must satisfy an additional property to come from a complex manifold structure. Note that every complex manifold $(M, \mathcal{O}_M)$ can be considered as a smooth real manifold

$(M, \mathcal{C}_M^{\mathbb{R},\infty})$ by considering not only holomorphic, but smooth real differentiable functions on the coordinate neighborhoods and gluing the thus obtained local sheaves. By defining the complex structures on the tangent spaces for holomorphic coordinates $z_k = x_k + iy_k$ on a complex manifold through

$$I \cdot \frac{\partial}{\partial x_k} = \frac{\partial}{\partial y_k} \quad \text{and} \quad I \cdot \frac{\partial}{\partial y_k} = -\frac{\partial}{\partial x_k} \tag{9.1}$$

one establishes a well-defined almost-complex structure I on M, which is called the *canonical almost-complex structure* on M.

For each almost-complex structure I, one defines a tensor field $N_I \in \mathcal{T}_M^{2,1}(M)$ by (cf. Remark 6.47)

$$\forall \mathfrak{X}, \mathfrak{Y} \in \mathcal{X}(M): \quad N_I(\mathfrak{X}, \mathfrak{Y}) := [I\mathfrak{X}, I\mathfrak{Y}] - I[I\mathfrak{X}, \mathfrak{Y}] - I[\mathfrak{X}, I\mathfrak{Y}] - [\mathfrak{X}, \mathfrak{Y}].$$

N_I is called the *Nijenhuis tensor* of I. The almost-complex structure I is called *integrable*, if its Nijenhuis tensor vanishes. The name is explained by the following non-trivial theorem, which is proven using analytical methods.

Theorem 9.10 (Newlander-Nirenberg) *Let M be a real manifold with an almost-complex structure I. Then the following statements are equivalent.*

(1) *I is integrable.*
(2) *M is the underlying real manifold of a complex manifold and I is its canonical almost-complex structure.*

For a symplectic manifold (M, ω), Darboux's Theorem 9.9 shows that one can define an almost-complex structure I on M for symplectic coordinates using the formula (9.1). It leaves ω invariant and through

$$\forall x \in M, \mathfrak{v}, \mathfrak{w} \in T_x(M): \quad g(x)(\mathfrak{v}, \mathfrak{w}) := \omega(x)(\mathfrak{v}, I(x)\mathfrak{w}) \tag{9.2}$$

a Riemannian metric on M is defined.

The question arises whether the almost-complex structures from (9.1) are integrable. This is generally not the case, see [CS01, § 17.3].

In contrast to the local indistinguishability of symplectic manifolds from Darboux's Theorem 9.9, there are invariants derived from the symplectic form that can distinguish symplectic manifolds. It is even possible to show that not every differentiable manifold of even dimension can carry a symplectic form, which is also a contrast to Riemannian geometry. Statements of this kind belong to the field of *symplectic topology*. To prove them, however, sophisticated analytical methods are needed to this day. Two concepts play an important role in this context, *I-holomorphic* or *pseudo-holomorphic curves* and the *Floer homology*, see [MS17, AD14]. Defining the Floer homology would go beyond the scope here, but

9.1 Special Tensors

I-holomorphic curves are mappings $\gamma : U \to M$ for open subsets $U \subseteq \mathbb{C}$, which fulfill the I-analogon to the Cauchy-Riemann differential equations:

$$\frac{\partial \gamma}{\partial s} + (I \circ \gamma) \frac{\partial \gamma}{\partial t} = 0.$$

To systematize the analytical methods of symplectic topology, the concept of a *polyfold* has been introduced, a kind of orbifolds enriched by families of Banach spaces (see [HWZ21]).

Kähler Manifolds

Kähler manifolds are complex manifolds which, when viewed as real manifolds, are both Riemannian and symplectic, with the two structures coupled via the complex structure. Kähler manifolds play an important role in both complex analysis and complex algebraic geometry. They also give rise to the introduction of new structures.

Let M be a complex manifold. Then the fibers $T_z(M)$ of the tangent bundle are complex vector spaces, and one can consider Hermitian inner products on each of these fibers. Let for each $z \in M$ such a Hermitian inner product $h(z) \colon T_z(M) \times T_z(M) \to \mathbb{C}$ be given. Then $g(z) := \operatorname{Re} \bigl(h(z) \bigr) T_z(M) \times T_z(M) \to \mathbb{R}$ is a Euclidean inner product, where $T_z(M)$ is considered as a real vector space. Because

$$h(z)(\mathfrak{v}, \mathfrak{w}) = \overline{h(z)(\mathfrak{w}, \mathfrak{v})}$$

$\omega := \operatorname{Im} \bigl(h(z) \bigr) T_z(M) \times T_z(M) \to \mathbb{R}$ is skew-symmetric. If $\mathfrak{v}_1, \ldots, \mathfrak{v}_n$ is a $h(z)$-orthonormal $\mathbb{C}$-basis for $T_z(M)$, then $\mathfrak{v}_1, \ldots, \mathfrak{v}_n, i\mathfrak{v}_1, \ldots, i\mathfrak{v}_n$ is a $\operatorname{Re} \bigl(h(z) \bigr)$-orthonormal $\mathbb{R}$-basis for $T_z(M)$, which satisfies

$$\forall j, k \in \{1, \ldots, n\} \colon \quad \operatorname{Im} h(z)(\mathfrak{v}_j, \mathfrak{v}_k) = 0 \text{ and } \operatorname{Im} h(z)(\mathfrak{v}_j, i\mathfrak{v}_k) = -\delta_{jk}.$$

Thus, ω is pointwise non-degenerate.

Note that for $\mathfrak{v}, \mathfrak{w} \in T_z(M)$

$$g(i\mathfrak{v}, \mathfrak{w}) = \operatorname{Re} i h(z)(\mathfrak{v}, \mathfrak{w}) = -\operatorname{Im} h(z)(\mathfrak{v}, \mathfrak{w}) = -\omega(\mathfrak{v}, \mathfrak{w}). \qquad (*)$$

So g can be obtained from ω and conversely ω can be obtained from g. In particular, h is completely determined by g or ω. This makes it easy to formulate the desired regularity properties.

Definition 9.11 (Hermitian Structure) A *Hermitian structure* h on a complex manifold $(M, \mathcal{O}_M)$ is given by a Riemannian metric g on the associated real manifold $(M, \mathcal{C}_M^{\mathbb{R},\infty})$, for which $h = g + i\omega$ holds with the 2-form ω defined by (∗) on $(M, \mathcal{C}_M^{\mathbb{R},\infty})$.

Definition 9.12 (Kähler Manifolds) A *Kähler manifold* is a complex manifold $(M, \mathcal{O}_X)$ together with a Hermitian structure h on M, for which $\omega := \operatorname{Im} h$ is closed.

The simplest example class for Kähler manifolds are the vector spaces $\mathbb{C}^n$ with their canonical Hermitian metrics. The most important class of example are the complex projective spaces and their closed complex submanifolds. In the following example, we explicitly write the Fubini-Study metric on $\mathbb{P}^n_{\mathbb{C}}$, but *ad hoc* in coordinates. It should be noted that the Fubini-Study form can be naturally obtained from the Kähler form on $\mathbb{C}^{n+1}$ through the process of symplectic reduction.

Example 9.13 (Complex Projective Spaces) We describe the *Fubini-Study metric* only on the coordinate neighborhood $U_0 = \{[x] \in \mathbb{P}^n_{\mathbb{C}} \mid x_0 \neq 0\}$, which we have described in Example 7.46. The coordinates are denoted by $z_1, \ldots, z_n$. By splitting into real and imaginary parts, one finds smooth coordinate functions $x_j = \operatorname{Re}(z_j)$ and $y_j = \operatorname{Im}(z_j)$ as well as smooth functions $\bar{z}_j := x_j - i y_j$. Then, one forms the 1-form $d\bar{z}_j = dx_j - i\, dy_j$ on $\mathbb{P}^n_{\mathbb{C}}$, which is no longer holomorphic, but should be considered as a section of the complexified tangent bundle of the real manifold $\mathbb{P}^n_{\mathbb{C}}$. Thus,

$$\omega_{\text{FS}} := \frac{1}{2i}\left(\left(1+\sum_{j=1}^n|z_j|^2\right)\sum_{j=1}^n dz_j \wedge d\bar{z}_j - \left(1+\sum_{j=1}^n|z_j|^2\right)^2 \sum_{j,k=1}^n z_j \bar{z}_k dz_j \wedge d\bar{z}_k\right)$$

becomes a Kähler form, whose associated Hermitian metric is given by

$$h_{\text{FS}} := \left(1+\sum_{j=1}^n|z_j|^2\right)\sum_{j=1}^n dz_j \otimes d\bar{z}_j$$

$$- \left(1+\sum_{j=1}^n|z_j|^2\right)^2 \sum_{j,k=1}^n z_j \bar{z}_k dz_j \otimes d\bar{z}_k$$

is given. □

Let M for the moment be a smooth real manifold with an almost-complex structure I. One can extend the endomorphisms $I(x)$ complex-linearly and thus obtain $\pm i$-eigenspace decompositions of $T_x(M)_{\mathbb{C}} := T_x(M) \otimes \mathbb{C}$. The $+i$-eigenspaces are combined to form the bundle $T^{1,0}(M)$, the $-i$-eigenspaces to $T^{0,1}(M)$ (not to be confused with the section sheaves $\mathcal{T}_M^{(r,s)}$ of the tensor bundles

9.1 Special Tensors

$\bigoplus_r^s TM$ from Secction 6.3). The decomposition $TM_\mathbb{C} = T^{1,0}(M) \oplus T^{0,1}(M)$ also induces decompositions

$$\bigwedge^k TM_\mathbb{C} = \bigwedge^k TM \otimes \mathbb{C} = \bigoplus_{p+q=k} \bigwedge^{p,q} TM$$

of the complexified k-form bundles from Section 6.4. Here, the bundles $\bigwedge^{p,q} TM$ are isomorphic to $\bigwedge^p T^{1,0}(M) \otimes \bigwedge^q T^{0,1}(M)$. The complex conjugation, which is induced by the complex conjugation on $\mathbb{C}$ on all these bundles, then provides the identity $\overline{\bigwedge^{p,q} TM} = \bigwedge^{q,p} TM$ and thus in particular $\dim_\mathbb{C}(\bigwedge^{p,q} TM) = \dim_\mathbb{C}(\bigwedge^{q,p} TM)$. Let $\Omega^k(M)_\mathbb{C}$ be the space of global sections of $\bigwedge^k TM_\mathbb{C}$, i.e., the complex differential forms of degree k. We denote the sections of $\bigwedge^{p,q} TM$ with $\Omega^{p,q}(M)$.

By complexifying the exterior derivative, one obtains a complex of $\mathbb{C}$-vector spaces, whose cohomology is the de Rham cohomology $H^k(M, \mathbb{C})$ with coefficients in $\mathbb{C}$. By combining the exterior derivative with the projections onto $\bigwedge^{p+1,q} TM$ and $\bigwedge^{p,q+1} TM$, one obtains the differentials

$$\partial : \Omega^{p,q}(M) \to \Omega^{p+1,q}(M) \quad \text{and} \quad \bar{\partial} : \Omega^{p,q}(M) \to \Omega^{p,q+1}(M).$$

The almost-complex structure I is integrable if and only if $d = \partial + \bar{\partial}$, see [Hu05, Prop. 2.6.15]. In this case, one can construct from the complexes $(\Omega^{p,\bullet}(M), \bar{\partial})$ the *Dolbeault cohomologies* $H^{p,q}(M)$ in analogy to the de Rham cohomology.

For compact Kähler manifolds, one can define L^2-inner products on the spaces of forms using the Hermitian structure. These in turn enable the construction of adjoint operators d^*, ∂^* and $\bar{\partial}^*$ to the differentials d, ∂ and $\bar{\partial}$. The associated *Laplace operators* are defined by

$$\Delta_d := dd^* + d^*d, \quad \Delta_\partial := \partial\partial^* + \partial^*\partial, \quad \Delta_{\bar{\partial}} := \bar{\partial}\bar{\partial}^* + \bar{\partial}^*\bar{\partial}.$$

These Laplace operators have good properties—they are, e.g., elliptic. Forms that are annihilated by the respective Laplace operators Δ are called Δ-*harmonic*. For Kähler manifolds, $\Delta_d = 2\Delta_\partial = 2\Delta_{\bar{\partial}}$, so the terms coincide in this case and one can simply speak of harmonic forms. We denote the space of harmonic elements of $\Omega^{p,q}(M)$ with $\mathcal{H}^{p,q}(M)$ and the space of harmonic elements of $\Omega^k(M)$ with $\mathcal{H}^k(M)$.

Theorem 9.14 (Hodge) *Let (M, ω) be a compact Kähler manifold. Then*

$$\mathcal{H}^k(M) \cong H^k(M, \mathbb{C}) = \bigoplus_{p+q=k} H^{p,q}(M) \cong \bigoplus_{p+q=k} \mathcal{H}^{p,q}(M),$$

where $H^{p,q}(M)$ can be read as Dolbeault cohomology space or as space of de Rham classes, which represent elements of $\Omega^{p,q}(M)$.

The Hodge theorem is one of the central results in [Vo16], which also contains various applications. Here we only go into the definition of a *Hodge structure*. The de Rham theorem in particular shows that the de Rham cohomology $H^k(M, \mathbb{C})$ can be written as $H^k(M, \mathbb{Z}) \otimes \mathbb{C}$, where $H^k(M, \mathbb{Z})$ is one of the usual cohomology theories with integer coefficients. For example, one can take the simplicial cohomology if M is homeomorphic to a simplicial complex. If one is familiar with sheaf cohomology, one can use the sheaf cohomology of M with coefficients in the locally constant sheaf defined by the ring $\mathbb{Z}$. The Hodge theorem then provides the decomposition

$$H^k(M, \mathbb{Z}) \otimes \mathbb{C} = \bigoplus_{p+q=k} H^{p,q}(M)$$

and thus an example of a Hodge structure.

Definition 9.15 (Hodge Structure) A *Hodge structure* of weight k consists of a finitely generated abelian group $H_\mathbb{Z}$ and $\mathbb{C}$-vector spaces $H^{p,q}$ with $p + q = k$, for which $\overline{H^{p,q}} = H^{q,p}$ and

$$H_\mathbb{Z} \otimes \mathbb{C} = \bigoplus_{p+q=k} H^{p,q}$$

holds.

Poisson Manifolds

Poisson manifolds are generalizations of symplectic manifolds, which play a role both in classical mechanics and in the description of quantum mechanical analogs of mechanical systems. The following definition does not fit the heading "special tensors", but it turns out that it is equivalent to the existence of a special tensor.

Definition 9.16 (Poisson Manifolds) Let M be a differentiable manifold. A Lie bracket $\{\cdot, \cdot\} : C^\infty(M) \times C^\infty(M) \longrightarrow \mathbb{R}$ (see Example 4.5) is called a *Poisson bracket*, if

$$\forall f_1, f_2, g \in C^\infty(M): \quad \{f_1 f_2, g\} = \{f_1, g\} f_2 + f_1 \{f_2, g\} \tag{9.3}$$

holds. The pair $(M, \{\cdot, \cdot\})$ is then called a *Poisson manifold*.

The skew symmetry of the Poisson bracket implies

$$\forall g_1, g_2, f \in C^\infty(M): \quad \{f, g_1 g_2\} = \{f, g_1\} g_2 + g_1 \{f, g_2\}.$$

9.1 Special Tensors

Thus, $\mathfrak{X}_f \colon C^\infty(M) \to C^\infty(M)$, $g \mapsto \{f, g\}$ for $f \in C^\infty(M)$ is a derivation, i.e., a vector field. It is called, as in the context of symplectic manifolds, the *Hamiltonian vector field* of f.

The announced special tensor field, the *Poisson tensor*, on a Poisson manifold is $\Lambda \in \bigwedge^2 \mathcal{T}_M(M) \subseteq \bigotimes_0^2 \mathcal{T}_M(M)$ given by

$$\forall f, g \in C^\infty(M): \quad \Lambda(x)\big(df(x), dg(x)\big) := \{f, g\}(x) \tag{9.4}$$

The Eq. (9.4) defines for each $\Lambda \in \bigwedge^2 \mathcal{T}_M(M)$ a Lie bracket on $C^\infty(M)$, which, however, does not necessarily also fulfill (9.3). To characterize the validity of this equation by a property of Λ, one first extends the contraction ι of a differential form with a vector field (see Exercise 6.16) to a contraction

$$\iota(\mathfrak{X}_1 \wedge \ldots \wedge \mathfrak{X}_k)\nu(\mathfrak{Y}_1, \ldots, \mathfrak{Y}_r) := \nu(\mathfrak{X}_1, \ldots, \mathfrak{X}_k, \mathfrak{Y}_1, \ldots, \mathfrak{Y}_r)$$

of differential forms with *multivector fields*, i.e., sections of $\bigwedge^k TM$, and then defines the *Schouten-Nijenhuis bracket*

$$[\cdot, \cdot] \colon C^\infty(M; \bigwedge^p TM) \times C^\infty(M; \bigwedge^q TM) \to C^\infty(M; \bigwedge^{p+q-1} TM)$$

via

$$\iota([P, Q])\omega := (-1)^{q(p+1)} \iota(P)d\big(\iota(Q)\omega\big) + (-1)^p \iota(Q)d\big(\iota(P)\omega\big) - \iota(P \wedge Q)d\omega.$$

Then one can show that (9.3) is equivalent to $[\Lambda, \Lambda] = 0$. One then also writes (M, Λ) instead of $(M, \{\cdot, \cdot\})$ when one wants to denote a Poisson manifold.

Example 9.17 (Symplectic Manifolds) Let (M, ω) be a symplectic manifold (smooth and real). Then $\omega^\flat \colon T(M) \to T^*(M)$, $\mathfrak{v} \mapsto \big(\mathfrak{w} \mapsto \omega(\mathfrak{w}, \mathfrak{v})\big)$ is an isomorphism of real vector bundles over M. We denote the inverse of $\omega^\flat$ by $\omega^\sharp$ and set

$$\Lambda(\alpha, \beta) := -\omega\big(\omega^\sharp(\alpha), \omega^\sharp(\beta)\big)$$

for $\alpha, \beta \in T_x^*(M)$ and $x \in M$. Then Λ defines a Poisson structure on M, and the Poisson bracket is given by

$$\{f, g\}(x) = \Lambda(x)\big(df(x), dg(x)\big) = -\omega\big(\omega^\sharp(df(x)), \omega^\sharp(dg(x))\big).$$

Let $\mathfrak{X}$ be the symplectic gradient of f in the sense of Definition 9.6. Then for every vector field $\mathfrak{Y}$ on M, we have $\omega(\mathfrak{X}, \mathfrak{Y}) = df(\mathfrak{Y})$, and we obtain $\omega^\sharp\big(df(x)\big) = \mathfrak{X}(x)$. So if $\mathfrak{Y}$ is the symplectic gradient of g, we get

$$-\omega\big(\omega^\sharp(df(x)), \omega^\sharp(dg(x))\big) = -\omega\big(\mathfrak{X}(x), \mathfrak{Y}(x)\big) = \omega\big(\mathfrak{Y}(x), \mathfrak{X}(x)\big) = (\mathfrak{X}g)(x).$$

Together, we get $\mathfrak{X} = \mathfrak{X}_f$, i.e., the Hamiltonian vector field for the function f coincides with the symplectic gradient of f. The double use of the name "Hamiltonian vector field" thus does not lead to any ambiguities. □

Example 9.18 (Dual Spaces of Lie Algebras) Let $(V, [\cdot, \cdot])$ be a finite-dimensional real Lie algebra. Then by

$$\forall f, g \in C^\infty(V^*): \quad \{f, g\}(x) := \langle x, [df(x), dg(x)] \rangle,$$

a Poisson bracket is defined on V^*. Note that V can be identified with $(V^*)^*$, because it was assumed to be finite-dimensional. Since linear mappings are smooth, V can thus be considered as a subspace of $C^\infty(V^*)$. Thus, the Poisson bracket $\{v, w\}$ of two elements $v, w \in V$ is defined. With this identification, we get $\{v, w\} = [v, w]$, i.e., the Poisson bracket is an extension of the Lie bracket on V.

To determine the Poisson tensor, we choose a basis $v_1, \ldots, v_n$ for V. Then by

$$\forall i, j \in \{1, \ldots, n\}: \quad [v_i, v_j] = \sum_{k=1}^n c_{ij}^k v_k$$

numbers $c_{ij}^k \in \mathbb{R}$ are defined, which are called the *structure constants* of V (with respect to the given basis). Let now $\alpha_1, \ldots, \alpha_n$ be the dual basis to $v_1, \ldots, v_n$ for $V^* = \mathrm{Hom}_\mathbb{R}(V, \mathbb{R})$. Then on V^*

$$\sum_{k=1}^n x_k \alpha_k = x \mapsto \Lambda_{ij}(x) := \sum_{k=1}^n x_k c_{ij}^k$$

defines a Poisson form $x \mapsto \Lambda(x) = (\Lambda_{ij}(x))_{i,j=1,\ldots,n}$, where $\Lambda(x)(v_j, v_j) = \Lambda_{ij}(x)$. Because of

$$\Lambda(x)(dv_i(x), dv_j(x)) = \Lambda(x)(v_i, v_j) = \Lambda_{ij}(x) = \sum_{k=1}^n c_{ij}^k x_k = \sum_{k=1}^n c_{ij}^k v_k(x)$$

$$= [v_i, v_j](x) = \{v_i, v_j\}(x),$$

Λ is the Poisson tensor for $\{\cdot, \cdot\}$. □

As we have seen in Example 9.8, the Poisson manifold V^*, when V is the Lie algebra of a Lie group G, can be foliated by symplectic manifolds. This is not just a special example. All Poisson manifolds have a natural *symplectic foliation*. The tangent spaces to the leaves are given by the linear span of the Hamiltonian vector fields. That is, if $B \subseteq M$ is the leaf of the symplectic foliation through $x \in M$, then

$$T_x(B) = \mathrm{Span}\{\mathfrak{X}_f(x) \mid f \in C^\infty(M)\}.$$

9.1 Special Tensors

The Poisson manifold is a symplectic manifold if and only if the Poisson tensor has full rank. In this case, there is only one leaf in the foliation (assuming the manifold is connected).

A typical example is the space $\mathbb{R}^3$, which is identified via a basis with the dual space of the Lie algebra $\mathfrak{so}_3$ consisting of the skew-symmetric real 3×3 matrices; for the resulting Poisson structure on $\mathbb{R}^3$, the leaves of the symplectic foliation are given by the concentric spheres. The sphere of radius 0, i.e., the origin, is also a leaf.

A significant difference in the mathematical modeling of classical and quantum mechanics is that the classical observables form commutative associative algebras, while the quantum mechanical observables form non-commutative associative algebras. We do not go into the subtle interplay of the two models here, but only briefly outline a method for constructing a non-commutative associative algebra from a Poisson manifold.

Definition 9.19 (Formal Star Product) Let (M, Λ) be a Poisson manifold. Let $C^\infty(M)[[\lambda]]$ be the space of formal power series in one variable λ with coefficients in $C^\infty(M)$, cf. Example 1.6. A *formal star product* is a $\mathbb{C}[[\lambda]]$-bilinear mapping

$$\star : C^\infty(M)[[\lambda]] \times C^\infty(M)[[\lambda]] \to C^\infty(M)[[\lambda]]$$

of the form

$$f \star g = \sum_{r=0}^{\infty} \lambda^r C_r(f, g),$$

where the $C_r : C^\infty(M) \times C^\infty(M) \to C^\infty(M)$ are $\mathbb{C}$-bilinear mappings, which are extended $\mathbb{C}[[\lambda]]$-bilinearly. Furthermore, the following properties are required from $\star$:

(a) $(C^\infty(M)[[\lambda]], \star)$ is an associative algebra with the constant function $1 \in C^\infty(M)$ as a unit.
(b) For all $f, g \in C^\infty(M)$, $C_0(f, g) = fg$ and $C_1(f, g) - C_1(g, f) = i\{f, g\}$.

Two star products $\star$ and $\star'$ are called *equivalent*, if there is a formal series $S = \mathrm{id} + \sum_{r=1}^{\infty} \lambda^r S_r$ of linear mappings $S_r : C^\infty(M) \to C^\infty(M)$ with $S_1 = 1$ and

$$\forall f, g \in C^\infty(M)[[\lambda]] : \quad f \star' g = S^{-1}(Sf \star Sg).$$

After star products had been constructed for various special cases over the decades, Maxim Kontsevich showed in [Ko03] that there are star products for every Poisson manifold, for which the C_r are given in both variables by differential operators of order r. He also provides a classification up to equivalence, which relies on the concept of *formal deformations* of Poisson structures. A central technical result in this context is Kontsevich's *formality theorem*, which connects the Schouten-Nijenhuis bracket with the so-called Hochschild cohomology of the algebra $C^\infty(M)$.

Literature: There is a large number of introductory and advanced books on Riemannian geometry. Very reader-friendly are, e.g., [dC92] and the first volume of [Sp99]. The selection of books on pseudo-Riemannian geometry is significantly smaller, but see [ON83] and [Ch11]. Specifically for the Lorentzian variant, [BE11] and [HE73] are two classics. For the physical interpretation of Minkowski space in the context of special relativity, see [Na12].

Traditionally, most books on symplectic geometry focus on the connection with classical mechanics. Examples of this are [LM87] and [AMR88]. Newer books like [CS01, MS17, Oh15] also take symplectic topology into account. Introductions to the analysis underlying symplectic topology are provided by [MS12, AD14, HWZ21]. In [AC21b], connections between symplectic geometry and methods of homological algebra are also discussed. The Newlander-Nirenberg theorem is explained in many books (e.g., [KN96, Hu05, MS17]), but only very few provide a complete (i.e., not just in the analytical case) proof. An example is the book [Ho73].

There are only a few books that deal exclusively with Kähler manifolds, such as [We58] or [Ba06]. However, they play an important role in most books on complex manifolds. Examples of this are [We79] and [Hu05] as well as [GH94], where in particular the connection to algebraic geometry is explained. The book [Vo16], which is an introduction to Hodge theory, also contains a detailed introduction to the theory of Kähler manifolds.

An early systematic presentation of the differential geometry of Poisson manifolds in the context of symplectic geometry and classical mechanics is [LM87]. Texts like [Va94] and [CLM21] focus exclusively on Poisson manifolds. In [LPV13], applications play a larger role. In particular, [LPV13, Chap. 13] is an introduction to Kontsevich's results on deformation quantization. Kontsevich's article in [AC21a] goes a step further. It describes, among other things, the role of deformation techniques and homological algebra in the so-called derived non-commutative geometry.

9.2 Connections and Fiber Bundles

In Ch. 6, we defined differentiable mappings between manifolds, and we know what a derivative is. We are thus familiar with higher derivatives in principle. We have also found that derivatives live on tangent spaces, and we have discovered how the tangent spaces of different points stick together in a differentiable way. This was done via charts, and because of the definition of charts and differentiable structures, this differentiable sticking together did *not* depend on the choice of the chart. It would be different if one tried to explain what it means when two vectors in *different* tangent spaces point in the same direction. Of course, in a chart (U, φ) on M or its associated chart, one can say that two directions are the same if the last n coordinates

9.2 Connections and Fiber Bundles

of the corresponding vectors are multiples of each other. But one can make other charts on M that contain the same points and for which these vectors *do not* (in this sense) point in the same direction. In fact, it is not at all possible to define a meaningful concept of the above-mentioned type globally, i.e., for arbitrary points without any further specification, to compare the affine structure of the tangent space. But even if one restricts the validity range of such a comparison process, one does not find a concept independent of the choice of charts, without first introducing an additional structure on the manifold.

Covariant Derivatives

The additional structure that is introduced is the ability to form directional derivatives of vector fields. With these directional derivatives, one can then compare tangent vectors $\mathfrak{v}$ and $\mathfrak{w}$ in two tangent spaces $T_p M$ and $T_q M$ by connecting p and q with a smooth curve γ, defining a vector field $\mathfrak{X}$ (at least on the image of γ) for which $\mathfrak{v} = \mathfrak{X}(p)$ and $\mathfrak{w} = \mathfrak{X}(q)$. Then one says that $\mathfrak{w}$ is the parallel shifted vector of $\mathfrak{v}$ along γ if the directional derivative of the vector field $\mathfrak{X}$ in the direction of $\gamma'(t)$ vanishes for all t.

Given our algebraic definition of the tangent space, which identified tangent vectors with directional derivatives, it may seem surprising that it should not be possible to define directional derivatives of vector fields without introducing new structures on a manifold. But these directional derivatives referred to functions, and when one tries to define these directional derivatives also for vector fields, e.g., by deriving the component functions of a vector field, one quickly realizes that one does not obtain a chart-independent result. For simplicity, we formulate the definitions only for smooth manifolds.

Definition 9.20 (Affine Connection) Let M be a $C^{\mathbb{R},\infty}$-manifold. An *affine connection* on M is a rule ∇, which assigns to each vector field $\mathfrak{X} \in \mathcal{T}_M(M)$ a linear mapping $\nabla_{\mathfrak{X}} : \mathcal{T}_M(M) \to \mathcal{T}_M(M)$ that has the following two properties:

(a) $\forall f, g \in C^\infty(M), \mathfrak{X}, \mathfrak{Y} \in \mathcal{T}_M(M): \quad \nabla_{f\mathfrak{X}+g\mathfrak{Y}} = f\nabla_{\mathfrak{X}} + g\nabla_{\mathfrak{Y}}$.
(b) $\forall f \in C^\infty(M), \mathfrak{X}, \mathfrak{Y} \in \mathcal{T}_M(M): \quad \nabla_{\mathfrak{X}}(f\mathfrak{Y}) = f\nabla_{\mathfrak{X}}(\mathfrak{Y}) + \big(\mathfrak{X}(f)\big)\mathfrak{Y}$.

The operator $\nabla_{\mathfrak{X}}$ is also called *covariant derivative* with respect to $\mathfrak{X}$.

The name *covariant derivative* obscures the fact that this is an additional structure. On the other hand, in the course of this section, we will discuss versions of connections that do not readily reveal their relationship to covariant derivatives. Therefore, we prefer the term covariant derivative here.

Example 9.21 (Hyperplanes) Let M be a hyperplane in $\mathbb{R}^{n+1}$, then $T_p M$ can be identified with a hyperplane of $\mathbb{R}^{n+1}$ for each $p \in M$. We assume that we have a smooth function $\eta : M \to \mathbb{R}^{n+1}$ such that $\eta(p)$ is always perpendicular to $T_p M$ (with respect to the usual scalar product on $\mathbb{R}^{n+1}$). For each $p \in M$, the projection from $\mathbb{R}^{n+1}$ along $\mathbb{R}\eta(p)$ onto $T_p M$ is denoted by Π_p. Note that in $\mathbb{R}^{n+1}$, directional derivatives of vector fields can be easily formed: For this, let $\mathfrak{X}$ and $\mathfrak{Y}$ be vector fields, then for $p \in M$ the directional derivative of $\mathfrak{Y}$ at p in the direction of $\mathfrak{X}(p)$ is given by

$$D_{\mathfrak{X}}(\mathfrak{Y})(p) := \lim_{h \to 0} \frac{1}{h}\Big(\mathfrak{Y}(p + h\mathfrak{X}(p)) - \mathfrak{Y}(p)\Big).$$

The vector $D_{\mathfrak{X}}(\mathfrak{Y})(p)$ generally does not lie in $T_p M$. According to our definition, however, the vector $\Pi_p\big(D_{\mathfrak{X}}(\mathfrak{Y})(p)\big)$ does indeed lie in $T_p M$. We define

$$\nabla_{\mathfrak{X}}(\mathfrak{Y})(p) := \Pi_p\big(D_{\mathfrak{X}}(\mathfrak{Y})(p)\big). \tag{$*$}$$

It can be shown that $\nabla_{\mathfrak{X}}(\mathfrak{Y})$ depends only on the values of $\mathfrak{X}$ and $\mathfrak{Y}$ on M. Furthermore, it can be shown that for every vector field on a neighborhood U in M, a vector field can be defined on a neighborhood V in $\mathbb{R}^{n+1}$, the restriction of which is the field on U. With these statements, it can be proven that the rule defined by $(*)$ is a covariant derivative.

Let a smooth vector field $\mathfrak{Y} \in \mathcal{T}_M(M)$ be fixed. Then the mapping $T : \mathcal{T}_M(M) \times \mathcal{T}_M^*(M) \to C^\infty(M)$, defined by $T(\mathfrak{X}, \omega) = \omega\big(\nabla_{\mathfrak{X}}(\mathfrak{Y})\big)$, is a tensor field of type $(1, 1)$. Thus, the value of the covariant derivative $\nabla_{\mathfrak{X}}(\mathfrak{Y})$ at the point $p \in M$ depends only on the value of $\mathfrak{X}$ at p. Therefore, the equation

$$\forall t \in I: \quad \nabla_{\gamma'(t)}(\mathfrak{Y})\big(\gamma(t)\big) = 0$$

for a smooth curve γ and a given vector field $\mathfrak{Y}$ makes sense. The solution of this equation is interpreted as the parallel transport of $\mathfrak{Y}$ along γ. A more detailed analysis of the differential equation shows that only the values of $\mathfrak{Y}$ on the image of γ play a role. Thus, the *parallel transport* along a curve $\gamma : I \to M$ from p to q in M becomes a linear isomorphism $\tau_{p,q,\gamma} : T_p(M) \to T_q(M)$. A curve is called a *geodesic* for ∇, if for any two times $t, s \in I$, it holds that

$$\tau_{\gamma(s),\gamma(t),\gamma}\big(\gamma'(t)\big) = \gamma'(s),$$

i.e., the curve γ has "constant speed". Thus, geodesics are something like straight lines in M. □

It turns out that on a pseudo-Riemannian manifold (M, g) there exists exactly one covariant derivative ∇ for which all parallel transports are isometric and for which the *torsion tensor* $T_\nabla \in \bigoplus_2^1 \mathcal{T}_M(M)$ defined by

$$\forall \mathfrak{X}, \mathfrak{Y} \in \mathcal{T}_M(M), \omega \in \Omega^1(M): \quad T_\nabla(\mathfrak{X}, \mathfrak{Y}, \omega) = \omega\big(\nabla_{\mathfrak{X}}(\mathfrak{Y}) - \nabla_{\mathfrak{Y}}(\mathfrak{X}) - [\mathfrak{X}, \mathfrak{Y}]\big)$$

vanishes. This covariant derivative is called the *Levi-Civita connection* for (M, g). In the case of Riemannian metrics, the geodesics for the Levi-Civita connection are (locally) distance-minimizing, i.e., they coincide with the geodesics defined in Section 9.1 for (M, g).

Curvature

Imagine the unit sphere in $\mathbb{R}^3$ with the covariant derivative from Example 9.21. The parallel transport along a great circle is then precisely the rotation about the corresponding axis of rotation of the sphere. If one follows the parallel transport for three successive quarter circles (first on the equator, then to the north pole, and finally back to the starting point), a tangential vector parallel to the equator will result in a tangential vector pointing to the north pole. In contrast, a vector does not change if it is transported along a closed curve in a Euclidean vector space. This suggests that one can discover *curvature* using parallel transport along closed curves. If one considers curves that arise for two vector fields $\mathfrak{X}$ and $\mathfrak{Y}$ by successively following the integral curves of $\mathfrak{X}$, $\mathfrak{Y}$, $-\mathfrak{X}$ and $-\mathfrak{Y}$ for a time ε each, and transports a third vector field $\mathfrak{Z}$ along this curve, then actually for $\varepsilon \to 0$ a pointwise defined quantity results. This quantity is a tensor of rank $(1, 3)$, which can be defined by

$$\forall \mathfrak{X}, \mathfrak{Y}, \mathfrak{Z} \in \mathcal{T}_M(M), \ \omega \in \Omega^1(M):$$
$$R_\nabla(\mathfrak{X}, \mathfrak{Y}, \mathfrak{Z}, \omega) := \omega\big((\nabla_\mathfrak{X} \nabla_\mathfrak{Y} - \nabla_\mathfrak{Y} \nabla_\mathfrak{X} - \nabla_{[\mathfrak{X},\mathfrak{Y}]})(\mathfrak{Z})\big).$$

It is called the *Riemannian curvature tensor*. This and derived tensors play an important role in many important differential equations of geometric analysis. Notably, in the Einstein field equations of general relativity (see e.g. [Sa22]) and in the Ricci flow, which played a central role in the solution of the Poincaré conjecture and its modern generalization, Bill Thurston's geometrization conjecture, by Grigori Perelman (see e.g. [MF10]).

Fiber Bundles

The concept of a covariant derivative, which we have introduced for the tangent bundle, can be easily generalized to general vector bundles in the sense of Definition 6.35. But one can go a step further and define connections on general fiber bundles that at first glance seem to have nothing to do with affine connections. We will not describe the connections here in more detail, but refer to Volume 2 of Spivak's five-volume work [Sp99], in which the different concepts of connections and their cross-connections are discussed in detail.

Definition 9.22 (Fiber Bundle) Let M and F be differentiable manifolds of type $C^{\mathbb{K},k}$. A $C^{\mathbb{K},k}$-*fiber bundle* with fiber F is a differentiable manifold E with a $C^{\mathbb{K},k}$-mapping $\pi : E \to M$, which has the following properties: For each $p \in M$ there is a neighborhood U of p in M and a diffeomorphism $\varphi_U : \pi^{-1}(U) \to U \times F$ with $\pi_1 \circ \varphi_U = \pi$ (π_1 is the projection onto the U-component), and for each $q \in U$, $\pi_2 \circ \psi_U|_{E_q} : E_q \to F$ is a $C^{\mathbb{K},k}$-isomorphism (where E_q is the fiber $\pi^{-1}(q)$ and π_2 is the projection onto the F-component). A *section* of (E, π) is a $C^{\mathbb{K},k}$-mapping $\sigma : M \to E$ with $\pi \circ \sigma = \mathrm{id}_M$.

The simplest example of a fiber bundle over M with fiber F is the product $M \times F$ with the projection π_1 onto the first component. It is called the *trivial F-bundle* over M. The definition of a fiber bundle thus includes that fiber bundles are locally trivial. A cover of the fiber bundle by open subsets of the form $\pi^{-1}(U)$ with the properties described in Definition 9.22 is called a *local trivialization* of the bundle.

The tensor bundles from Section 6.3 are examples of fiber bundles. For these bundles, F is a vector space and the fibers themselves have vector space structures, for which the fiber-wise diffeomorphisms become isomorphisms of vector spaces. These are exactly the vector bundles from Definition 6.35. If the fiber of a vector bundle has dimension 1, it is also referred to as a *line bundle*.

Example 9.23 (Orientation Bundle) In Section 6.5 we saw that the integration of n-forms on an n-dimensional differentiable manifold could only be justified from the transformation theorem if the manifold was oriented. Orientation is also an additional structure. The existence of an orientation can be characterized by the existence of a nowhere vanishing n-form. Another characterization is found using the *orientation bundle* $\mathrm{Or}(M)$. This bundle is defined by considering the trivial line bundle $U \times \mathbb{R}$ on each chart neighborhood $(U_\alpha, \varphi_\alpha)$ and gluing these trivial bundles on $U_\alpha \cap U_\beta$ using

$$(x, r) \mapsto \left(x, \mathrm{sign}\det\left(\varphi_\alpha \circ \varphi_\beta^{-1}\right)'(x)\right)r\right)$$

(see Exercise 6.13). The manifold is orientable if and only if the orientation bundle is *trivial*, i.e., if there is a diffeomorphism $\varphi : M \times \mathbb{R} \to \mathrm{Or}(M)$ that maps the fibers $\{m\} \times \mathbb{R}$ linearly isomorphic to the fiber over m in $\mathrm{Or}(M)$. An orientation is then given as a cover of M by charts $(U_\alpha, \varphi_\alpha)$ for which the transition functions do not take values in the group $\{\pm 1\}$, but only in the smaller group $\{1\}$. □

If one considers geometric objects such as the space of all bases (possibly with additional properties) for each fiber of a tensor bundle, one encounters the following type of fiber bundle.

9.2 Connections and Fiber Bundles

Definition 9.24 (Principal Fiber Bundle) Let G be a Lie group. A fiber bundle (E, π) is called a *principal fiber bundle* with *structure group* G if there is a smooth right action $E \times G \to E$ with

$$\forall e \in E, g \in G: \quad \pi(e \cdot g) = \pi(g),$$

i.e., it preserves the fibers $E_m := \pi^{-1}(m)$, and which on the fibers is *free* ($e \cdot g = e \implies g = 1$) as well as *transitive* ($\forall e_1, e_2 \in E$ with $\pi(e_1) = \pi(e_2)$ $\exists g \in G$ such that $e_1 \cdot g = e_2$).

Example 9.25 (Homogeneous Spaces) The simplest example of a principal fiber bundle is the quotient mapping $\pi : G \to G/H$ of a Lie group G with respect to a closed subgroup H. Each fiber of this mapping is a coset gH, on which the group H acts freely and transitively from the right. Since G/H naturally carries a smooth manifold structure, with respect to which all mappings and actions are smooth, (G, π) thus becomes a principal fiber bundle over G/H with structure group H. □

Example 9.26 (Frame Bundle of a Vector Bundle) Let (E, π) be a vector bundle over M with fiber $\mathbb{R}^n$. We set

$$\mathrm{GL}(E) := \bigcup_{x \in M} \{\varphi \in \mathrm{Hom}(\mathbb{R}^n, E_x) \mid \varphi \text{ invertible}\},$$

and consider the mapping

$$\widetilde{\pi} : \mathrm{GL}(E) \longrightarrow M, \quad \varphi \longmapsto x \text{ for } \varphi \in \mathrm{Hom}(\mathbb{R}^n, E_x).$$

Then $(\mathrm{GL}(E), \widetilde{\pi})$ is a principal fiber bundle with structure group $\mathrm{GL}(n, \mathbb{R})$. To see this, one can reduce to the case of a trivial vector bundle, i.e.

$$E = U \times \mathbb{R}^n, \quad E_x = \{x\} \times \mathbb{R}^n.$$

assume. Then one can identify $\{\varphi \in \mathrm{Hom}(\mathbb{R}^n, E_x) \mid \varphi \text{ invertible}\}$ with $\mathrm{GL}(n, \mathbb{R})$ and finds $\mathrm{GL}(E) \cong U \times \mathrm{GL}(n, \mathbb{R})$. Thus, $(\mathrm{GL}(E), \widetilde{\pi})$ becomes a fiber bundle with fiber $\mathrm{GL}(n, \mathbb{R})$. For $g \in \mathrm{GL}(n, \mathbb{R})$ and $\varphi \in \mathrm{Hom}(\mathbb{R}^n, E_x)$,

$$(\varphi \cdot g)(v) = \varphi(g \cdot v)$$

defines the structure of a principal fiber bundle on $\mathrm{GL}(E)$. This bundle is often called the *frame bundle* of E. □

Principal fiber bundles with structure group G are examples of G-structures on fiber bundles.

Definition 9.27 (*G*-**Structures on Fiber Bundles**) Let $E \to M$ be a $C^{\mathbb{K},k}$-fiber bundle with fiber F and G a group. Furthermore, let $(U_\alpha, \psi_\alpha)_{\alpha \in A}$ be a family of local trivializations of the fiber bundle E, which cover E. A *G-structure* on E consists of a group action $G \times F \to F$, $(g, f) \mapsto g \cdot f$ and a family of *transition functions* $g_{\alpha\beta} \colon U_\alpha \cap U_\beta \to G$, which satisfy

$$\forall \alpha, \beta, \gamma \in A, x \in U_\alpha \cap U_\beta \cap U_\gamma : \quad g_{\alpha\beta}(x) g_{\beta\gamma}(x) = g_{\alpha\gamma}(x)$$

and

$$\forall \alpha \in A, x \in U_\alpha : \quad g_{\alpha\alpha}(x) = 1 \in G$$

as well as

$$\forall \alpha, \beta \in A, x \in U_\alpha \cap U_\beta, f \in F : \quad \psi_\alpha \circ \psi_\beta^{-1}(x, f) = (x, g_{\alpha\beta}(x) \cdot f). \qquad (*)$$

In Example 9.23 we see the *reduction* of a structure group, which describes an additional structure.

Definition 9.28 (**Reduction of the Structure Group**) Let $E \to M$ be a $C^{\mathbb{K},k}$-fiber bundle with *G*-structure and *H* a subgroup of *G*. Then an *H*-structure, which results from a covering subfamily $(U_\alpha, \psi_\alpha)_{\alpha \in A'}$, for which

$$\forall \alpha, \beta \in A', x \in U_\alpha \cap U_\beta : \quad g_{\alpha\beta}(x) \in H$$

holds, is called a *reduction* of the structure group.

Example 9.29 (**Metrics from Structure Group Reductions**) Every vector bundle E with fiber V has a natural $GL(V)$-structure. The transition functions are given by

$$g_{\alpha\beta} := \left(\pi_2 \circ \psi_\alpha|_{E_x}\right) \circ \left(\pi_2 \circ \psi_\beta|_{E_x}\right)^{-1} \colon V \to V$$

Now let V be equipped with an inner product $\langle \cdot, \cdot \rangle_V$ and H the subgroup of $GL(V)$ formed by the isometries of this inner product. Suppose we have a reduction of the structure group $GL(V)$ to H on a vector bundle $\pi \colon E \to M$ with fiber V. Then, due to $(*)$ in Definition 9.27, on each fiber E_x with $x \in U_\alpha, \alpha \in A'$, an inner product can be defined by

$$\forall \mathfrak{v}, \mathfrak{w} \in E_x : \quad \langle \mathfrak{v}, \mathfrak{w} \rangle_x := \langle \pi_2 \circ \psi_\alpha(\mathfrak{v}), \pi_2 \circ \psi_\alpha(\mathfrak{w}) \rangle_V$$

which does not depend on the choice of local trivialization. From the definitions, it is immediately clear that the mapping

$$\{(\mathfrak{v}, \mathfrak{w}) \in E \times E \mid \pi(\mathfrak{v}) = \pi(\mathfrak{w})\} \to \mathbb{K}, \quad (\mathfrak{v}, \mathfrak{w}) \mapsto \langle \mathfrak{v}, \mathfrak{w} \rangle_{\pi(\mathfrak{v})}$$

has the same degree of differentiability as the vector bundle E.

9.2 Connections and Fiber Bundles

In particular, this example shows that by reducing the structure group of $GL(n, \mathbb{R})$ of the tangent bundle of a real n-dimensional manifold to the orthogonal group $O(n, \mathbb{R})$, a Riemannian metric is specified. If there is even a reduction to the special orthogonal group $SO(n, \mathbb{R})$, we obtain an oriented Riemannian manifold.

□

Analogous to Example 9.29, for a complex manifold, the reduction of the structure group $GL(n, \mathbb{C})$ to the unitary group $U(n)$ provides a Hermitian structure. The construction can also be transferred to other types of bilinear forms. For example, from the reduction of the structure group of $GL(n, \mathbb{R})$ of the tangent bundle of a real n-dimensional manifold to the symplectic group $Sp(n, \mathbb{R})$, a non-degenerate 2-form is obtained. We see that a number of the additional structures discussed in Section 9.1 through tensors can also be described by reduction of structure groups.

Instead of reducing a structure group G to a subgroup H, one can consider group homomorphisms $\rho : H \to G$ and view an H-structure $(h_{\alpha\beta})_{\alpha,\beta \in A}$ as a *lift* of a G-structure $(g_{\alpha\beta})_{\alpha,\beta \in A}$, when

$$\forall \alpha, \beta \in A, x \in U_\alpha \cap U_\beta: \quad g_{\alpha\beta}(x) = \rho(h_{\alpha\beta}(x)).$$

Lifts of G-structures are also additional structures. In geometry and mathematical physics, e.g., the *spin structures* (and the associated *Dirac operators*) play an important role, which are given by the lift of the $SO(n, \mathbb{R})$-structure of an oriented Riemannian manifold to a $Spin(n)$-structure (see [LM89]). The group $Spin(n)$ is a compact group with a two-element normal subgroup, for which the quotient group is isomorphic to $SO(n, \mathbb{R})$. Topologically, $Spin(n)$ is the simply connected covering group of $SO(n, \mathbb{R})$.

In Example 9.26, a principal fiber bundle was constructed from a vector bundle. Conversely, for a principal fiber bundle with structure group G over M and a differentiable group action $G \times F \to F$, a fiber bundle over M with fiber F can always be assigned.

Example 9.30 (Associated Bundles) Let (E, π) be a principal fiber bundle with structure group G over M and $G \times F \to F$ a differentiable group action. Then G acts on the right on $E \times F$ by $(e, f) \cdot g := (e \cdot g, g^{-1} \cdot f)$. Let $E \times_G F$ be the set of G-orbits of this action and $[e, f] := (e, f) \cdot G$. Then

$$\pi_F : E \times_G F \to M \quad [e, f] \mapsto \pi(e)$$

is a fiber bundle over M with fiber F.

□

On the principal fiber bundles from Example 9.25, there is a left action of G that commutes with the right action of H. This also gives a left action of G on all associated bundles $G \times_H F$. The sections of such an associated bundle can be identified with functions $f : G \to F$ that satisfy

$$\forall h \in H, g \in G : \quad f(gh) = h^{-1} \cdot f(g).$$

Note the by no means accidental similarity to the induced modules from Example 4.49. This explains why associated bundles of homogeneous spaces play a central role in the theory of induced representations of Lie groups (see e.g. [Vo87]).

Example 9.31 (Pullback of Fiber Bundles) Let (E, π) be a $C^{\mathbb{K},k}$-fiber bundle over M with fiber F, N a $C^{\mathbb{K},k}$-manifold and $\varphi : N \to M$ a $C^{\mathbb{K},k}$-mapping. Then

$$\varphi^* E := \{(y, e) \in N \times E \mid \varphi(y) = \pi(e)\} \to N, \quad (n, e) \mapsto n$$

is a $C^{\mathbb{K},k}$-fiber bundle over N with fiber F. The bundle $\varphi^* E$ is called the *pullback* of E with φ. We summarize the construction in the following commutative diagram (cf. Example 4.36)

$$\begin{array}{ccc} \varphi^* E & \xrightarrow{\pi^*\varphi} & E \\ \varphi^*\pi \downarrow & & \downarrow \pi \\ N & \xrightarrow{\varphi} & M \end{array}$$

and note that the restrictions of $\pi^*\varphi$ to fibers are each a diffeomorphism onto a fiber of E.

□

Connections on Fiber Bundles

Let (E, π) be a fiber bundle over M. A *vertical vector* of E is a tangential vector $X \in T_e(E)$ with $X \in T_e(E_{\pi(e)})$, i.e., $\pi'(X) = 0$, where $\pi' : TE \to TM$ is the derivative of π. With the canonical projection $\pi_E : TE \to E$ and $VE := \{X \in TE \mid X \text{ vertical}\} \subseteq TE$ then (VE, π_E) is a vector bundle, which is called the *vertical bundle* of E is called. We note that the fiber $V(E)_e$ of $V(E)$ in $e \in E$ is equal to $T_e(E_{\pi(e)})$.

Definition 9.32 (Connection on a Fiber Bundle) Let (E, π) be a fiber bundle over M. A VE-valued 1-form Φ on E, i.e., a smooth section of $\bigotimes_1^1 TE = TE^* \otimes TE = \text{Hom}(TE, TE)$ with values in VE is a *connection*, if $\Phi \circ \Phi = \Phi$ and $\text{Image}(\Phi(e)) = VE_e$, i.e., if Φ is a projection from TE onto the sub-bundle VE. The kernel HE of Φ is called the *horizontal bundle* of E to Φ.

9.2 Connections and Fiber Bundles

If one now has a smooth curve $\gamma : I \to M$, then one can pull back the bundle E with γ to a bundle $\gamma^* E$ over I according to Example 9.31. One can also pull back the vector-valued form Φ. To do this, one must first show that the vertical sub-bundle of a pulled back bundle is the pullback of the vertical sub-bundle. The vertical bundle of $\gamma^* E$ is given by

$$V(\gamma^* E) = \{Z \in T(\gamma^* E) \mid (\gamma^* \pi)'(Z) = 0\}$$

For $Z \in V(\gamma^* E)$, we thus have

$$\pi' \circ (\pi^* \gamma)'(Z) = \gamma' \circ (\gamma^* \pi)'(Z) = 0,$$

i.e., $(\pi^* \gamma)'(Z) \in VE$. This gives us the commutative diagram

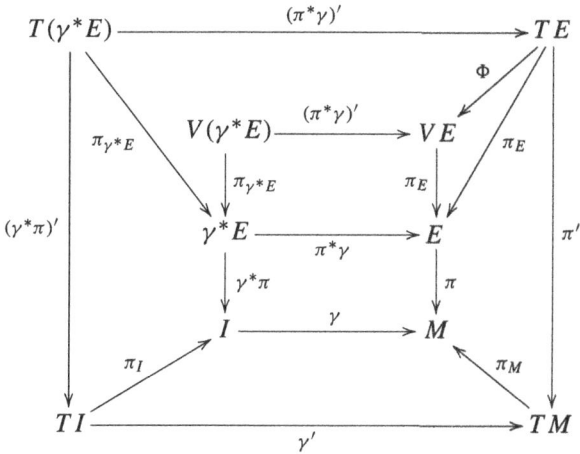

and we want to see that there is a projection $\gamma^* \Phi : T(\gamma^* E) \to V(\gamma^* E)$ with

$$\Phi \circ (\pi^* \gamma)' = (\pi^* \gamma)' \circ \gamma^* \Phi$$

Since $\pi^* \gamma |_{(\gamma^* E)_{\gamma^* \pi(t,e)}} : (\gamma^* E)_{\gamma^* \pi(t,e)} \to E_{\pi(e)}$ is a diffeomorphism for each $(t, e) \in \gamma^* E$, $\varphi_{t,e} := (\pi^* \gamma)' |_{V(\gamma^* E)_{(t,e)}} : V(\gamma^* E)_{(t,e)} \to V(E)_e$ is a linear isomorphism. Thus, for each $(t, e) \in \gamma^* E$, we can define a mapping

$$\gamma^* \Phi : T(\gamma^* E)_{(t,e)} \to V(\gamma^* E)_{(t,e)}, \quad Z \mapsto \varphi_{t,e}^{-1} \circ \Phi \circ (\pi^* \gamma)'(Z)$$

These mappings are projections and can be combined into a connection $\gamma^* \Phi : T(\gamma^* E) \to V(\gamma^* E)$ that completes the above diagram.

Note that for the construction of the pullback of a connection, the special properties of I and γ are not needed at all. It would suffice to assume $\gamma : I \to M$ as a differentiable mapping between manifolds. Next, however, we consider the horizontal bundle $H(\gamma^*E)$ and want to use that it is *integrable*, i.e., that there is a foliation of γ^*E whose leaves have the fibers of $H(\gamma^*E)$ as tangent spaces. This is not the case for general horizontal bundles. But for our γ^*E this is possible, because the fibers of $H(\gamma^*E)$ are 1-dimensional and the integrability follows from the Picard-Lindelöf theorem on the unique solvability of ordinary differential equations.

We choose a fixed point $t_0 \in I$. Then the curve γ can be lifted for each "starting point" $e_0 \in E_{\gamma}(t_0)$ to a horizontal curve $\widetilde{\gamma}_{e_0}$ in γ^*E with $\widetilde{\gamma}_{e_0}(t_0) = (t_0, e_0)$ and for $t \in I$ the resulting mapping

$$\tau_{\gamma(t_0),\gamma(t),\gamma} : E_{\gamma(t_0)} \to E_{\gamma(t)}, \quad e_0 \mapsto \pi^*\gamma(\widetilde{\gamma}_{e_0}(t))$$

is the desired parallel transport along the curve.

Curvature properties can also be read from the parallel transport along closed curves for this parallel transport. For each $x_0 \in M$ one gets a group of diffeomorphisms of E_{x_0}, which one calls the *holonomy group* of the connection in x_0. We do not want to go into the holonomy groups in more detail, but rather illuminate another approach to curvature by introducing the curvature form of a connection.

We denote the smooth sections of $\bigwedge^k T^*M \otimes TM$ with $\Omega^k(M, TM)$ and call them *vector-valued k*-forms. The vector-valued differential forms

$$\Omega^{\bullet}(M, TM) := \bigoplus_{k=0}^{\infty} \Omega^k(M, TM)$$

form a $\mathbb{Z}_2$-graded Lie algebra (see Example 4.5) with respect to the *Frölicher-Nijenhuis bracket*, which we simply denote by $[\cdot, \cdot]$, but only want to define for arguments in $\Omega^1(M, TM)$ (a complete discussion can be found in [Mi08, § 16]).[1] For $K, L \in \Omega^1(M, TM)$ and $\mathfrak{X}, \mathfrak{Y} \in \mathcal{X}(M)$ we set

$$[K, L](\mathfrak{X}, \mathfrak{Y}) := [K\mathfrak{X}, L\mathfrak{Y}] - [K\mathfrak{Y}, L\mathfrak{X}] - L([K\mathfrak{X}, \mathfrak{Y}] - [K\mathfrak{Y}, \mathfrak{X}])$$
$$- K([L\mathfrak{X}, \mathfrak{Y}] - [L\mathfrak{Y}, \mathfrak{X}]) + (LK + KL)[\mathfrak{X}, \mathfrak{Y}],$$

where on the right side the brackets are the usual commutator brackets of vector fields. It should be noted that vector-valued 1-forms are sections of $\mathrm{Hom}(TM, TM)$, i.e., they map vector fields to vector fields.

The *curvature form* of a connection $\Phi \in \Omega^1(E, V(E))$ is defined as $R := \frac{1}{2}[\Phi, \Phi] \in \Omega^2(E, TE)$. Since the vertical bundle VE is always integrable, R has values in VE. On the other hand, the horizontal bundle is integrable exactly when $R = 0$ holds, i.e., the curvature vanishes. If the fiber bundle (E, π) has symmetries, e.g., if it is a principal fiber bundle with structure group G, then one can also ask for symmetry properties of a connection on the bundle. Since for principal fiber bundles the vertical bundle is trivializable with fiber $\mathfrak{g} = \mathrm{Lie}(G)$, one is lead to the concept of a $\mathfrak{g}$-valued connection. For such a G-invariant connection $\Phi \in \Omega^1(E, VE)$, one finds an associated connection form $\omega \in \Omega^1(E, \mathfrak{g})$ and to its also G-invariant curvature form $R \in \Omega^2(E, VE)$ a curvature form $\Omega \in \Omega^2(E, \mathfrak{g})$ i.e., equivariant with respect to the G actions on E and $\mathfrak{g}$. From such G-invariant connections, connections on associated bundles can be constructed and ultimately, if it is a vector bundle, covariant derivatives.

The exterior product of differential forms can be generalized to $\mathfrak{g}$-valued forms by considering forms with values in the symmetric algebra of $\mathfrak{g}$. Using the natural mapping $S(\mathfrak{g}^*) \to S(\mathfrak{g})^*$, such forms can be linked with $f \in S(\mathfrak{g}^*)$ and made into ordinary differential forms. They become G-invariant when f is chosen in the space $S(\mathfrak{g}^*)^G$ of G-invariant polynomial functions on $\mathfrak{g}$. If this principle is applied to the k-th power of the curvature form Ω of a principal fiber bundle E, one obtains a G-invariant $2k$-form $\mathrm{CW}_E(f)$ on E. This form, like Ω, is *horizontal*, i.e., as a multilinear form it vanishes as soon as one of the arguments is a vertical vector field. But the pullback with π is an isomorphism between the $\Omega(M)$ and the G-equivariant horizontal differential forms on E. Thus, one obtains a differential form $\mathrm{CW}_{M,\pi}(f) \in \Omega^{2k}(M)$ with $\mathrm{CW}_E(f) = \pi^*(\mathrm{CW}_{M,\pi}(f))$. This form does not depend on the choice of the G-invariant connection and is called *Chern-Weil form* of f. One can verify that the Chern-Weil forms are closed and therefore define (de Rham) cohomology classes. The resulting cohomology classes (of even degree) are also called *characteristic classes*. They play an important role in algebraic topology and geometry. Particularly prominent is their appearance on the topological side of the *Atiyah-Singer index theorem*, which expresses the analytic index of elliptic differential operators on compact manifolds by topological quantities, namely integrals of certain characteristic classes.

Literature: Connections are the basic structure for the approach of [KN96] to differential geometry. The different variants of the description of connections and their relevance for curvature concepts are described and compared in volume 2 of [Sp99]. Covariant derivatives, Levi-Civita connections, and Riemannian curvature are of course also explained in any book on Riemannian geometry. Bundles are extensively treated in [KN96]. Because of their role in gauge theory and other areas of theoretical physics, they also find their way into physics textbooks like [Na15]. But they can also be found in many more specialized texts like [LM89] or [MS12]. The approach to connections on general fiber bundles outlined here, up to a formulation of the Atiyah-Singer index theorem, can be found in [Mi08, Chap. IV]. Another approach, which among other things makes massive use of Dirac operators, can be found in [BGV92].

9.3 Structured Analysis?

So far, analysis has only been mentioned in passing in this text. Implicitly through the differential calculus in Ch. 6 on manifolds and as a kind of termination criterion when content would have required the introduction of specific analytic methods that do not correlate well with structures.

It would not be correct to say that structures play no role in analysis. Even the differential and integral calculus of Newton and Leibniz can be interpreted as algebraizations of analytic content of Greek mathematics. Functional analysis as a combination of linear algebra and topology was invented as a structural framework for integral equations and later expanded to become an indispensable tool for dealing with differential equations. And basically, the Kolmogorov axioms of probability theory are also an algebraization of an analytic problem.

There has been no shortage of attempts in later times to further algebraize analysis, but these attempts have not prevailed among analysts. An example is the *non-standard analysis*, which, motivated not least by Euler's handling of Leibniz's infinitesimal mathematics, introduces "infinitely small" and "infinitely large" numbers and calculates with them (see e.g., [Ro96]). Bourbaki works in [Bo67] with a differential calculus based on metric fields. Even more general is the calculus presented in [Be11], which operates on the basis of topological rings.

A very common approach in analysis is to work with scales of topological vector spaces tailored to the respective problem (see [St23] for current examples). Classic examples of such scales are Lebesgue's L^p-spaces with $1 < p \leq \infty$ or the Sobolev spaces H_s with $s \in \mathbb{R}$. There are attempts at systematic representations of such scales of spaces (see e.g., the four-volume work [Tr83]), but so far little capital seems to have been made from such a system.

In specific situations, attempts are also made to solve analytic problems by defining structures, originally associated with finite-dimensional vector spaces, such as manifolds, on the basis of infinite-dimensional vector spaces and using them to solve the problems. An example is the treatment of partial differential evolution equations, which are dealt with using ordinary differential equations on infinite-dimensional spaces (see e.g., [EN00, Am19, AK21]).

Many examples could be given of specific situations in which structural considerations flow into the solution of analytic problems. However, structures within which general theorems and lemmas are proven, which are repeatedly referred to, play a subordinate role in analysis. When analysts are asked about this, they usually emphasize that in analysis *principles* are at the center, which need to be understood and which allow solution methods to be adapted to given problems.

Algebraic Analysis

At this point, I would like to highlight and describe in more detail an example of a profound algebraization of analysis. It goes back to Mikio Sato and his students Masaki Kashiwara and Takahiro Kawai, see [SKK73, KKK86]. This *algebraic analysis*, which makes massive use of sheaf theory and homological algebra, is the actual origin of microlocal analysis, but in its original form it has found little distribution. Only the variant developed by Lars Hörmander, in which the topological and algebraic components are reduced to a minimum (see [Ho83]), is by now established as a frequently used tool by analysts. The starting point of algebraic analysis according to Sato are the *hyperfunctions*. These are the most general among

9.3 Structured Analysis?

the so-called "generalized functions". The best-known example class of generalized functions are the *distributions* on open subsets U of $\mathbb{R}^n$. They are defined as continuous linear functionals on the space $\mathcal{C}_c^\infty(U)$ of smooth functions with compact support on U. Every smooth function $f \in \mathcal{C}^\infty(U)$ on U defines a distribution by $\varphi \mapsto \int_U f(x)\varphi(x)\,dx$. In this way, one embeds $\mathcal{C}^\infty(U)$ into the space $\mathcal{D}'(U)$ of distributions on U, which explains the term generalized functions. If one restricts a distribution $\varphi \in \mathcal{D}'(U)$ to the space $\mathcal{A}(U)$ of all real analytic functions, one obtains a linear functional on $\mathcal{A}(U)$, but not every such functional can be extended to a distribution. These ideas can be transferred to real analytic manifolds with volume form. The volume form is used to be able to consider functions as distributions via integrals (otherwise, one must replace the compactly supported smooth functions by smooth sections of the density bundle). On a compact real analytic manifold M, one can actually define hyperfunctions as linear functionals on the space of real analytic functions and then obtains the inclusion chain

$$\mathcal{A}(M) \hookrightarrow \mathcal{C}^\infty(M) \hookrightarrow \mathcal{D}'(M) \hookrightarrow \mathcal{B}(M),$$

where $\mathcal{B}(M)$ denotes the space of hyperfunctions on M.

For non-compact manifolds, this approach does not work. In the case of one-dimensional manifolds M, however, there is another simple approach. In this case, one considers a complexification $M_\mathbb{C}$ of M and holomorphic functions on the complement of M in $M_\mathbb{C}$. Locally, this complement looks like two half-spaces. If the function is defined and holomorphic on all of $M_\mathbb{C}$, then it is uniquely determined by its restriction to M due to the principle of analytic continuation. This restriction can then be considered as the boundary value of each of the two restrictions to the two half-spaces. If the function vanishes on one of the two half-spaces, but can be continuously extended from the other half-space to M, the boundary value is something like a jump between the functions on the two half-spaces. These observations lead to considering the quotient $\mathcal{O}(M_\mathbb{C} \setminus M)/\mathcal{O}(M_\mathbb{C})$ as objects on M and the cosets $\varphi + \mathcal{O}(M_\mathbb{C})$ as "boundary values" of the $\varphi \in \mathcal{O}(M_\mathbb{C} \setminus M)$. However, these considerations were purely local, i.e., one should actually work with only locally defined functions. It is appropriate here to work with the sheaves of holomorphic functions and thus obtain the sheaf $\mathcal{B}_M := \mathcal{O}_{M_\mathbb{C} \setminus M}/\mathcal{O}_{M_\mathbb{C}}$ of hyperfunctions.

In higher dimensions, this approach cannot be adopted unchanged, because functions in $\mathcal{O}(M_\mathbb{C} \setminus M)$ can always be holomorphically extended to all of $M_\mathbb{C}$ according to Hartogs' theorem, so the above quotient always vanishes. The solution is to consider M locally as an open subset of $\mathbb{R}^n \subseteq \mathbb{C}^n$ in holomorphic functions on the set of points whose coordinates have non-vanishing imaginary parts in all components. This leads to families of holomorphic functions and the appropriate subspace to be singled out is of cohomological nature. As a result, one obtains a definition of the sheaf $\mathcal{B}_M$ of hyperfunctions as the part of n-cohomology of $M_\mathbb{C}$ with coefficients in $\mathcal{O}_{M_\mathbb{C}}$, which is supported in M (tensored with the orientation sheaf, i.e., the sheaf associated with the presheaf of all sections of the orientation bundle). With this general definition, one then obtains the embeddings at the sheaf level

$$\mathcal{A}_M \hookrightarrow \mathcal{C}_M^\infty \hookrightarrow \mathcal{D}'_M \hookrightarrow \mathcal{B}_M.$$

Generalized functions are an extension of the mathematical framework of functions that allow problems for functions in this extended framework to be solved and the solutions to be analyzed in more detail. For example, if one wants to solve linear differential equations, it is shown that differential operators also act on distributions. In some situations, one can then find distributional solutions to the differential equation, so-called *weak solutions*. For special differential operators, one can also prove "regularity theorems", which state that every weak solution must have been a smooth function. An example of such a special property of differential operators $D : C^\infty(M) \to C^\infty(M)$ is *ellipticity*. It can be read off the principal symbol $\sigma \in C^\infty(T^*M)$ of the differential operator D. If D is given in local coordinates by $D = \sum_{\alpha \in \mathbb{N}_0^n} c_\alpha \partial_x^\alpha$, then σ is given in the associated coordinates of T^*M by $\sigma(x, \xi) = \sum_{|\alpha|=k} c_\alpha \xi^\alpha$, where k is the order of the highest non-vanishing derivative in D. Ellipticity means that $\sigma(x, \xi)$ does not vanish for $\xi \neq 0$.

Hyperfunctions are an even wider framework for functions than distributions and there are problems that can be better handled in this framework than for distributions. The price one pays is the high conceptual effort already for the definitions, but also for the (sheaf- and module-theoretical) tools that one can use for working in this framework. This effort is shunned by almost all analysts and they try to reformulate the problems so that they can be handled within the framework of distributions using methods of functional and harmonic analysis. This is not always successful. In particular, the construction of boundary value mapping in the framework of distributions often only works under additional hypotheses.

The next step in Sato's algebraic analysis after the introduction of hyperfunctions is the construction of a suitable algebra of operators. Suitable here means that it contains the linear differential operators $D = \sum_{\alpha \in \mathbb{N}_0^n} c_\alpha \partial_x^\alpha$ (finite sum) with real analytic coefficient functions c_α and many elements are invertible up to residues which are easy to handle. As a starting point, one takes the *full symbols* $\sum_{\alpha \in \mathbb{N}_0^n} c_\alpha(x) \xi^\alpha$ defined on coordinate transformations and considers the summands as homogeneous holomorphic functions on $T^*M_\mathbb{C}$, where $M_\mathbb{C}$ is a "small" complex neighborhood of M. Then one follows an approach similar to polynomials, which are generalized to power series. One considers not necessarily finite formal sums of holomorphic functions P_ζ on $T^*M_\mathbb{C}$. These functions should be homogeneous in the fiber variable of degree $\zeta \in \mathbb{C}$, i.e., P_ζ is an eigenfunction of the Euler operator $\sum_j \xi_j \partial_{\xi_j}$ with eigenvalue ζ. Furthermore, it is assumed that there is a principal value $\lambda \in \mathbb{C}$ for the homogeneity and all other summands are homogeneous of degree $\lambda - j$ with $j \in \mathbb{N}$. In addition, there is a technical estimate in the fiber variables. In this way, one obtains for each $\lambda \in \mathbb{C}$ a sheaf $\mathcal{E}_{M_\mathbb{C}}(\lambda)$, whose sections are called *microdifferential operators*. It is not obvious on which spaces the microdifferential operators should naturally act. One possibility is to introduce

9.3 Structured Analysis?

the sheaf $\mathcal{C}_M$ of microfunctions on $iT^*M := \coprod_{x \in M} iT_xM \subseteq T^*M_{\mathbb{C}}|_M$, which is obtained from the sheaf $\mathcal{B}_M$ of hyperfunctions. For this, one needs the concept of the *singular set* $\mathrm{SS}(u) \in iT^*M$ of a hyperfunction. This is the complement of the set of all $(x, i\xi) \in iT^*M$, for which u is *microanalytic* in a neighborhood of $(x, i\xi)$, which essentially means that u is represented by a sum of holomorphic functions on open cones around $[0, \epsilon[i\xi$. In this way, one finds a presheaf

$$U \mapsto \mathcal{B}_M(M)/\{u \in \mathcal{B}_M(M) \mid \mathrm{SS}(u) \cap V = \emptyset\}$$

on iT^*M, whose associated sheaf is the sheaf of *microfunctions*. Microfunctions can be multiplied, pulled back, and integrated. Microdifferential operators act on microfunctions and the algebra of microdifferential operators has good properties. Together, microfunctions and microdifferential operators provide a powerful framework in which many problems for differential operators with holomorphic coefficient functions can be elegantly solved.

Hörmander developed a variant of this algebraic analysis for distributions, in which the role of analytic continuation is taken over by the Fourier transformation. Regularity of distributions is described using decay conditions of the Fourier transformation, making it accessible for classical methods of analysis, which are centrally concerned with inequalities. In this variant, the singular sets become *wavefront sets*, the microdifferential operators become *pseudo-differential operators* and the *quantized contact transformations*, which are obtained from coordinate transformations in Sato, become *Fourier integral operators*. These are the essential building blocks of *microlocal analysis*, which has been continuously developed over the last decades and is now a standard tool, especially in the spectral theory of differential operators.

One could say that microlocal analysis is a collection of principles that are constantly varied according to the mathematical preferences of contemporary analysts to solve a wide range of analytic problems. Its origin in a structural theory is hardly recognizable anymore. It does not seem impossible to me that the role of structures in microlocal analysis and analysis in general may at some point (again) gain importance.

Literature: The original source for algebraic analysis according to Sato is [SKK73]. The book [KKK86] is a somewhat easier-to-read introduction. There are various introductions specifically to hyperfunctions. The book [Mo91] is dedicated to this topic, but also includes a chapter on microfunctions. Both [Sch84] and [Ho83] contain introductory chapters on hyperfunctions, aimed at readers familiar with the analysis of distributions. The book [Sch85] is an introduction to the theory of microdifferential operators. [Bj79, Bj93] consider microdifferential operators in the context of general rings of differential operators and their modules. The sheaf-theoretical tools used in the theory are described in [KS94]. A standard source for microlocal analysis and pseudodifferential operators is [Ho83]. More recent texts are [Zw12] and [Iv19].

References

[AMR88] Abraham, R., Marsden, J., & Ratiu, T. (1988). *Manifolds, tensor analysis, and applications*. Springer.
[AHW18] Alldridge, A., Hilgert, J., & Wurzbacher, T. (2018). Superorbits. *Journal of the Institute of Mathematics of Jussieu, 17*, 1065–1120.
[Am19] Amann, H. (2019). *Linear and quasilinear parabolic problems II*. Birkhäuser.
[AC21a] Anel, M., & Catren, G. (2021). *New spaces in mathematics – Formal and conceptual reflections*. Cambridge University Press.
[AC21b] Anel, M., & Catren, G. (2021). *New spaces in physics – Formal and conceptual reflections*. Cambridge University Press.
[Ap75] Apostol, T. (1975). *Calculus* (vols. 1–2, 2nd edn.). Wiley.
[AK21] Arnold, V.I., & Khesin, B.A. (2021). *Topological methods in hydrodynamics* (2nd edn.). Springer.
[Ar23] Artin, M. (2023). *Algebra* (2nd edn.). Pearson.
[AD14] Audin, M., & Damian, M. (2014). *Morse theory and floer homology*. Springer.
[Ba06] Ballmann, W. (2006). *Lectures on kähler manifolds*. European Mathematical Society.
[BE11] Beem, J., & Ehrlich, P. (1981). *Global lorentzian geometry*. Marcel Dekker.
[BGV92] Berline, N., Vergne, M., & Getzler, E. (1992). *Heat kernels and dirac operators*. Springer.
[Be11] Bertram, W. (2011). *Elementary topological differential calculus*. Calvage & Mounet.
[Bj79] Björk, J.-E. (1979). *Rings of differential operators*. North-Holland.
[Bj93] Björk, J.-E. (1993). *Analytic $\mathcal{D}$-modules and applications*. Kluwer.
[Bo91] Borel, A. (1991). *Linear algebraic groups* (2nd edn.). Springer.
[Bo67] Bourbaki, N. (1967). *Variétés différentielles et analytiques – Fascicule de résultats*. Hermann.
[Bo70] Bourbaki, N. (1970). *Algebrè I*. Hermann.
[Br97] Bredon, G. E. (1997). *Sheaf theory*. Springer.
[Bu96] Bumb, D. (1996). *Automorphic forms and representations*. Cambridge University Press.
[CS01] Cannas da Silva, A. (2001). *Lectures on symplectic geometry*. Springer.
[CK05] Capiński, M., & Kopp, E. (2005). *Measure, integral and probability* (2nd edn.). Springer.
[CCF11] Carmeli, C., Caston, L., & Fioresi, R. (2011). *Mathematical foundations of supersymmetry*. European Mathematical Society.
[Ch11] Chen, B.-Y. (2019). *Pseudo-Riemannian geometry, δ-invariants and applications*. World Scientific.
[CS22] Clausen, D., & Scholze, P. (2022). Condensed mathematics and complex geometry. https://people.mpim-bonn.mpg.de/scholze/Complex.pdf
[Co94] Connes, A. (1994). *Noncommutative geometry*. Academic Press.

[Co95] Coutinho, S. C. (1995). *A primer of algebraic D-modules*. Cambridge University Press.
[CLM21] Crainic, M., Loja Fernandez, R., & Mărcuţ, I. (2021). *Lectures on poisson geometry*. American Mathematical Society.
[DM99] Deligne, P., & Morgan, J. (1999). Notes on supersymmetry (following J. Bernstein). In P. Deligne et al. (Eds.), *Quantum fields and strings: A course for mathematicians* (vol. I, pp. 41–98). American Mathematical Society.
[Di69] Dieudonné, J. (1969). *Foundations of modern analysis*. Enlarged and corrected printing. Academic Press.
[Di74] Dieudonné, J. (1974). *Cours de Géométrie Algébrique 2*. Presses Universitaires de France.
[dC92] do Carmo, M. P. (1992). *Riemannian geometry*. Birkhäuser.
[EH07] Eisenbud, D., & Harris, J. (2007). *The geometry of schemes*. Springer.
[EN00] Engel, K.-J., & Nagel, R. (2000). *One-parameter semigroups for linear evolution equations*. Springer.
[GM03] Gelfand, S. I., & Manin, Y. I. (2003). *Methods of homological algebra* (2nd edn.). Springer.
[Go73] Godement, R. (1973). *Théorie des Faisceaux*. Hermann.
[GW10] Görtz, U., & Wedhorn, T. (2010). *Algebraic geometry I*. Vieweg+Teubner.
[Go11] Gowers, T., et al. (2011). *The Princeton companion to mathematics*. Princeton University Press.
[GVF01] Gracia-Bondía, J., Várilly, J., & Figueroa, H. (2001). *Elements of noncommutative geometry*. Birkhäuser.
[Gr08] Grätzer, G. (2008). *Universal algebra*. Springer.
[GH94] Griffith, P., & Harris, J. (1994). *Principles of algebraic geometry*. Wiley.
[Ha77] Hartshorne, R. (1977). *Algebraic geometry*. Springer.
[Ha02] Hatcher, A. (2002). *Algebraic topology*. Cambridge University Press.
[HE73] Hawking, S., & Ellis, G. (1973). *The large scale structure of space-time*. Cambridge University Press.
[HS73] Herrlich, H., & Strecker, G. E. (1973). *Category theory*. Allyn and Bacon.
[HN12] Hilgert, J., & Neeb, K.-H. (2012). *Structure and geometry of lie groups*. Springer.
[HWZ21] Hofer, H., Wysocki, K., & Zehnder, E. (2021). *Polyfold and fredholm theory*. Springer.
[Ho12] Holme, A. (2012). *A royal road to algebraic geometry*. Springer.
[Ho73] Hörmander, L. (1973). *An introduction to complex analysis in several variables* (2nd edn.). North-Holland.
[Ho83] Hörmander, L. (1983). *The analysis of linear partial differential operators I-IV*. Springer.
[Hu05] Huybrechts, D. (2005). *Complex geometry*. Springer.
[Iv86] Iversen, B. (1986). *Cohomology of sheaves*. Springer.
[Iv19] Ivrii, V. (2019). *Microlocal analysis, sharp spectral asymptotics and applications*. Springer.
[KS94] Kashiwara, M., & Shapira, P. (1994). *Sheaves on manifolds*. Springer.
[KS06] Kashiwara, M., & Shapira, P. (2006). *Categories and sheaves*. Springer.
[KKK86] Kashiwara, M., Kawai, T., & Kimura, T. (1986). *Foundations of algebraic analysis*. Princeton University Press.
[Ke95] Kempf, G. (1995). *Algebraic structures*. Vieweg.
[Kn06] Knapp, A. W. (2006). *Basic algebra*. Birkhäuser.
[KV95] Knapp, A. W., Vogan, D. A. (1995). *Cohomological induction and unitary representations*. Princeton University Press.
[KN96] Kobayashi, S., & Nomizu, K. (1996). *Foundations of differential geometry*. Wiley.
[Ko97] Koch, H. (2000). *Number theory*. American Mathematical Society.
[Ko03] Kontsevich, M. (2003). Deformation quantization of Poisson manifolds. *Letters in Mathematical Physics, 66*, 157–216.
[La21] Land, M. (2021). *Introduction to infinity-categories*. Springer.
[La67] Lang, S. (1967). *Algebraic structures*. Addison-Wesley.

[La93] Lang, S. (1993). *Algebra*. Addison-Wesley.
[LPV13] Laurent-Gengoux, C., Pichereau, A., & Vanhaecke, P. (2013). *Poisson structures*. Springer.
[LM89] Lawson, B., & Michelsohn, M.-L. (1989). *Spin geometry*. Princeton University Press.
[Le14] Leinster, T. (2014). *Basic category theory*. Cambridge University Press.
[LS23] Le Stum, B. (2023). An introduction to condensed mathematics. https://perso.univ-rennes1.fr/bernard.le-stum/bernard.le-stum/Enseignement_files/CondensedBook.pdf
[LM87] Libermann, P., & Marle, C.-M. (1987). *Symplectic geometry and analytical mechanics*. D. Reidel.
[LS20] Lins Neto, A., and Scárdua, B. (2020). *Complex algebraic foliations*. De Gruyter.
[ML98] Mac Lane, S. (1998). *Categories for the working mathematician*. Springer.
[MT97] Madsen, I., & Tornehave, J. (1997). *From calculus to cohomology – De Rham cohomology and characteristic classes*. Cambridge University Press.
[MS12] McDuff, D., & Salamon, D. (2012). *J-holomorphic curves and symplectic topology* (2nd edn.). American Mathematical Society.
[MS17] McDuff, D., & Salamon, D. (2017). *Introduction to symplectic topology* (3rd edn.). Oxford University Press.
[Mi08] Michor, P. W. (2008). *Topics in differential geometry*. American Mathematical Society.
[Mi05] Milne, J. S. (2005). *Algebraic geometry: V5.0*. Taiaroa Publishing.
[Mi17] Milne, J. S. (2017). *Algebraic groups*. Cambridge University Press.
[MM03] Moerdijk, I., & Mrčun, J. (2003). *Introduction to foliations and Lie groupoids*. Cambridge University Press.
[MF10] Morgan, J. W., & Fong, F. (2010). *Ricci flow and geometrization of 3-manifolds*. American Mathematical Society.
[Mo91] Morimoto, M. (1991). *An introduction to sato's hyperfunctions*. American Mathematical Society.
[Mu18] Munkres, S. (2018). *Topology* (2nd edn.). Pearson.
[Na12] Naber, G. L. (2012). *The geometry of minkowski spacetime – An introduction to the mathematics of the special theory of relativity*. Springer.
[Na15] Nakahara, M. (2015). *Differential geometry, topology and physics*. Springer.
[Oh15] Oh, Y.-G. (2015). *Symplectic topology and floer homology*. Cambridge University Press.
[ON83] O'Neill, B. (1983). *Semi-Riemannian geometry*. Academic Press.
[Pe95] Perrin, D. (1995). *Algebraic geometry*. InterEditions.
[PR94] Platonov, V., & Rapinchuk, A. (1994). *Algebraic groups and number theory*. Academic Press.
[Po17] Pohl, A. (2017). The category of reduced orbifolds in local charts. *Journal of the Mathematical Society of Japan, 69,* 755–800.
[Ra04] Ramanan, S. (2004). *Global calculus*. American Mathematical Society.
[Re90] Reid, M. (1990). *Undergraduate algebraic geometry*. Cambridge University Press.
[Re95] Reid, M. (1995). *Undergraduate commutative algebra*. Cambridge University Press.
[Ro96] Robinson, A. (1996). *Non-standard analysis*. Reprint of the second (1974) edition. With a foreword by Wilhelmus A. J. Luxemburg. Princeton University Press.
[Sa22] Sasane, A. (2022). *A mathematical introduction to general relativity*. World Scientific.
[Sa56] Satake, I. (1956). On a generalization of the notion of manifold. *Proceedings of the National Academy of Sciences of the United States of America, 42,* 359–363.
[SKK73] Sato, M., Kashiwara, M., & Kawai, T. (1973). Microfunctions and pseudo-differential equations. In *Hyperfunctions and pseudo-differential equations. Lecture notes in mathematics* (vol. 287, pp. 265–529). Springer.
[Sch85] Schapira, P. (1985). *Microdifferential systems in the complex domain*. Springer.
[Sch84] Schlichtkrull, H. (1984). *Hyperfunctions and harmonic analysis on symmetric spaces*. Birkhäuser.
[Sp66] Spanier, E. (1966). *Algebraic topology*. McGraw-Hill.
[Sp71] Spivak, M. (1971). *Calculus on manifolds*. Addison-Wesley.

[Sp99] Spivak, M. (1999). *A comprehensive introduction to differential geometry* (3rd edn.). Publish or Perish.
[Sp88] Springer, T. (1988). *Linear algebraic groups* (2nd edn.). Springer.
[St23] Street, B. (2023). *Maximal subellipticity*. De Gruyter.
[Ta02] Taylor, J.L. (2002). *Several complex variables with connections to algebraic geometry and lie groups*. American Mathematical Society.
[Te75] Tennison, B.R. (1975). *Sheaf theory*. Cambridge University Press.
[SP] The Stacks project – an open source textbook and reference work on algebraic geometry. https://stacks.math.columbia.edu
[Tr83] Triebel, H. (1983–2020). *Theory of function spaces I–IV*. Birkhäuser.
[Tu04] Tuynman, G. M. (2004). *Supermanifolds and supergroups*. Kluwer Academic Publishers.
[Va94] Vaisman, I. (1994). *Lectures on the geometry of Poisson manifolds*. Birkhäuser.
[Vo87] Vogan, D. A. (1987). *Unitary representations of reductive Lie groups*. Princeton University Press.
[Vo16] Voisin, C. (2016). *Hodge theory and complex algebraic geometry*. The Société Mathématique de France.
[Wa83] Warner, F. W. (1983). *Foundations of differentiable manifolds and lie groups*. Springer.
[We15] Wedhorn, T. (2015). *Manifolds, sheaves, and cohomology*. Springer.
[We94] Weibel, C. (1994). *An introduction to homological algebra*. Cambridge University Press.
[We58] Weil, A. (1958). *Introduction to the study of Kähler manifolds*. Hermann.
[We79] Wells, R. O. (1979). *Differential analysis on complex manifolds*. Springer.
[Zw12] Zworski, M. (2012). *Semiclassical analysis*. American Mathematical Society.

Index

Symbols
Ab, category of abelian groups, 99
$\mathbf{Alg}_R^{\text{alt}}$, category of alternating R-algebras, 126
$\mathbf{Alg}_{\mathbb{K}}^{\text{fg}}$, category of finitely generated $\mathbb{K}$-algebras, 244
$\mathbf{Alg}_R$, category of associative R-algebras, 100
$\mathbf{Alg}_{R,1}$, category of associative R-algebras with identity, 110
$\text{Alt}_R(M^q, N)$, alternating mappings, 86
alt, alternation, 87
Ann, annulator, 268
$\mathbf{AS}_{\mathbb{K}}$, category of $\mathbb{K}$-algebraic sets, 245
Aut $R(M)$, module automorphisms, 31
$\mathcal{B}_{X,R}$, presheaf of bounded functions, 140
$\mathcal{C}^{\mathbb{K},k}(U; E)$, sections of E over U, 198
$\mathcal{C}_X^{\mathbb{K},k}$, structure sheaf of the manifold X, 171
$\mathcal{C}_X^{\mathbb{K},k} \otimes V$, locally free module sheaf, 169
$\mathbf{CAlg}_R$, category of commutative R-algebras, 100
$\mathbf{C}_{\mathbb{C}-\text{Vect}}^{\text{diff}}$, category, 101
$\mathbf{C}_{\mathbb{C}-\text{Vect}}^k$, category, 101
χ_φ, characteristic polynomial, 51
$\mathbf{C}_{\mathbb{K}-\text{Vect}}^\omega$, category, 101
$\mathcal{C}_{X,V}^{\mathbb{K},\omega}$, sheaf of analytic V-valued functions, 140
$\mathcal{C}_{X,V}^{\mathbb{K},k}$, sheaf of differentiable V-valued functions, 140
CMet, category of complete metric spaces, 124
$\mathbf{C}_{\Phi,\Gamma}$, category of an algebraic structure, 100
$\mathbf{C}_{R-\text{Vect}}^{\text{diff}}$, category, 101
$\mathbf{C}_{R-\text{Vect}}^k$, category, 101
$\mathbf{CRing}_1$, category of commutative rings with identity, 99
$\mathcal{C}_{X,R}$, sheaf of continuous functions, 140
df, differential of f, 196

$D(f)$, standard open set in the maximal spectrum, 258
$D(f)$, standard open set in the spectrum, 266
$D_{\vec{x}}$, directional derivative, 304
$Df(p)$, Jacobian matrix of f, 192
$\left(\frac{\partial y_I}{\partial x_I}\right)$, Jacobian matrix in multi-index notation, 211
$\frac{\partial}{\partial x_j}$, basis field, 205
$\frac{\partial}{\partial x_j}|_p$, derivation in coordinates, 185
$\mathcal{D}(U)$, differential operators on U, 9
∂M, boundary of M, 223
∂^α, higher partial derivatives, 8
∂_j, partial derivatives, 8
Δ_X, diagonal in X, 262
$d\alpha$, exterior derivative of α, 212
d_g, distance function, 291
df_p, derivative of f, 190
dx_I, differential forms in multi-index notation, 210
$dx_j(p)$, φ-basis for T_p^*M, 196
$\text{Der}(A)$, derivations of A, 206
$\text{Der}(A, B)$, derivations from $A \subseteq B$, 206
$\text{Der}_\Phi(A, B)$, Φ-derivations from A to B, 206
$\mathcal{D}er(\mathcal{C}_M^{\mathbb{K},k})$, derivation sheaf on M, 207
$\mathcal{D}er(\mathcal{C}_M^{\mathbb{K},k}, \mathcal{C}_M^{\mathbb{K},k-1})$, derivation sheaf on M, 207
$\dim_{\mathbb{K}}(X)$, dimension of X, 173
$D_V(f)$, standard open set, 252
$dx_1(p)$, φ-basis for the cotangent space, 197
$\deg(f)$, degree of a polynomial, 8
E_p, fiber of a vector bundle, 198
$\acute{\text{E}}(X, \mathbf{C})$, category of étalé spaces, 156
$\mathcal{E}_\pi$, sheaf of sections, 199
$\acute{\text{E}}(\mathcal{F})$, étalé space of a sheaf, 155

Index

ÉMod$_{\mathcal{O}_X}$, étalé-category of $\mathcal{O}_X$-modules, 167
End $R(M)$, module endomorphisms, 31
End**C**(X), endomorphisms of a category, 98
ev, evaluation functor, 129
ev, evaluation mapping, 92
ev$_a$, evaluation mapping, 240
ev$_x$, evaluation, 11
ev$_\varphi$, evaluation homomorphism, 48
$F(E)$, free module over E, 41
$_R F(E)$, free left-R-module over E, 41
Field, category of fields, 100
$\gcd(a_1,\ldots,a_k)$, gcd for rings, 20
$\mathbb{G}_{k,n}$, Grassmann manifold, 178
$_G\mathbf{Mod}_R$, category of G-modules, 125
Grp, category of groups, 99
Grp$_{\text{top}}$, category of topological groups, 101
$H^\bullet$, representing functor, 128
$H_\bullet$, representing functor, 129
H_x, stalk functor in x, 145
$\mathbb{H}$, quaternions, 6
H^X, represented by X functor, 106
H_X, represented by X functor, 106
$\mathcal{H}ol_{X,V}$, sheaf of holomorphic V-valued functions, 140
$\text{Hom}_R(M, N)$, module homomorphisms, 29
Hom**C**(X, Y), morphisms of a category, 98
Hom$_\mathbf{A}$, Hom-functor, 107
$\mathcal{H}om_{\mathcal{O}_X}(\mathcal{F},\mathcal{G})$, $\mathcal{O}_X$-modules of homomorphisms, 168
$I(X)$, vanishing ideal of a set, 241
$\iota_{\mathfrak{x}}\alpha$, contraction, 215
I, ideal for the alternating algebra, 84
ID, category of integral domains, 100
ID$_m$, subcategory of integral domains, 122
$\text{im}(\varphi)$, image of a module homomorphism, 30
Ind_H^G, induction functor, 125
J, ideal for the symmetric algebra, 77
J_q, q-part of J, 82
$\ker(\varphi)$, kernel of a module homomorphism, 30
$\ell_g(\gamma)$, curve length, 290
$L(V_1,\ldots,V_s; V_0)$, bundle of s-linear mappings, 203
$L_{\mathfrak{x}}\alpha$, Lie derivative of a differential form, 215
$\mathcal{L}_X^1$, presheaf of integrable functions, 140
$\bigwedge T^*M$, form bundle, 211
$\bigwedge^k TM^*$, k-form bundle, 210
$\Lambda(M)$, exterior algebra, 84
$\Lambda(\psi)$, exterior algebra functor, 85
$\Lambda_q(M)$, antisymmetric tensors, 81
$\Lambda^q(M)$, q-th exterior power of M, 86
lcm, least common multiple, 112
LieAlg$_R$, category of Lie algebras over R, 127
$\varprojlim$, categorical limit, 113

$\varinjlim$, categorical colimit, 113
$L_R(M_1,\ldots,M_n; P)$, multilinear mappings, 56
$L_R(M_\lambda; P)$, multilinear mappings, 56
$\mathfrak{m}_a$, vanishing ideal of a, 246
Man$_{\mathbb{K},k}$, category of manifolds, 172
Mat$_\mathbb{K}$, category of $\mathbb{K}^n$'s, 136
Met, category of metric spaces, 124
Mod$_{\mathcal{C}_M^{\mathbb{K},k}}^{\text{lff}}$, category of locally free module sheaves, 200
Mod$_R$, category of right-R-modules, 100
Mod$_{\mathcal{O}_X}$, category of $\mathcal{O}_X$-modules, 166
$_R\mathbf{Mod}$, category of left-R-modules, 100
$\nabla_{\mathfrak{x}}$, covariant derivative, 303
ob(**C**), objects of a category, 98
Ω_M^ℓ, sheaf of ℓ-forms, 199
$\Omega^r(M)$, differential forms of degree r, 210
Ω_M, sheaf of differential forms, 199, 210, 212
Or(M), orientation bundle, 306
$P(a, \varrho)$, open polycylinder, 235
$P(a, \varrho)^-$, closed polycylinder, 235
$\mathbb{P}_R^n$, projective space over R, 272
PS$(X, \mathbf{C})$, category of presheaves, 141
φ-basis, 184, 187
φ_{ij}, transition map, 173
$\mathbb{P}_\mathbb{K}^n$, projective space, 178
$\mathbb{P}_\mathbb{K}^n$, projective space, 178
PVar$_\mathbb{K}$, category of prevarieties, 258
$Q(R)$, quotient field, 15
$\mathbb{Q}_p$, p-adic numbers, 118
rad(I), radical of an ideal, 245
Res_H^G, restriction functor, 125
rest$_{|c,d|}$ restriction, 11
ρ_U, section functor, 149
$\rho_{V,U}$, restriction mappings, 140
Ring, category of rings, 99
Ring$_1$, category of rings with identity, 99
RSp$_R$, category of R-ringed spaces, 160
RSp$_{\text{loc}, R}$, category of locally R-ringed spaces, 164
S, section functor, 200
$(S\varphi)_U$, section functor, 156
$S(E, X, \pi)$, section functor, 156
S$(X, \mathbf{C})$, category of sheaves, 144
Sch, category of schemes, 266
Sch^{aff}, category of affine schemes, 266
Set, category of sets, 98
Sheaf, sheafification functor, 148
sign(σ), signature of a permutation, 81
$S_{k,n}$, Stiefel manifold, 178
$S(M)$, symmetric algebra, 77
$\mathbb{S}^n$, n-sphere, 175
Spec(R), spectrum of a ring, 161
$S(\varphi)$, symmetric algebra functor, 78

Index

Spm(A), maximal spectrum, 258
S_q, symmetric group on q letters, 79
$S^q(M)$, q-th symmetric power of M, 79
$S_q(M)$, symmetric tensors, 81
$\mathrm{Sym}_R(M^q; N)$, symmetric mappings, 79
sym, symmetrization mapping, 81
TM, tangent bundle, 193
Tf, derivative of f, 195
$T(a, \varrho)$, determining surface of a polycylinder, 235
T^*M, cotangent bundle, 196
T_pM, geometric tangent space, 183
T_p^*M, cotangent space, 196
$T_p^{\mathrm{alg}}M$, algebraic tangent space, 185
$T_p^{\mathrm{geo}}M$, geometric tangent space, 183
$T_p^{\mathrm{phy}}M$, physicist's tangent space, 184
$\mathrm{T}(M)$, tensor algebra, 73
$\mathrm{T}(\varphi)$, tensor algebra functor, 74
$\mathrm{T}^n(M)$, tensors of n-th level, 72
$\mathcal{T}_M^*$, cotangential sheaf of M, 199
$\mathcal{T}_M$, tangential sheaf of M, 199
Top, category of topological spaces, 100
$T_\Phi(X)$, term algebra, 92
$\mathrm{U}(L)$, universal enveloping algebra, 127
$\mathrm{Unit}(R)$, units of R, 4
$V(I)$, vanishing set in $\mathrm{Spec}(R)$, 161
$\mathbf{V}(I)$, vanishing set, 258
$\mathcal{V}(I)$, zero set of an ideal, 240
$V(f)$, vanishing set in the spectrum, 266
$v_\varphi^{(j)}$, φ-basis, 184
$\mathrm{Var}_\mathbb{K}^{\mathrm{aff}}$, category of affine varieties, 259
$\mathrm{Var}_\mathbb{K}$, category of varieties, 262
$\mathrm{VB}_M^{\mathbb{K},k}$, category of vector bundles over M, 199
$\mathrm{Vect}_\mathbb{K}^{\mathrm{fin}}$, category of finite-dimensional $\mathbb{K}$-vector spaces, 102
$\mathcal{X}(M)$, vector fields on M, 205
$\mathbb{Z}_p$, p-adic integers, 118
$\mathbb{Z}[i]$, Gaussian integers, 5
$\mathbb{Z}/n\mathbb{Z}$, residue class ring, 5
$\langle E \rangle$, submodule generated by E, 32
$\{\cdot, \cdot, \cdot\}$, Jordan triple product, 94
$R[X_1, \ldots, X_k]$, polynomial ring, 8
$R[X_1, \ldots, X_k]_d$, homogeneous polynomials of degree d, 8
$R[[X_1, \ldots, X_k]]$, formal power series, 7
$[\cdot, \cdot, \cdot]$, Lie triple product, 94
$[\cdot, \cdot]$, Lie bracket, 93
$[\gamma]_p$, equivalence class of curves through p, 183
$[\gamma_j]_p$, φ-basis, 183
$[a, b]$, Lie bracket, 127
$[m]$, equivalence class, 95
$[s]$, germs of a section, 155
$[x]$, coset, 12
$[\mathbf{A}, \mathbf{B}]$, functor category, 127
$\mathbb{K}[V]$, coordinate ring of V, 241
$\mathbb{K}[V]_f$, localization of $\mathbb{K}[V]$ at f, 252
$\mathbb{K}[\varphi]$, subalgebra of $\mathrm{End}_\mathbb{K}(V)$, 48
$\sigma \cdot t$, permutation of a tensor, 81
$g \cdot m$, group action, 96
$G \circ F$, functor composition, 107
$X \cong_\mathbf{C} Y$, isomorphism in a category, 103
$\cong$, isomorphism, 13
$A_X(U)$, locally constant A-valued functions on U, 143
E_x, fiber in étalé space, 154
$\mathcal{X}_x$, value of $\mathcal{X}$ at x, 205
φ_U, presheaf morphism, 141
φ_x, induced morphism of stalks in x, 145
s_x, germ of s in x, 145
$\mathcal{F}_\pi(U)$, section sheaf of an étalé space, 154
$\mathcal{F}_x$, stalk in x, 145
1_X, identity (morphism), 98
$\varphi^\flat$, sheaf morphism, 153
$\widetilde{D}(i)$, diagram presheaves, 150
$f^{-1}(\mathcal{F}')$, inverse image of a sheaf, 151
$f^{<-1>}(\mathcal{F}')$, inverse presheaf image, 151

$$\mathbf{A} \underset{G}{\overset{F}{\rightleftarrows}} \bot\Gamma\ \mathbf{B}$$

, adjoint functors, 122

$$\mathbf{A} \underset{F'}{\overset{F}{\rightrightarrows}} \Downarrow\Phi\ \mathbf{B}$$

, natural transformation, 127
$F \dashv G$, adjoint functors, 122
$I \trianglelefteq R$, ideal in a ring, 10
$\mathbf{S} \hookrightarrow \mathbf{C}$, subcategory, 100
$\mathbf{S} \overset{v}{\hookrightarrow} \mathbf{C}$, full subcategory, 100
$\bigoplus_{\lambda \in \Lambda} M_\lambda$, direct sum of modules, 36
$M \otimes_R N$, tensor product over R, 58
$M^{\otimes n}$, tensor power of M, 72
$\bigotimes_r^s TM$, tensor bundle, 202
$\bigotimes_r^s T_pM$, tensor space, 202
$\bigotimes_r^s \mathcal{T}_M^{(r,s)}$, tensor sheaf of level (r, s), 203
$\varphi \otimes \psi$ tensor product of homomorphisms, 59
$\mathcal{F} \otimes_{\mathcal{O}_X} \mathcal{G}$, tensor product of $\mathcal{O}_X$-modules, 168
$\bigotimes_{\lambda \in \Lambda} M_\lambda$, tensor product of the M_λ, 65
$f(\mathfrak{m})$, evaluation of a ring element, 258
$\mathbf{C}(X)$, endomorphisms of a category, 98
$\mathbf{C}(X, Y)$, morphisms of a category, 98
$x + I$, coset, 12

$\bigsqcup_{i \in I} X_i$, categorical sum, 109
$\prod_{i \in I} X_i$, categorical product, 108
$\prod_{\lambda \in \Lambda} M_\lambda$, direct product of modules, 36
$\psi^\sharp$, sheaf morphism, 153
$F \simeq F'$, naturally isomorphic functors, 128
$M_\sim$, space of equivalence classes, 95
$\sim_x$, stalk relation in x, 145
M/N, quotient module, 31
$f_*\mathcal{F}$, direct image of a presheaf, 142
$f'(p)$, derivative of f, 190
$\mathbf{A} \times \mathbf{B}$, product category, 103
$(\widetilde{U}, \widetilde{\varphi})$, associated chart on TM, 195
$(\widetilde{U}, \widetilde{\varphi})$, associated chart on T^*M, 197
$(\widetilde{U}, \widetilde{\varphi})$, associated chart on $\bigotimes_r^s TM$, 202
$\widetilde{\mathfrak{X}}$, vector field as differential operator, 206
$E \times_X E'$, fiber product, 158
$X_1 \times \cdots \times X_n$, categorical product, 108
F^{op}, opposing functor, 113
M^0, the set $\{\emptyset\}$, 91
M^G, G-invariants, 125
M^{adj}, adjoint matrix, 247
$S^{-1}R$, localization, 17
$\mathbf{C}^{\mathrm{op}}$, opposite category, 102
$\mathcal{F}^+$, sheafification of $\mathcal{F}$, 145
$M^\vee$, dual module, 85, 107
$\varphi^\vee$, dual morphism, 107
$m \vee m'$, infimum of m and m', 112
$\mathcal{F}^\vee$, dual module sheaf, 169
$^\vee$, duality functor, 107
$|\alpha|$, length of the multi-index α, 8
$\wedge$, exterior product for differential forms, 212
$\wedge$, exterior product, 84
$m \wedge m'$, supremum of m and m', 112

A

Abelian
   category, 281
Action, 97
   trivial, 97
Addition, 3
   pointwise, 4
Adjoint action, 292
Adjoint matrix, 247
Affine
   connection, 303
   scheme, 266
   subset, 260
   variety, 258
Algebra
   associative, 71
   commutative, 71
   homomorphism, 72
   over a commutative ring, 71
   over a ring, 246
   symmetric, over a module, 76
Algebraic
   analysis, 314
   field extension, 249
   group, 284
   set, 240
   structure, 92
   topology, 279
Algebraically
   closed, 51
   independent, 247
Almost-complex structure, 293
   canonical, 294
   integrable, 294
Alternating, 126
   forms, 210
   mapping, 86
Analytic mappings, 174
Antisymmetric
   mapping, 86
   tensor, 81
Arrows of a category, 98
Associated
   chart
      of the cotangent bundle, 197
      of the tangent bundle, 195
   elements of a ring, 20
Atiyah, Michael (1929–2019), 313
Atiyah-Singer index theorem, 313
Atlas, 173
   oriented, 217, 221
   for a manifold with boundary, 224
Automorphism in a category, 103

B

Basis
   field, 205
   of a module, 33
   of a topology, 171
Bimodule, 60
   homomorphismus, 61
Boundary of a manifold, 223
Bundle
   associated, 309
   horizontal, 310
   of k-forms, 210
   trivial, 306
   vertical, 310

Index

C

Cartan, Henri (1904–1986), 215
Cartan-identity, 215
Categorical
    coproduct, 109
    product, 108
    sum, 109
Category, 98
    abelian, 281
    derived, 283
    exact, 281
    locally small, 128
    opposite, 102
    preadditive, 281
    small, 102
Cauchy, Augustin Louis (1789–1857), 229
    integral
        formula, 234
        formula, in several variables, 235
        theorem, 231
    product, on formal power series, 7
Cauchy-Riemann differential equations, 229
Center of a polycylinder, 235
Chain complex, 282
Characteristic
    class, 313
    polynomial, 51
Chart
    of a manifold, 172
    neighborhood, 172
Chern, Shiing-shen (1901–2004), 313
Chern-Weil
    form, 313
Chevalley, Claude (1909–1984), 53
Chinese remainder theorem, 12
Class of a mapping, 172
Clausen, Dustin, 283
Closed differential form, 213
Co-adjoint action, 292
Cofunctor, 107
Cokernel of a morphism, 281
Colimit, 113
Commutator product, 127
Complex structure, 293
Composition of morphisms, 98
Condensed mathematics, 283
Congruence relation, 95
Connected, 179
    component, 180
    topological space, 173
Connection, 310
Constant sheaf, 143
Contraction of a differential form, 215
Contravariant functor, 107
Coordinate
    neighborhood, 172
    ring of an algebraic set, 241
Coprime, 20
Coproduct, 109
Coset, 5, 96
Cotangent
    bundle, 196
    sheaf, 199
    space, 196
Covariant
    derivative, 303
    functor, 107
Curvature, 305
    form, 312
Cyclic
    module, 26
    vector, 50

D

Darboux, Gaston (1842–1917), 294
D-cocone, 113
D-colimit, 113
D-cone, 112
Degree
    of a differential form, 210
    function, 19
    of a polynomial, 8
De Rham cohomology, 281
De Rham, Georges (1903–1990), 209
Derivation, 185
    of an algebra, 206
Derived category, 283
Determinant, 86
Determining surface of a polycylinder, 235
Diagram
    commutative, 35
    of the form **I**, 112
Differentiable structure, 174
Differential
    form, 85, 199, 210
    of a mapping, 190
    operators
        with constant coefficients, 9
        with smooth coefficients, 9
Dimension of a manifold, 173
Direct
    image of presheaves, 142
    image of sheaves, 150
    product of modules, 37
    sum of modules, 37
Distributions, 315
Distributivity, 3

Division ring, 6
Division with remainder in a Euclidean ring, 19
Divisor, 20
D-limit, 113
Dolbeault cohomology, 297
Dolbeault, Pierre (1924–2015), 297
Dual module, 107

E
Ellipticity, 316
Epimorphism, 103
Equationally defined class, 92
equivalence of categories, 134
Essentially surjective, 134
étalé space, 153
Euclid (ca. 325–265 BC), 19
Euclidean
  ring, 19
  space, 290
Evaluation of a ring homomorphism, 49
Exact
  category, 281
Exterior
  algebra, 83
  derivative, 212
  power of a module, 86

F
Factor module, 31
Faithful functor, 131
Fiber
  bundle, 306
  product, 115
  of a vector bundle, 198
Field
  of fractions, 15
Finitely
  generated ideal, 255
  generated R-algebra, 246
Finite R-algebra, 246
Floer, Andreas (1956–1991), 294
Floer homology, 294
Foliation, 286
Formal
  deformation, 301
  star product, 301
Formality theorem, 301
Frölicher, Alfred (1927–2010), 312
Frölicher-Nijenhuis bracket, 312
Frame bundle
  of a vector bundle, 307

Free group action, 307
Fubini, Guido (1897–1943), 296
Fubini-Study metric, 296
Full
  functor, 131
  subcategory, 99
Fully faithful functor, 132
Functor
  category, 127
  contravariant, 107
  covariant, 107
  fully faithful, 132
  opposite, 113
Functoriality, 59
Fundamental theorem
  of algebra, 52
  of arithmetic, 21
  of calculus, 225
  on finitely generated abelian groups, 45

G
G-action, 97
Galois, Evariste (1811–1832), 279
Galois theory, 279
Gauß, Carl Friedrich (1777–1855), 5
Gaussian integers, 5
Geodesic, 291, 304
Germ of a presheaf, 145
Gluing, 116
  data, 144
  of sheaves, 144
G-module, 125
Gradient, 291
  symplectic, 292
Grassmann, Hermann (1809–1877), 178
Grassmann manifolds, 178
Greatest common divisor, 20
Group, 93
  algebraic, 284
  object, 284
  structure, 93
Groupoid, 286
G-set, 96
G-structure, 308

H
Hamilton, William Rowan (1805–1865), 7
  equations of motion, 292
  function, 291
  vector field, 292, 299
Harmonic, 297
Hermite, Charles (1822–1901), 295

Hermitian structure, 296
Hilbert, David (1862–1943), 246
  basis theorem, 256
  Nullstellensatz, 246
Hodge structure, 298
Hodge, William (1903–1975), 297
Hörmander, Lars (1931–2012), 314
Holonomy group, 312
Homogeneous polynomial, 8
Homological algebra, 280
Homology, 279
Homomorphism
  of local rings, 164
  of modules, 29
  of rings, 9
Homotopy, 279
  category, 283
  of chain complexes, 283
Hopf, Heinz (1894-1971), 291
Horizontal
  bundle, 310
  form, 313
Hyperfunctions, 314
Hyperplane, 177

I
Ideal, 10
  in an algebra, 72
  finitely generated, 255
  generated by a set, 11
  maximal, 17
  prime, 17
Identity
  as functor, 104
  of a group, 5
  in a ring, 6
I-holomorphic curve, 294
Image of a module homomorphism, 30
Inclusion of a categorical sum, 109
Independence in a module, 33
Induction, 125
Inductive
  limit, 116
  system, 116
Infimum in a partially ordered set, 112
Integrable sub-bundle, 312
Integral domain, 14
Integral of a differential form, 222
Internal direct sum, 40
Invariance of basis length, 35
Inverse
  additive, 6
  image

  of an étalé space, 159
  of a sheaf, 151
  multiplicative, 6
Irreducible
  components of an algebraic set, 272
  element of an integral domain, 23
  topological space, 272
Isomorphism
  of algebraic sets, 242
  of categories, 103, 134
  of modules, 29
  of rings, 10
Isomorphism theorem for modules, 32

J
Jacobi, Carl-Gustav (1804–1851), 93
Jacobi identity, 93
Jordan
  algebra, 94
  block, 51
  identity, 94
  normal form, 52
  triple system, 94
Jordan, Camille (1838–1922), 51
Jordan-Chevalley decomposition, 53
Jordan, Pascual (1902–1980), 94

K
Kähler, Erich (1906–2000), 296
Kähler manifold, 296
Kashiwara, Masaki (*1947), 314
Kawai, Takahiro (*1945), 314
Kernel
  of a module homomorphism, 30
  of a morphism, 281
  of a ring homomorphism, 10
Kontsevich, Maxim (*1964), 301
$\mathbb{K}$-rational points, 274
$\mathbb{K}$-valued points, 274

L
Laplace operators, 297
Laplace, Pierre-Simon (1749–1827), 297
Lattice, 112
Laurent, Pierre (1813–1854), 271
Laurent polynomials, 271
Leading coefficient, 8
Left ideal, 26
Level
  of a tensor bundle, 202
  of a tensor field, 202

Levi-Civita connection, 305
Levi-Civita, Tullio (1873–1941), 305
Lie, Sophus (1842–1899), 93
    algebra, 93
        of vector fields, 208
        derivative of differential forms, 215
        groups, 284
        supergroups, 285
        triple system, 94
Lift of a G-structure, 309
Limit, 113
    direct, 116
    inductive, 116
    inverse, 116
    projective, 116
Line
    bundle, 306
    integral, 229
Linear combination in a module, 32
Line integral
    complex, 231
Local
    ring, 163
    ring homomorphism, 164
    trivialization, 306
Locally
    closed, 263
    finite covering, 220
    ringed space, 163
Lorentz, Hendrik (1853–1928), 290
    manifold, 290
    metric, 290

M

Manifold, 171
    with boundary, 223
    differentiable, 171
    oriented, 218
Microdifferential operators, 316
Microfunctions, 317
Minimal polynomial, 49
Minkowski, Hermann (1864–1909), 290
Minkowski space, 290
Module, 26
    cyclic, 26
    dual, 107
    finitely generated, 33
    free, 33
    homomorphism, 29
    left, 26
    right, 28
    sheaf, 166
Möbius, August Ferdinand (1790–1868), 218

Möbius strip, 218
Monic polynomial, 8
Monomorphism, 103
Morphism, 160, 164
    of a category, 98
    of étalé spaces, 156
    of locally ringed spaces, 164
    of manifolds, 172
    of module sheaves, 166
    of presheaves, 141
    of ringed spaces, 160
    of sheaves, 144
    of vector bundles, 199
Multi-index, 7, 8
Multilinear
    form, 56
    mapping, 56
Multiplication, 3
    pointwise, 4
Multivector field, 299

N

n-ary operation, 91
Natural
    isomorphism, 128
    transformation, 127
Newlander, August, 294
Nijenhuis, Albert (1926–2015), 312
Nijenhuis tensor, 294
Nirenberg, Louis (1925–2020), 294
Noether
    normalization, 248
    property, 255
    ring, 255
    spaces, 257
Noether, Emmy (1882–1935), 255
Noether, Max (1844–1921), 248
Non-commutative geometry, 286
Non-standard analysis, 314
Normalized polynomial, 8
Normal subgroup, 96
Nullstellensatz, 246

O

Object of a category, 98
$\mathcal{O}_X$-module
    dual, 169
    free, 169
    locally free, 169
        of finite type, 169
1-form, 85, 198

Index   331

operation
   n-ary, 91
Opposite
   category, 102
   functor, 113
Orbifold, 286
Orbit, 97
Orientation, 222
   bundle, 306
   induced, 225

P

p-adic
   integers, 118
   numbers, 118
Paracompact, 220
Parallel transport, 304
Parametrization of a manifold, 172
Partition of unity, 218
   subordinate to a cover, 218
Perelman, Grigori (*1966), 305
Poincaré, Henri (1854–1912), 181
Pointwise
   addition, 4
   multiplication, 4
Poisson, Siméon (1781–1840), 289
   bracket, 298
   manifold, 298
   tensor, 299
Polycylinder, 235
Polyfold, 295
Polynomial, 8
   division, 18
   function, 241
   homogeneous, 8
   mapping, 242
   normalized, 8, 49, 52
Polyradius, 235
Power series, 7
Preadditive, 281
Presheaf, 140
Prevariety, 255, 258
Prime element, 20
Principal
   fiber bundle, 307
   ideal, 11
Principal ideal
   domain, 19
Product
   categorical, 108
   category, 103
   of manifolds, 177

   in a module, 33
   topology, 109
Projection
   canonical, on submodules, 40
   of a categorical product, 108
   stereographic, 175
Projective
   limit, 116
   space, 178
      over a ring, 272
   system, 116
Pseudo-differential operators, 317
Pseudo-holomorphic curve, 294
Pseudo-Riemannian
   manifold, 290
   metric, 290
Pullback, 115
   of differential forms, 216
   of a fiber bundle, 310
Pushout, 116

Q

Quaternions, 7
Quotient
   group, 26
   module, 31
   ring, 12
   structure, 95

R

Radical, 245
   ideal, 245
Radius of a polycylinder, 235
Rank
   of a module, 36
   of a vector bundle, 198
Rational normal form of a linear mapping, 50
Reduced, 241
Reduction of the structure group, 308
Refinement of a cover, 220
Regular function, 259
Regular mapping
   of affine varieties, 260
   of prevarieties, 260
Residue
   class, 5
      ring, 5
   field of a local ring, 163
Restriction, 140
Riemann, Bernhard (1826–1866), 229
   surface, 180

Riemannian
  curvature, 305
  manifold, 290
  metric, 290
Right ideal, 30
Ring, 3
  commutative, 4
  Euclidean, 19
  extension, 29
  factorial, 21
  with identity, 4
  Noetherian, 255
Ringed space, 160
Rinow, Willi (1907–1979), 291

S

Sato, Mikio (1928–2023), 314
Scheme, 266
  affine, 266
  over a ring, 274
Scholze, Peter (*1987), 283
Schouten, Jan (1883–1971), 299
Schouten-Nijenhuis bracket, 299
Section
  of a fiber bundle, 306
  functor, 149
  of a presheaf, 140
  sheaf, 154
  of a vector bundle, 198
Separated prevariety, 262
Sheaf, 143
  of sections, 199
Short exact sequence, 43
Singer, Isadore (1924–2021), 313
Singular set, 317
Skew-field, 6
Skew-symmetric
  mapping, 86
Skyscraper sheaf, 144
Small category, 102
Smooth function, 174
  on singular sets, 223
Solution curve, 27
Spectrum of a ring, 161
Sphere, 175
Spin structure, 309
Stack, 286
Stalk
  functor, 145
  of a presheaf, 145
Standard open set, 252
  of an affine scheme, 266
Stiefel, Eduard (1909–1978), 178

Stiefel manifold, 178
Stokes, George Gabriel (1819–1903), 225
  theorem, 225
Structure
  algebraic, 92
  constants of a Lie algebra, 300
  group of a principal fiber bundle, 307
  local, 137
  sheaf of a ringed space, 160
Study, Eduard (1862–1930), 296
Subalgebra, 72
Subcategory, 99
  full, 99
Submanifold, 178
Submodule, 29
  generated by a set, 32
Subring, 29
Sum
  categorical, 109
  topological, 110
Supremum in a partially ordered set, 112
Symbol, 316
Symmetric
  mapping, 79
  power of a module, 79
  tensor, 81
Symplectic
  coordinates, 293
  foliation, 300
  form, 291
  gradient, 292
  manifold, 291
  topology, 294
  vector space, 293

T

Tangent
  bundle, 193
  functor, 196
  sheaf, 199
  space
    algebraic, 185
    geometric, 183
    physicist's, 184
  vector, 184
Tensor, 199
  algebra over a module, 72
  bundle, 202
  field, 75, 202
  product
    of algebras, 110
    of modules, 57, 64
    of $\mathcal{O}_X$-modules, 168

Terminal object of a category, 284
Theorem
    of Darboux, 293
    of de Rham, 283
    of Hodge, 297
    of Hopf-Rinow, 291
    of Newlander-Nirenberg, 294
Thurston, Bill (1946–2012), 305
Topological sum, 110
Transition function, 201, 308
Transitive group action, 307
Type of a manifold, 171

U
Unital ring homomorphisms, 122
Unit of a ring, 4
Universal algebra, 32
Universal property
    direct sum of modules, 37
    exterior algebra, 84
    free modules, 34
    product of modules, 37
    symmetric algebra, 76
    tensor algebra, 72
    tensor product, 57, 64
Universe, 128

V
Vanishing ideal, 241
Variety, 262
    affine, 258
    projective, 265
    quasiprojective, 265

Vector
    bundle, 198
    field, 27, 203, 205
    subbundle, 209
Vector-valued differential forms, 312
Vertical
    bundle, 310
    vector, 310
Volterra, Vito (1860–1940), 181

W
Wavefront set, 317
Weak solution, 316
Weil, André (1906–1998), 313
Weyl algebra, 9
Weyl, Hermann (1885–1955), 9

Y
Yoneda, Nobuo (1930–1996), 130
    embedding, 132
    lemma, 130

Z
Zariski, Oscar (1899–1986), 161
Zariski topology, 161, 240, 242, 258
Zero
    of an additive group, 5
    divisor, 14
    object, 281
Zorn, Max (1906–1993), 33
Zorn's lemma, 33

SPRINGER NATURE

GPSR Compliance

The European Union's (EU) General Product Safety Regulation (GPSR) is a set of rules that requires consumer products to be safe and our obligations to ensure this.

If you have any concerns about our products, you can contact us on ProductSafety@springernature.com

In case Publisher is established outside the EU, the EU authorized representative is:

Springer Nature Customer Service Center GmbH
Europaplatz 3
69115 Heidelberg, Germany

The manufacturer's authorised representative in the EU is Springer Nature Customer Service Centre GmbH, Europaplatz 3, 69115 Heidelberg, Germany. If you have any concerns regarding our products, please contact ProductSafety@springernature.com

Printed and bound by CPI Group (UK) Ltd, Croydon, CR0 4YY
25/03/2026
02078172-0012